GLOBAL COMPETITION IN TRANSPORTATION MARKETS: ANALYSIS AND POLICY MAKING

RESEARCH IN TRANSPORTATION ECONOMICS

Series Editor: Martin Dresner

Volume 1–6:	Research in Transportation Economics – Edited by B. Starr McMullen
Volume 7:	Railroad Bankruptcies and Mergers from Chicago West 1975 – 2001: Financial Analysis and Regulatory Critique – Michael Conant
Volume 8:	Economic Impacts of Intelligent Transportation Systems: Innovations and Case Studies – Evangelos Bekiaris and Yuko Nakanishi
Volume 9:	Road Pricing: Theory and Evidence – Georgina Santos
Volume 10:	Transportation Labor Issues and Regulatory Reform – James Peoples and Wayne K. Talley
Volume 11:	Interurban Road Charging for Trucks in Europe – José Manuel Viegas
Volume 12:	Shipping Economics – Kevin Cullinane

RESEARCH IN TRANSPORTATION ECONOMICS VOLUME 13

GLOBAL COMPETITION IN TRANSPORTATION MARKETS: ANALYSIS AND POLICY MAKING

EDITED BY

ADIB KANAFANI

University of California, Berkeley, USA

KATSUHIKO KURODA

Kobe University, Kobe, Japan

2005

ELSEVIER

JAI

Amsterdam – Boston – Heidelberg – London – New York – Oxford
Paris – San Diego – San Francisco – Singapore – Sydney – Tokyo

ELSEVIER B.V.
Radarweg 29
P.O. Box 211
1000 AE Amsterdam
The Netherlands

ELSEVIER Inc.
525 B Street, Suite 1900
San Diego
CA 92101-4495
USA

ELSEVIER Ltd
The Boulevard, Langford
Lane Kidlington
Oxford OX5 1GB
UK

ELSEVIER Ltd
84 Theobalds Road
London
WC1X 8RR
UK

First edition 2005

British Library Cataloguing in Publication Data
A catalogue record is available from the British Library.

ISBN: 0-7623-1204-1
ISSN: 0739-8859 (Series)

The paper used in this publication meets the requirements of ANSI/NISO Z39.48-1992 (Permanence of Paper).
Printed in The Netherlands.

CONTENTS

LIST OF CONTRIBUTORS

Adib Kanafani	University of California, Berkeley, California, USA
Anming Zhang	University of British Columbia, Vancouver, BC, Canada
Atsushi Koike	Tottori University, Tottori, Japan
David Gillen	University of British Columbia, BC, Canada
Eddy E. M. Van de Voorde	University of Antwerp, Antwerp, Belgium
Emine Yetiskul	Kyoto University, Kyoto, Japan
Gang Gong	Boston University Center for Transportation Studies, Boston, MA, USA
Hajime Inamura	Tohoku University, Sendai, Japan
Hideyuki Kita	Tottori University, Tottori, Japan
Hilde M. A. Meersman	University of Antwerp, Antwerp, Belgium
Hitoshi Ieda	University of Tokyo, Tokyo, Japan
Kakuya Matsushima	Kyoto University, Kyoto, Japan
Katsuhiko Kuroda	Kobe University, Kobe, Japan
Katsuhiro Yamaguchi	Ministry of Land Infrastructure and Transport, Tokyo, Japan
Kazuhiko Ishiguro	Kobe University, Kobe, Japan
Kazuyuki Tsuchiya	Mitsubishi Research Institute, Inc., Tokyo, Japan
Keishi Tanimoto	Tottori University, Tottori, Japan
Kenneth Button	George Mason University, VA, USA

Kerati Kijmanawat	University of Tokyo, Tokyo, Japan
Kiyoshi Kobayashi	Kyoto University, Kyoto, Japan
Mikio Takebayashi	Kobe University, Kobe, Japan
Naoto Ida	Muroran Institute of Technology, Hokkaido, Japan
Ryuichi Shibasaki	National Institute for Land and Infrastructure Management, Ministry of Land and Infrastructure, Kanagawa, Japan
Takayuki Morikawa	Nagoya University, Nagoya, Japan
Takayuki Ueda	Tokyo Institute of Technology, Tokyo, Japan
Tatsuo Takase	Shinshu University, Nagano, Japan
Tohru Tamura	Muroran Institute of Technology, Hokkaido, Japan
Tomihiro Watanabe	Ministry of Land and Infrastructure, Tokyo, Japan
Toshiaki Tsuji	Nomura Research Institute, Tokyo, Japan
T.R. Lakshmanan	Boston University Center for Transportation Studies, Boston, MA, USA
William P. Anderson	Boston University Center for Transportation Studies, Boston, MA, USA

PREFACE

These proceedings cover an international symposium held at the University of Kobe in December 2002. The purpose of the symposium, entitled "*International Symposium on Global Competition in Transportation Markets – Analysis and Policy Making*" was to bring together an international group of researchers to explore the question of how transportation markets will behave in the era of global trade and of the emerging transportation networks that support it. The papers presented at the symposium covered two modes that might appear at first sight to be too different to form the topic of a single symposium. But, as many of these papers showed, if at least implicitly, these two modes share important features that underlie the behavior of the transportation sector, especially at the global level. As such, they provide a coherent basis for a discourse on the question of competition in transportation markets.

While the title of the symposium might promise a direct treatment of competition within the transportation sector *at the global level*, the fact is that the global dimension is addressed by extending to the global level knowledge based on the study of how transportation systems engage in competition at a more local or network-regional level. Such is the state of the art of the study of global competition in transportation: relatively little has been done directly on this topic and much of what is known about it is based on extrapolation from the more local or network-regional level. While it is now accepted that there are many features of transportation markets that are robust to geographical scale, it is still hoped that this symposium and these proceedings will encourage researchers to direct attention to addressing transportation problems at the truly global level. This will not be easy, since conducting research at such a level has its difficulties, not the least of which are practical issues such as information acquisition and management.

It is on the other hand encouraging for those who want to do global-level research that some fundamental attributes of transportation are independent of the geographical scale at which the system is observed. These are clearly recognized and dealt with in the papers assembled in these proceedings. We mention some of these here in the hope of motivating the reader of these proceedings to look for them and to make a comparison of the insights to be found in the various papers presented here.

First, there are the economies of scale that arise in transportation. Of particular interest to the topic at hand are the economies that arise in networks, such as economies of scope and density. Perhaps the most important manifestation of these arises in air transportation in the form of hub dominance. In these proceedings, the reader will find a number of papers that touch on this. These include the paper by Kita, Koike and Tanimoto; the paper by Yetiskul, Matsushima and Kobayashi; and the paper by Zhang. The issue also arises in maritime transportation where the economies of vessel scale exert a very strong influence on network structure, and where port dominance has an influence not too dissimilar from that of airline hub dominate. Here the reader will find the papers by Kuroda, Takebayashi and Tsuji, and by Shibasaki, Ieda and Watanabe addressing this important subject.

Another fundamental feature of transportation system that figures prominently in the work presented here is network externalities. These externalities arise when effects, usually beneficial ones, accrue to elements of the systems from actions intended for other elements. Positive synergies arise in transportation networks between local (OD) traffic and connecting (hub-and-spoke) traffic. This is addressed in the paper by Kita, Koike and Tanimoto. Network-regional externalities also arise from airport capacity expansion. The paper by Ueda, Koike, Yamaguchi and Tsuchiya illustrate that in the case of Haneda airport. Externalities resulting from adjustments in the regulatory framework of transportation markets have also received attention. The paper by Ida and Tamura looks at a negative effect: the loss of regional air transport service; and the paper by Ishiguro, Inamura and Higuchi looks at the positive regional economic effects of trade flow liberalization.

Finally, but not least important, the notion of uncertainty arises here and plays an important role when seeking appropriate policies. Uncertainty in the behavior of the transportation market results from the inherent principle that many of the equilibriums to be found in these markets are not stable. The same economies and externalities mentioned above combined with the utility maximizing behavior of travelers and shippers result in the elusiveness of stable equilibrium. This means that the behavior of the system should be predicted as a dynamic, continually changing quest for equilibrium rather than a demonstrative, stable and predictable solution. Many of the papers presented here recognize this either as volatility in forecasting, as e.g. in the paper by Meersman or as instability of choice model solutions as in the case of air passenger choice with the paper by Takase and Morikawa; or in the analysis of port choice in the paper by Kuroda, Takebayashi and Tsuji.

The papers presented at the Kobe symposium represent a rich and diverse set of reports on insight research into some of the fundamental aspects of the behavior of transportation markets. It is hoped that the symposium will inspire additional work that would this research one step further toward addressing the global nature of interactions that arise in transportation markets.

PART I:
AIR TRANSPORT

COMPETITION IN A DEREGULATED MARKET FOR AIR TRAVEL: THE U.S. DOMESTIC EXPERIENCE AND LESSONS FOR GLOBAL MARKETS

William P. Anderson, Gang Gong and T. R. Lakshmanan

ABSTRACT

This paper presents evidence on the nature of competition in the post-deregulation U.S. air travel market. It begins with an historical review of the U.S. market since deregulation. An econometric study of domestic air fares for 1995 and 2000 is used to address the following questions: How does the level of competition on a particular route or in a particular airport affect fares? Does competition from low-cost airlines have an especially large effect on fares? Do hub-and-spoke systems make air travel cheaper for the consumer?

Global Competition in Transportation Markets: Analysis and Policy Making
Research in Transportation Economics, Volume 13, 3–25

ISSN: 0739-8859/doi:10.1016/S0739-8859(05)13001-7

INTRODUCTION

It is now almost 25 years since the U.S. federal government deregulated domestic air travel. This period has seen major changes in the airline industry, including the entry of low cost carriers, waves of mergers among the major carriers, rapid growth in the number of air travelers, general decline in airfares, increased variability in fares across markets, and the emergence of hub-and-spoke (H-S) systems. As the global air travel industry gradually moves toward a more market-oriented regime, much can be learned from America's experience of deregulation.

The purpose of this paper is to provide information about the experience of deregulated air travel markets in the U.S. It begins with an historical review of deregulation and a synopsis of current views on its success or failure. It also presents some new empirical results based on very large data sets for 1995 and 2000 that test hypotheses about the variation in air fares in the deregulated market. Particular emphasis is placed on the effect of competition and hubs on fares.

Finally, we attempt to draw some general lessons from the U.S. experience that can be applied to the case of deregulation in global markets.

DEREGULATION AND COMPETITION IN THE U.S.

The Civil Aeronautics Act of 1938 instituted formal regulation of the airline industry in the U.S. It was part of a larger scale regulatory movement that started in the railroad, motor carrier, telephone and banking industries. Since these industries are important "public interest" industries and there is a natural tendency for oligopoly, the U.S. government decided to intervene for the purposes of protecting either the general public, the industry itself, or both.

The Act created the Civil Aeronautics Authority (later reorganized as the Civil Aeronautics Board (CAB)) and authorized it to control route entry and exit of air carriers, regulate fares, award subsidies, and control mergers and inter-carrier agreements (Bailey, Graham, & Kaplan, 1985). Regulation of the airlines was deemed necessary to prevent "destructive competition" from damaging a fledgling industry vital to the future security and commerce of the nation (Sampson et al., 1990).

Due to the inherent lack of efficiency of government regulation, economic studies burgeoned in the 1970s to question the desirability of continuing regulation. They also argued that the airline industry had become more mature and competitively balanced since regulation began in 1938, so it was time to remove regulatory restraints.

In 1978, Congress passed and President Carter signed into law the Airline Deregulation Act, which signified the beginning of airline deregulation. The Act took the regulatory power away from the CAB, allowing carriers that were "fit, willing, and able" to serve any route and charge fares to any level that they deemed appropriate (Morrison & Winston, 1986).

In the first few years after the Act, new entry boomed and prices fell substantially on most routes, especially long distance routes. On the shorter routes, which had been cross-subsidized under regulation, real prices did not fall as much and even increased in some cases. The new entrants weakened the dominance of incumbent airlines. The top 10 trunk airlines' market share declined from 87% in 1978 to 75% in 1983 (Goetz & Sutton, 1997). As the deep recession of the early 1980s set in, entry of new airlines came to a halt. Instead, a spate of mergers, acquisitions, and bankruptcies contributed to a period of industrial consolidation whereby the top four airlines controlled nearly 70% and the top 10 airlines accounted for 97% of domestic revenue passenger kilometers by 1991. Then from 1992 there was another wave of new entrants. The four-firm concentration ratio declined to 56% and top 10 airlines market share dropped to just over 90% by 2000 (Goetz, 2002).

So generally, deregulation has led to an attenuation of competition through consolidation and bankruptcy (Graham, 1993), and this was reinforced by the market power exercised by the dominant airlines at their hubs. However, even though competition decreased in general, route level competition increased with a larger proportion of routes having more effective competitive carriers. Moreover, longer-haul markets had significantly more effective competitors than shorter-haul markets.

One of the most important developments of the post-deregulation airline industry has been the emergence of a low cost entrant called Southwest Airlines as one of the largest carriers in the market. Southwest follows a strategy of cost containment that includes flying a fleet of almost identical planes and in many cases offering service from smaller airports, avoiding congested and hub-dominated airports. Because of its aggressive pricing strategies, the entrance of Southwest into a market has in many cases forced incumbent carriers to cut their fares significantly. Thus, Southwest is credited by some as having single-handedly driven a large share of the post-deregulation reduction in fares (Morrison, 2001).

Overall, deregulation has been widely hailed as a success, because the rate of growth in air travel has far exceeded population growth (the number of enplanements has more than doubled since 1978), and passengers have enjoyed a larger network of services, greater service choices and lower fares on average. According to the U.S. Department of Transportation (USDOT) (2001), domestic consumers pay a third less (in constant dollars) than they paid 20 years ago. The U.S. Government Accounting Office (USGAO) (1996) showed that among the largest 112 airports, average fares declined over a range of 8–11% (in constant dollars) during the 1979–1994 period. The Transportation Research Board (TRB) (1999) indicated a 25% decline from 1990 to 1998 (Goetz, 2002).

At the same time, most of these studies also raised concerns that the average fare figures mask large differences among market segments. For example, the USGAO (1996) found that, even though fares have declined overall for each airport size classification, fares at small- and medium-sized community airports have been consistently higher than fares at large community airports. In terms of geographic patterns, cities in the Southwest enjoyed lower fares while the highest fares were found in the Southeast, Northeast, and Upper Midwest regions (Goetz, 2002). Some research has suggested that a key factor explaining much of the fare differences over the last two decades is the level of carrier competition at airports and city-pair market levels.

Some recent research focused on several problems in the airline industry, which include high fares at the fortress hubs, pockets of pain, and increasing industry consolidation (USGAO, 1999; Goetz & Vowles, 2000). The principal explanation for these problems is the lack of competition from low-cost airlines such as Southwest or one of the new entrants, and the dominance of one or two higher-cost airlines (Goetz, 1993; USGAO, 1996; Goetz & Sutton, 1997; Goetz & Vowles, 2000). But the deeper reasons, Goetz (2002) argues, lie in flaws in the theoretical foundation for deregulation.

There are three key theoretical pillars to the structure of the airline industry (Goetz, 2002): economies of scale are non-existent; barriers to entry are insignificant; and airline markets are contestable. These theoretical pillars have been used to argue that the industry would not become dominated by an oligopoly of carriers and deregulation could achieve desired results.

However, by the late 1980s, some studies began to suggest that not only did airlines possess economies of scope and traffic density but that, contrary to previous findings, they were also found to have significant economies of scale (Antoniou, 1991). Economies of scope in the airline industry have been reinforced through the airlines' use of computerized reservation systems

(CRS) and loyalty marketing schemes such as frequent flyer programs, travel agent commission overrides, and corporate discounts (Levine, 1987; Goetz, 2002).

For many economists and policy-makers, a cornerstone of support for airline deregulation was contestability theory. While contestability theory in its pure form suggests that the number of actual competitors should have no effect on prices, many studies have found that the number of airlines actually competing on a route has a significant effect on the price level (Bailey et al., 1985; Call & Keeler, 1985; Morrison & Winston, 1987; Borenstein, 1989; Hurdle et al., 1989). If sunk costs are non-trivial, albeit small, and an incumbent can respond in price and quantity as quickly as a new competitor can enter, then the incumbent has little incentive to respond in advance of actual entry (Stiglitz, 1987). Advertising and the short-run losses associated with inauguration of service on a new route seem to be sufficient sunk costs to inhibit contestability in the airline industry (Borenstein, 1992).

Hubs

H-S networks appear to be a natural result of deregulation. Upon the removal of regulatory restraints, most carriers[1] adopted H-S route structures to accommodate larger volumes of traffic from an increased number of city-pairs.

For the airlines, there are many advantages to switching to an H-S network. Kanafani and Ghobrial (1985) pointed out that this system enables airlines to take advantage of economies of aircraft size, Toh and Higgins (1985) found an increase in airline profitability, and McShan and Windle (1989) reported significant cost savings. Thus, the H-S leads to cost reductions that could be passed on to consumers in the form of lower fares. However, it is difficult to disentangle the cost efficiency aspects of hubbing from the strategic aspects. Oum et al. (1995) noted that hubbing is a dominant strategy in an oligopoly and can deter entry by other airlines. Thus, whether passengers are better or worse off as a result of H-S networks is a point of some debate.

Morrison and Winston (1986) found that passengers have benefited from airline deregulation mainly through increased flight frequency, which is a direct result of the increase in H-S activity. Butler and Huston (1990) examined service to small non-hub airports (those most at risk to lost service) and discovered that most passengers using these airports have benefited from more frequent flights to a hub airport,[2] even though service to non-hub facilities has been curtailed. Due to this change, passengers originating from small non-hub airports can reach many more destinations on a one-stop-or-less flight than before. Barnett et al. (1992) discovered that a 1989 traveler was far more likely to find a timely non-stop jet flight than was her 1977 counterpart, even after adjusting for the growth in passenger traffic since 1977.

While the net benefits to passengers from the adoption of H-S networks have been substantial, not all passengers have fared equally. Travel times for those who lost non-stop service have increased and passengers originating from hub airports or facilities with severe capacity constraints face higher fares than they otherwise would.

Recent research has documented an association between high fares or yields and hub dominance by one or two firms. It has been reported that an increase in a hub's airline-enplanement share of 10% results in a 4.3% increase in average fares at that hub (Borenstein, 1989).

As for the far periphery, average fares tend to be higher than those in other centers. The USGAO (1991) found that passengers in small communities paid 3% more than passengers flying from major airports and 6% more at concentrated small-city airports than at small-city airports. It also found that passengers flying from small-city airports to major airports paid 34% more if the major airport was concentrated and 42% more if both the small-city and the major airport were concentrated (Goetz & Sutton, 1997).

In addition to the question of whether passengers fare better under H-S networks is the related question of whether the establishment of a hub at a particular airport yields economic benefits to the region that airport serves. Overall, it seems that cities hosting concentrated domestic hubs have benefited from increased airline employment, flight departures, and passenger flow, but not from lower air fares.

The dynamics of competition in the deregulated market for domestic air travel have given rise to a complex structure of fares (Anderson et al., 2002). It has been generally agreed that a key factor that explains much of the differences in fares is the level of carrier competition at the airport and city-pair market levels.

Numerous econometric or descriptive studies have attempted to test whether the observed dominance and concentration at airports or city-pair markets confers any pricing power to the dominant carriers (Bailey et al., 1985; Call & Keeler, 1985; Hurdle et al., 1989; Morrison & Winston, 1989; Bailey & Williams, 1989; Baker & Pratt, 1989; Strassman, 1990; Evans & Kessides, 1993; Anderson et al., 2002). The major analytical technique in these studies involves a regression of some measure of airfares as a function of competition and concentration, route characteristics, and airport characteristics.

Many, but by no means all of these studies conclude that competitive structure has an important impact on fares. A contrary conclusion is reached by Evans and Kessides (1993) who conclude that once they "appropriately controlled for inter- and intra-route heterogeneity in prices," they detect no significant correlation between route market share and price, and the effect of route concentration, though statistically significant, is quantitatively unimportant.

A recent study by Morrison (2001) provides empirical support for the unique role of Southwest Airlines in fare reductions. By including variables to explicitly account for actual or potential competition by Southwest in each market, he measures a "Southwest effect" that he concludes accounts for about 50% of the fare reduction since deregulation!

Fig. 1, which compares air travel costs by city of origin, provides visual evidence of the heterogeneity of fares across different markets. The fare per mile is used in order to control for variations in the average distance traveled by people from different cities. Here, we see that the cost per mile is about three times as high in the most expensive city (Pittsburgh) as in the least expensive (Ft. Lauderdale). Some of this variation can be explained by the fact that fares increase with distance at a decreasing rate, so that fare per mile is lower on long flights than short flights. It is not surprising therefore that cities in the west and in Florida, which have high proportions of long flights, have lower fare per mile than cities in the Northeast (New York, Boston, Washington, etc.) which have high proportions of short flights. It is also noteworthy, that the highest costs are in two cities (Charlotte and Pittsburgh) whose airports serve as "fortress hubs" for a single airline (U.S. Airways).

EMPIRICAL ANALYSIS

In order to test hypotheses about the impact of competitive structure on the cost of air travel, we regressed airfares against a number of pertinent variables. Our general hypotheses concerning factors affecting airfares are as follows:

- Airfares increase with distance traveled, but at a decreasing rate. This reflects the fact that terminal costs are fairly high relative to line-haul costs.

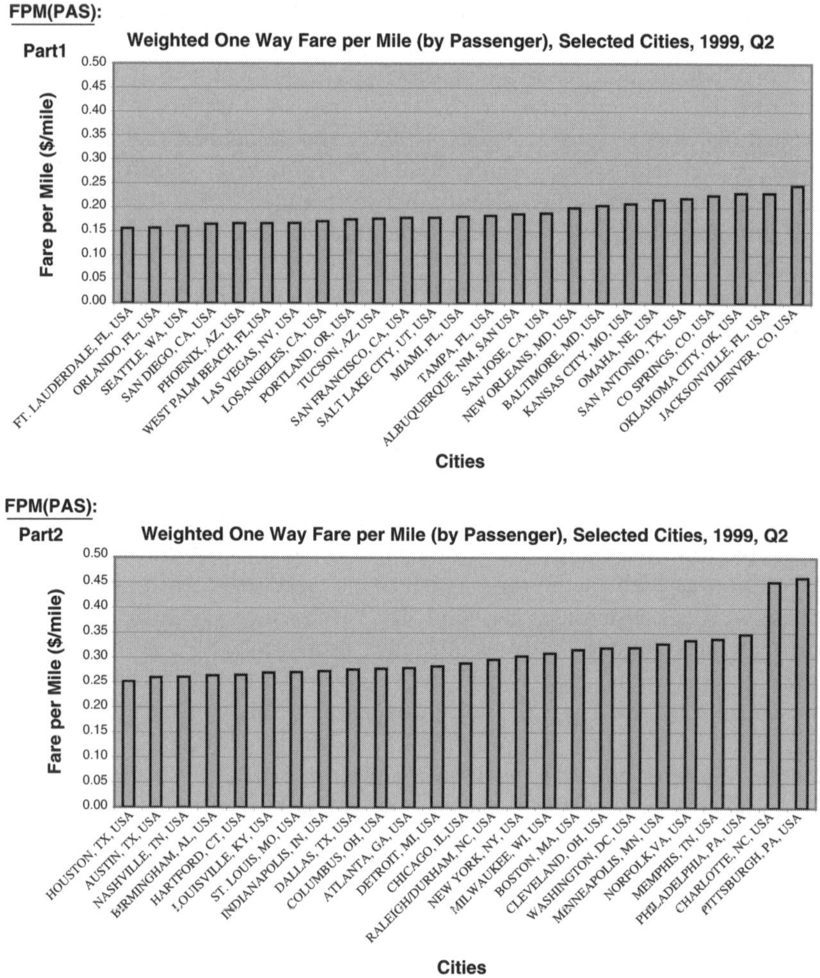

Fig. 1. Average fares per mile weighted by observed number of passengers from the Domestic Airline Fares Consumer Report (2nd Quarter, 1999).

- Airfares are lower in heavily traveled markets.[3] This may be true for two reasons: first, heavily traveled markets have higher load factors and second, larger planes may be used in more heavily traveled markets, yielding scale economies.

- Airfares are lower in markets with more competition. The truth of this hypothesis depends naturally on the presence or absence of a "contestability" effect. As we will see, testing this hypothesis is not simple because there are a number of different ways to define competition.
- The actual or potential presence of Southwest Airlines in a market has a negative impact on fares.
- Airfares to and from hubs are lower, after controlling for the level of competition.[4] This arises from higher load factors, use of larger planes, economies of scale in ground operations etc., which yield cost savings which may be passed on to the consumer in a competitive market. Different airlines have different hubs, so we expect each airline to offer lower fares from and to its hubs than it would to other locations.
- Round trip fares are lower than one-way fares.

Data Resources

Our analysis is based on data from the *Origin and Destination Survey* provided by the Office of Airline Information of the Bureau of Transportation Statistics. This survey comprises a 10% sample of all domestic airline tickets. Each record represents an individual ticket (or several tickets purchased together) with information on carrier, fare, route (origin-destination with intermediate stops), class, and restriction category. Our analysis is limited to records representing the most common class/restriction category: "restricted coach class".[5] We also limited our analysis to the top 1,000 domestic origin-destination pairs. We screened the data set for very low fares that were obviously indicative of frequent-flyer or employee purchases and eliminated them. The reason for this is that such fares are likely to fall disproportionately on certain vacation-oriented routes and might therefore bias estimates of the impact of certain variables such as distance on the pricing decisions of the airlines. As discussed below, it is important to take the elimination of these records into account when interpreting the results of our analysis.

We developed our data set from two files comprising all *Origin and Destination Survey* records for 1995 and 2000. The total number of observations in the data set were over 4.5 million for 1995 and over 5.2 million for 2000. To create the observations used in our regression analysis, we aggregated records within routes and carriers. That is, each observation represents service provided by a single carrier for a single origin destination pair.[6] Since the average number of carriers providing services in any given market is about five and we consider only the 1,000 largest origin-destination pairs, our estimations are based on about 5,000 observations in each year.

Our dependent variable is *average fare* $\bar{f}_{ijc}$ for carrier c in the market defined by origin i and destination j. Independent variables include the *total passengers* at the origin and destination airports p_i and p_j, and the *average distance* d_{ijc}. The passenger values are entered as proxies for market densities.[7] The distance is measured as total air travel miles, which may vary across records where different intermediate stops are included. Therefore the average may vary across carriers with different hub locations.

The degree of competition is inversely related to market concentration, which may be measured in a number of ways. The market position of any particular airline may be measured by its market share. Presumably, an airline with a very high share is able to charge something approaching monopoly prices. However, the share of a single airline does not necessarily give a good indication of the overall level of concentration in the market. The Herfindahl index, which is better for this purpose, is defined:

$$I_{ij} = \sum_{c} s_{ijc}^2$$

where s_{ijc} is the share of carrier c in the ij market. We include both the market share and the Herfindahl index in the regression equations as the former represents a carrier-specific concentration measure while the latter represents a general measure of concentration for the market.

Competitive conditions may be relevant both within markets and at airports. For example, some carriers may seek strategic advantage by establishing a dominant share of trips originating and ending in a particular airport. This is reflected in Fig. 1, where cities such as Charlotte and Pittsburg whose airports are dominated by a single carrier have some of the highest fares.

The specific effect of Southwest Airlines is captured by two dummy variables. $W^a_{ij} = 1$ if Southwest offers service in the ij market. To capture the effect of *potential* competition, $W^p_{ij} = 1$ if Southwest does not offer service in the market but has a presence in airports i and j such that it could easily enter the ij market and therefore represents a threat to other carriers in the market.

In order to capture a full range of competitive effects we include eight measures of competition in the regression equations: $s_{ijc}, s_{ic}, s_{jc}, I_{ij}, I_i, I_j$, W^a_{ij} and W^a_{ij}, where s_{ic} is the share of carrier c in all trips originating at origin i.

Table 1. Weighted Herfindahl Index.

Year	Airports	Markets
1995	0.424	0.561
2000	0.360	0.515

Table 1 provides a rough indication of trends in concentration from 1995 to 2000 based on weighted average Herfindahl indices at airports and in origin-destination markets. Not surprisingly, these indices are higher in markets than at airports. What is more surprising is that the indices decline by a fairly significant amount between 1995 and 2000. Despite the popular perception, airline activity is becoming less rather than more concentrated. The current wave of mergers and the demise of small carriers in the aftermath of September 11, however, may reverse this trend.

In order to capture the effect of hubs on air fares, we define a simple measure of "hubness" based on the ratio of connecting to total flights in the origin and destination airports i and j and a number of other factors (see the appendix for details). We identify each airport as either a hub or a spoke airport for each airline. Based on this measure we define two dummy variables: $H_{ic} = 1$ if the origin i is a hub for carrier c and $H_{jc} = 1$ if the destination j is a hub for carrier c. (Both equal zero otherwise.) Our hub assignments are shown in Table 2. One additional variable, R_{ijc}, represents the proportion of carrier c passengers on round trip flights in the ij market.

Parameters were estimated via ordinary least squares for a linear and linear-in-logarithms specification in each of the 2 years. In the linear version, the square of the average distance was added to make it possible to test the hypothesis that fares increase with distance but at a decreasing rate. In the logarithmic version the same hypothesis may be tested by observing whether the parameter on distance is less than one. ε_{ijc} is a residual term.

Linear specification

$$\bar{f}_{ijc} = \alpha + \beta_1 d_{ijc} + \beta_2 d_{ijc}^2 + \beta_2 p_i + \beta_3 p_j + \beta_4 s_{ijc} + \beta_5 s_{ic} + \beta_6 s_{jc} + \beta_7 I_{ij} + \beta_8 I_i + \beta_9 I_j + \gamma_1 W_{ij}^a + \gamma_2 W_{ij}^p + \gamma_3 H_{ic} + \gamma_4 H_{jc} + \gamma_5 R_{ijc} + \varepsilon_{ijc}$$

In the linear-in-logarithms specification, the dependent variable is the logarithm of the average fare and the right-hand side variables are the

Table 2. National Hubs.

1995		2000	
Carrier	Hubs	Carrier	Hubs
Trans World Express	St. Louis	Trans World Express	St. Louis
American Airlines	Dallas	American Airlines	Dallas
	Chicago		Chicago
Continental Airlines	Greensboro	Continental Airlines	Houston
	Houston		
Delta Airlines	Atlanta	Delta Airlines	Atlanta
	Cincinnati		Cincinnati
	Dallas		Dallas
	Salt Lake City		Salt Lake City
America West Airlines	Las Vegas	America West Airlines	Las Vegas
	Phoenix		Phoenix
Midway Airlines	Raleigh	Midway Airlines	Raleigh
		American Eagle	Dallas
			Chicago
Northwest Airlines	Memphis	Northwest Airlines	Memphis
	Minneapolis		Minneapolis
	Detroit		Detroit
		Comair	Cincinnati
		Horizon Air	Seattle
			Portland
Continental Express	Cleveland	Continental Express	Cleveland
	Houston		Houston
Trans World Airlines	St. Louis	Trans World Airlines	St. Louis
United Airlines	Denver	United Airlines	Denver
	Chicago		Chicago
US Airways	Charlotte	US Airways	Charlotte
	Pittsburgh		Pittsburgh
			Philadelphia

logarithms of all the dependent variables, except percent round trip and the hub and Southwest dummies.[8] η_{ijc} is a residual term.

Linear-in-logarithms specification:

$$\begin{aligned}\ln \bar{f}_{ijc} = {} & \delta + \delta_1 \ln d_{ijc} + \delta_2 \ln p_i + \delta_5 \ln p_j + \delta_4 \ln s_{ijc} + \delta_5 \ln s_{ic} + \delta_6 \ln s_{jc} \\ & + \delta_7 \ln I_{ij} + \delta_8 \ln I_i + \delta_9 \ln I_j + \mu_1 W^a_{ij} + \mu_2 W^p_{ij} + \mu_3 H_{ic} \\ & + \mu_4 H_{jc} + \mu_5 D_{ijc} + \eta_{ijc}\end{aligned}$$

RESULTS

Estimation results for linear and linear-in-logarithms specifications for 1995 are included in Table 3a and b, respectively. (See "Key to Variables" mentioned as footnote to Table 3 to read these tables.) Turning first to the linear specification, we see that the standard results of airline economics are confirmed: Fares increase with distance, but at a decreasing rate, as indicated by the negative sign on distance squared. The results do not, however, indicate the expected market density effect as the number of passengers at the origin and destination have unexpected positive signs which are statistically significant only in the case of the origin. One possible explanation is that the busiest airports have higher congestion and therefore it is more, rather than less, expensive for the airlines to serve them.

As for the measures of concentration, the results are more mixed. Both the market share and Herfindahl index are inverse measures of competitiveness – i.e., the higher their values the less competitive the market. We therefore expect these measures to have positive effects on fares. For the

Table 3. Estimation Results for Linear and Linear-in-Logarithms Specifications for 1995.

Variable	Parameter Estimate	Standard Error	t	PROB>$\|t\|$
(a) *Linear regression 1995: average fare*				
INTERCEP	−9.3995	4.3202	−2.1760	0.0296
DIST_AVE	0.1262	0.0035	35.7980	0.0001
DIST_SQR	−0.0000202	0.0000	−17.2780	0.0001
ORIG_SH	37.2792	4.6377	8.0380	0.0001
ORIG_H	−2.3159	3.6625	−0.6320	0.5272
DEST_SH	35.9050	4.1708	8.6090	0.0001
DEST_H	9.3378	3.4340	2.7190	0.0066
MARKET_SH	−5.9917	3.6783	−1.6290	0.1034
MARKET_H	43.9426	3.7376	11.7570	0.0001
PCT_ROUD	11.7854	3.3728	3.4940	0.0005
PSG_I	0.00003752	0.0000	8.5100	0.0001
PSG_J	0.00000128	0.0000	0.2940	0.7688
ORIG_HUB	16.6946	2.6606	6.2750	0.0001
DEST_HUB	17.7459	2.6907	6.5950	0.0001
SW_A	−65.9575	3.7042	−17.8060	0.0001
SW_P	−27.5778	1.3331	−20.6870	0.0001

$R^2 = 0.686$

Table 3. *(Continued)*

Variable	Parameter Estimate	Standard Error	t	PROB>|t|
(b) *Linear-in-logarithm regression 1995: average fare*				
INTERCEP	0.4928	0.1232	4.0000	0.0001
DIST_LN	0.5402	0.0076	71.1610	0.0001
O_SH_LN	0.0419	0.0045	9.3090	0.0001
O_H_LN	0.0107	0.0088	1.2120	0.2255
D_SH_LN	0.0395	0.0039	10.0920	0.0001
D_H_LN	0.0386	0.0083	4.6560	0.0001
M_SH_LN	−0.0189	0.0043	−4.3590	0.0001
M_H_LN	0.1128	0.0119	9.4450	0.0001
PCT_ROUD	0.0797	0.0234	3.4000	0.0007
PSG_I_LN	0.0638	0.0068	9.4030	0.0001
PSG_J_LN	0.0187	0.0070	2.6890	0.0072
ORIG_HUB	0.1188	0.0179	6.6360	0.0001
DEST_HUB	0.1056	0.0184	5.7490	0.0001
SW_A	−0.6062	0.0250	−24.2900	0.0001
SW_P	−0.2050	0.0090	−22.8640	0.0001
Adjusted $R^2 = 0.759$				

Key to Variables

FARE_AVE	average fare (for class X only)
FARE_LN	logarithm of FARE_AVE
DIST_AVE	average flight distance (over both direct and indirect flights)
DIST_LN	logarithm of DIST_AVE
DIST_SQR	square of DIST_AVE
ORIG_SH	market share of the airline at the origin airport
O_SH_LN	logarithm of ORIG_SH
ORIG_HI	Herfindahl index at the origin airport
O_HI_LN	logarithm of ORIG_HI
DEST_SH	market share of the airline at the destination airport
D_SH_LN	logarithm of DEST_SH
DEST_HI	Herfindahl index at the destination airport
D_HI_LN	logarithm of DEST_HI
MARKET_SH	market share of the airline on the route
M_SH_LN	logarithm of MARKET_SH
MARKET_H	Herfindahl index of the route
M_H_LN	logarithm of MARKET_H
SW_A	actual competition from Southwest
SW_P	potential competition from Southwest
PCT_ROUD	round trip flights percentage on the route
PSG_I	total passengers from airport I
PSG_J	total passengers to airport J
PSG_I_LN	logarithm of PSG_I
PSG_J_LN	logarithm of PSG_J
ORIG_HUB	dummy variable, = 1 if the origin is the airline's hub
DEST_HUB	dummy variable, = 1 if the destination is the airline's hub.

most part this is true. In those cases where the parameter estimate is negative we cannot reject the hypothesis that the true value is 0 with a high degree of confidence in the linear specification. The only perverse sign that is significantly different from zero is on the market share in the linear-in-logarithms specification. The fact that most of these parameters have the expected positive effect attests to importance of measuring competitiveness on a number of different dimensions.

The percentage of round trips has an unexpected positive effect on fares in the 1995 result. We believe this is because in 1995 Southwest Airlines classified all of its reported tickets as one-way tickets. Since Southwest fares are low this may have given a false indication that one-way fares are lower. In 2000 (see below), Southwest reported a large proportion of its tickets as round trips and the perverse effect was not present in our results.

The Southwest competition variables have large, significant negative values. In effect, the result in the linear model implies that fares in a particular market are lower by $66 if Southwest offers service in that market. Just the threat of competition from Southwest reduces fares by $27.[9]

The positive sign on the parameters of the hub dummies, a result that contradicts our prior hypothesis, repeats itself throughout our analyses. Theory tells us that costs are lower for trips to and from hubs, but this result suggest that trips whose origin or destination is a hub for the carrier in question have higher fares. We will discuss this result at some length below.

The linear-in-logarithms specification for the most part echoes the results of the linear specification. The fact that fares increase with distance at a decreasing rate is confirmed here by the fact that the parameter estimate for distance is less than one. Since the parameters in this specification can be read as elasticities, we get a better feel for the magnitude of the effects. For example, the elasticities of most of the concentration measures (with the exception of the market Herfindahl index) are fairly low. All the concentration measures take the expected positive signs, but those that take negative signs in the linear specification are not significantly different from 0 in the linear-in-logarithms specification. The unexpected round trip and hub effects are consistent with the linear specification.

The results for 2000, which are presented in Table 4a and b, are broadly consistent with the results for 1995, with a few noteworthy differences. This is not surprising given that both years are within the envelope of the same period of economic expansion in the national economy and relative prosperity in the airline industry. Distance effects are consistent with 1995. The negative Southwest effects and positive hub effects are still present and still highly statistically significant. The perverse positive round trip

Table 4. Estimation Results for Linear and Linear-in-Logarithm Regression for 2000.

Variable	Parameter Estimate	Standard Error	t	Prob>$\|t\|$
(a) *Linear regression 2000: average fare*				
INTERCEP[a]	68.3502	4.7744	14.3160	0.0001
DIST_AVE	0.0903	0.0031	28.7410	0.0001
DIST_SQR	−0.0000086	0.0000	−9.3710	0.0001
ORIG_SH	21.7189	5.3767	4.0390	0.0001
ORIG_H	−8.6731	4.4911	−1.9310	0.0535
DEST_SH	9.4615	4.7392	1.9960	0.0459
DEST_H	14.4954	4.7118	3.0760	0.0021
MARKET_SH	25.7126	4.1351	6.2180	0.0001
MARKET_H	16.8657	4.1369	4.0770	0.0001
PCT_ROUD	−33.2592	3.6538	−9.1030	0.0001
PSG_I	0.00001547	0.0000	3.5490	0.0004
PSG_J	−0.00001505	0.0000	−4.5710	0.0001
ORIG_HUB	13.8666	2.6479	5.2370	0.0001
DEST_HUB	17.7754	2.6422	6.7280	0.0001
SW_A	−48.3077	2.5495	−18.9480	0.0001
SW_P	−25.3592	1.3521	−18.7560	0.0001
$R^2 = 0.579$				
(b) *Linear-in-logarithm regression 2000: average fare*				
INTERCEP	1.2816	0.1091	11.7490	0.0001
DIST_LN	0.4927	0.0068	71.9970	0.0001
O_SH_LN	0.0080	0.0039	2.0560	0.0399
O_H_LN	0.0075	0.0082	0.9140	0.3606
D_SH_LN	−0.0102	0.0033	−3.1200	0.0018
D_H_LN	0.0266	0.0083	3.2080	0.0013
M_SH_LN	0.0404	0.0037	11.0410	0.0001
M_H_LN	0.0555	0.0101	5.4970	0.0001
PCT_ROUD	−0.2085	0.0207	−10.0720	0.0001
PSG_I_LN	0.0466	0.0060	7.8170	0.0001
PSG_J_LN	0.0029	0.0056	0.5210	0.6022
ORIG_HUB	0.0999	0.0150	6.6700	0.0001
DEST_HUB	0.1312	0.0151	8.6770	0.0001
SW_A	−0.3282	0.0140	−23.4360	0.0001
SW_P	−0.1463	0.0076	−19.1350	0.0001
Adjusted $R^2 = 0.687$				

[a]Refer Table 3 for "Key to Variables."

effect found in the 1995 result has changed to the expected negative effect. The results for the concentration measures is once again mixed with most signs positive as expected.

For a number of concentration measures we cannot reject the hypothesis that their effect on fares is zero. This does not necessarily mean, however, that their impact on fares is very large. For example, based on the results in Table 3b, if a carrier's share in a market increases by 10%, its fare increases by only 0.5%. Naturally, such an increase would trigger sympathetic but smaller increases in the origin and destination shares and in the Herfindahl indices, but the total effect would still be less than a 1% increase. Thus, the results indicate that the effect of general competition is significant but not extremely large. The effect of competition specifically from Southwest Airlines, however, appears to be large in all of our results.

Discussion of the Hub Effect

Our prior expectation was that an airline will provide lower cost fares to and from its hub than other airlines can provide to the same airport. For example, if Delta Airlines and U.S. Airways were to offer service from Boston to Atlanta (which is Delta's hub), Delta could enjoy greater scale economies by combining Boston–Atlanta passengers with Boston–New Orleans passengers, and Boston–Memphis passengers with Boston–Houston passengers, etc. Since U.S. Airways does not have a hub in Atlanta, it can only serve Boston–Atlanta passengers on that route.

The results, however, indicate the opposite: After controlling for other factors, fares to and from an airport are higher if the carrier maintains a hub there. At first glance, one might conclude that this reflects the competitive position of the hub carrier in that market. Also, the effect may be attributable to the fact that Southwest Airlines generally avoids hub airports. However, after controlling for competitive effects in a number of ways, the hub effect is still evident. At least two other explanations seem plausible.

The first is that there are qualitative differences between services provided by the hub carrier and non-hub carriers for which travelers are willing to pay a premium. For example, the hub carrier may offer more frequent service or service at more convenient times. Since there is no schedule data in the *Origin and Destination Survey* data base, we are not able to test this hypothesis explicitly.

The second, which applies more to the origin hub effect than the destination hub effect, is that travelers who regularly use a hub airport will benefit from frequent-flyer programs if they consistently use the hub airline. The hub airline can charge a premium for this brand loyalty which in part is offset by the cost of providing a large number of frequent flyer trips. From the traveling public's perspective, the higher fares at the hub are offset by frequent-flyer benefits, so their effective cost of air travel is not necessarily higher if they frequent hub airports.[10]

In assessing the role of hubs on fares, it is also important to bear in mind that in this relatively short time window (1995–2000) we may be observing a strategic process of competition among airlines rather than a competitive equilibrium. Large airlines may be able to seize competitive advantage in the short run by employing H-S networks, but in the future when (and if) the hub strategy becomes universal the results may be very different and more of the efficiency gains may be passed on to the traveling public.

Whatever the reason, the results suggest that establishment of a hub is a mixed blessing for regular users of the airport in question. The presence of the hub ensures a high degree of connectivity, including a relatively large number of direct flights and may make it easier for travelers to take advantage of frequent-flyer benefits. The fares those passengers pay, however, may not be any lower than at a non-hub airport. In fact, our results suggest the opposite is true.

LESSONS FROM THE U.S. EXPERIENCE

Both the historical narrative of the post-deregulation U.S. airline industry and the statistical analysis presented here provide some useful lessons that may be applied to liberalization of the global air transportation industry. The most significant lessons come in two interrelated areas: the dynamics of competition and the emergence of H-S networks.

The greatest fear expressed by critics of deregulation is that bankruptcies and mergers will lead to rapid industry concentration which in turn will lead to higher fares. The American experience is mixed on the link between deregulation and industrial concentration. Since deregulation, the industry has gone through at least two periods of significant concentration and may be about to embark on another. There have also been periods, however, when the industry became less concentrated due the entrance of low cost airlines, some serving niche markets and others challenging the industry leaders. Our data indicate that the most recent trend (1995–2000) has been of decreased concentration, as indicated by Table 1. One can hardly conclude from the American experience that concentration is an inevitable outcome of deregulation.

The impact of opening up global markets, however, may be different because of the presence of a number of national flagship carriers that only survive because of subsidies and national access restrictions. Because of the small size of their national and international markets, these carriers may be unable to achieve the economies of scale and scope to make them competitive, especially with airlines from large countries or airlines that have already established a strong position in international markets. Free trade in global air services may therefore have a more profound culling effect than did deregulation in the U.S.

As for the question of whether concentration necessarily leads to higher fares, there is still considerable debate in the U.S. Our statistical results suggest that the effect of concentration on fares is real, but it is not so large that it is likely to swamp the effect of efficiency improvements in a liberalized market.

Truly multilateral liberalization of international air travel markets, including rights of cabotage, may make it possible for low cost airlines to operate on a regional or even global scale. (The rise of low cost carriers in the European Union is evidence of this possibility.) Based on the U.S. experience, this new element of competition may have much greater implications for the future cost of air travel than the general level of concentration.

One of the most critical issues for international deregulation is how and where hub airports will emerge. There is already some important work examining the likely location of international hubs (Takebayashi & Kuroda, 2002). While it is beyond the scope of this paper to suggest where international hubs will be located, we can say something about the economic implications for the countries in which they locate. Hubs will confer a high level of connectivity and the convenience of frequent and direct flights on travelers in their home regions. They will also make those regions more accessible at a global scale, which may be a boon to business. There is no evidence, however, that they will make air travel any cheaper for those regions – in fact they may have the opposite effect. For those countries and regions that do not get hubs, the future is more uncertain. Some may actually benefit from more frequent service via a hub in another country, while others may lose a number of direct services.

NOTES

1. There is one important exception: Southwest Airlines has no prominent hub airline.

2. The Federal Aviation Administration (FAA) has the following size classification scheme: a large hub is a city and its metropolitan area that enplanes more than 1% of the total number of U.S. domestic airline passengers per year; a medium hub enplanes between 0.25 and 0.999%, a small hub enplanes between 0.05–0.249%, and a non-hub enplanes less than 0.05%. Although this definition is not based on hub's function, it has been used in several analyses such as in Goetz and Sutton (1997) and Button (2002). Some researchers have tried to develop other measures of "hubness" based on the hub's functionality. For example, Toh and Higgins (1985) calculate a hub index "by dividing the number of outlying cities served by the hub by the number of spokes radiating from it," and McShan and Windle (1989) provide an index which equals the proportion of an airline's total departures leaving from the 3% most utilized airports in its network. More recently, Bania et al. (1998) try to develop an index using cluster analysis on a value calculated from route connections. The USGAO (1990, 1993) defined a concentrated market as one where one airline accounted for at least 60% of the enplaning passengers or two airlines had at least 85%. The USDOT (1990) used a one-firm market share of 70% to identify concentrated hub fare premiums.

3. We refer to each origin–destination fare as a "market." This is different from a "route," which implies a specific way of getting from an origin to a destination (direct or through one or more intermediate airports).

4. While this may seem at odds with empirical findings described earlier, the key phrase here is "after controlling for competition." Hub dominance may lead to a high degree of concentration, but by controlling for this we may be able to identify a positive efficiency effect.

5. Southwest Airlines classifies its lowest priced fares as "restricted business class" rather than "restricted coach." We regard this as a semantic distinction only, and include those tickets in the data set.

6. We elected not to conduct our analysis on the individual traveler records for two reasons. First, most of the variation across fares offered by a single carrier within a market depends on time-of-day and day-of-week variations. Since the *Origin and Destination Survey* contains no schedule information, we have no basis for explaining this deviation. Second, the values of the independent variables would be identical for all observations within the same carrier/market combination.

7. Entering densities directly in the fare equation would imply a simultaneous equation bias.

8. Logarithms for these variables could not be taken since they have zero values for some observations.

9. This is consistent with the results of Morrison (2001), who also found that Southwest service to a nearby airport also reduces fares.

10. As noted earlier, frequent flyer flights have been eliminated from our data set.

REFERENCES

Anderson, W. P., Gong, G., & Lakshmanan, T. R. (2002). Geographical variation in the cost of air travel: Analysis of the Domestic Airline Fare Consumer Report. *Transportation Research Record, 1788*, 13–18.

Antoniou, A. (1991). Economies of scale in the airline industry: The evidence revisited. *Logistics and Transportation Review, 27*, 159–184.

Bailey, E. E., Graham, D. R., & Kaplan, D. P. (1985). *Deregulating the Airlines*. Cambridge, MA: MIT Press.

Bailey, E. E., & Williams, J. R. (1989). Sources of economic rent in the deregulated airline industry. *Journal of Law and Economics, 31*, 173–202.

Baker, S. H., & Pratt, J. B. (1989). Experience as a barrier to contestability in the airline markets. *The Review of Economics and Statistics, 71*, 352–356.

Bania, N., Bauer, P. W., & Zlatoper, T. R. (1998). US air passenger service: A taxonomy of route networks, hub locations, and competition. *Transportation Research, 34E*, 53–74.

Barnett, A., Curtis, J., Goranson, J., & Patrick, A. (1992). Better than ever: Nonstop jet service in an era of hubs and spokes. *Sloan Management Review, 33*, 49–54.

Borenstein, S. (1989). Hubs and high fares: Dominance and market power in the US airline industry. *Rand Journal of Economics, 20*, 344–365.

Borenstein, S. (1992). The evolution of U.S. airline competition. *Journal of Economic Perspectives, 6*, 45–73.

Butler, R. V., & Huston, J. H. (1990). Airline service to non-hub airports ten years after deregulation. *Logistics and Transportation Review, 26*, 3–16.

Button, K. (2002). Debunking some common myths about airport hubs. *Journal of Air Transport Management, 8*, 177–188.

Call, G. D., & Keeler, T. E. (1985). Airline deregulation, fares, and market behavior: Some empirical evidence. In: A. F. Daughety (Ed.), *Analytic studies in transport economics* (pp. 221–247). Cambridge, UK: Cambridge University Press.

Evans, W. N., & Kessides, I. N. (1993). Localized market power in the US airline industry. *The Review of Economics and Statistics, 75*, 66–75.

Goetz, A. R. (1993). Geographic patterns of air service frequencies and pricing at US cities. *Journal of the Transportation Research Forum, 33*, 56–72.

Goetz, A. R. (2002). Deregulation, competition, and antitrust implications in the US airline industry. *Journal of Transport Geography, 10*, 1–19.

Goetz, A. R., & Sutton, C. J. (1997). The geography of deregulation in the US airline industry. *Annals of the Association of American Geographers, 87*, 238–263.

Goetz, A. R., & Vowles, T. M. (2000). 'Pockets of Pain' across the deregulated landscape: The geography of US airline fares and service in the 1990s. Paper presented at annual meeting of the Association of American Geographers, Pittsburgh.

Graham, B. J. (1993). The regulation of deregulation: A comment on the liberalization of the UK's scheduled airline industry. *Journal of Transport Geography, 1*, 125–131.

Hurdle, G. J., Johnson, R. L., Jaskow, A. S., Werden, G. J., & Williams, M. A. (1989). Concentration, potential entry, and performance in the airline industry. *Journal of Industrial Economics, 38*, 119–139.

Kanafani, A., & Ghobrial, A. A. (1985). Airline hubbing – some implications for airport economics. *Transportation Research, 19A*, 15–27.

Levine, M. E. (1987). Airline competition in deregulated markets: Theory, firm strategy, and public policy. *Yale Journal on Regulation*, *29*, 393–494.

McShan, S., & Windle, R. (1989). The implications of hub-and-spoke routing for airline costs and competitiveness. *Logistics and Transportation Review*, *25*, 209–230.

Morrison, S. (2001). Actual, adjacent and potential competition: Estimating the full effect of Southwest Airlines. *Journal of Transport Economics and Policy*, *35*, 239–256.

Morrison, S., & Winston, C. (1986). *The economics effects of airline deregulation.* Washington, DC: The Brookings Institution.

Morrison, S., & Winston, C. (1987). Empirical implications and tests of the contestability hypothesis. *Journal of Law and Economics*, *30*, 53–66.

Morrison, S., & Winston, C. (1989). Enhancing the performance of the deregulated air transportation system. *Brookings Papers on Economic Activity: Microeconomics*, 61–112.

Oum, T. H., Zhang, A., & Zhang, Y. (1995). Airline network rivalry. *Canadian Journal of Economics*, *28*, 836–857.

Sampson, R. J., Farris, M. T., & Schrock, D. L. (1990). *Domestic transportation: Practice, theory and policy* (6th ed.). Boston, MA: Houghton-Mifflin Co.

Stiglitz, J. E. (1987). Technological change, sunk costs, and competition. *Brookings papers on economic activity: Microeconomics*, 883–937.

Strassman, D. L. (1990). Potential competition in the deregulated airlines. *The Review of Economics and Statistics*, *72*, 696–702.

Toh, R. S., & Higgins, R. G. (1985). The impact of hub and spoke network centralization and route monopoly on domestic air profitability. *Transportation Journal*, *24*, 16–27.

U.S. Department of Transportation. (1990). *Airports, airtraffic control, and related concerns. Office of the secretary of transportation, secretary's task force on competition in the U.S. domestic airline industry*, Washington, DC.

U.S. General Accounting Office. (1990). *Airline competition: Higher fares and reduced competition at concentrated airports.* Washington, DC.

U.S. General Accounting Office. (1993). *Airline competition: Higher fares and less competition continue at concentrated airports.* Washington, DC.

U.S. General Accounting Office. (1996). *Airline deregulation: Changes in airfares, services, and safety at small, medium-sized, and large communities.* Washington, DC.

U.S. General Accounting Office. (1999). *Airline deregulation: Changes in airfares, service quality and barriers to entry.* Washington, DC.

APPENDIX: IDENTIFYING HUBS

In the process of identifying hubs for airlines, we first looked at the ratio of connection passengers to the total passenger volume and used it as a base indicator. The cut-off point we selected is set at 30%, which is quite low. However, since this is only the first step to find candidate hubs, a higher percentage might prematurely disqualify some airports.

After defining a list of hub candidates for each airline, we next looked at the passenger volume by airline at airports. An airline should have a relatively large number of passenger volumes at its hubs. Therefore, an airport with a large connecting passenger ratio for an airline may not be that airline's hub. For example, in 1995, Northwest had a 60% connecting passenger ratio at Palm Springs, CA, but with a total passenger volume of only 5, Palm Springs could not be a hub for Northwest.

When comparing the passenger volume across airports by airline, distinctive gaps helped distinguish national from regional hubs. For example, for American Airlines in 1995, Nashville and Miami had relatively high connecting passenger ratios, but there was a big gap between the volumes at these airports and the volumes at Dallas-Ft. Worth and Chicago, so we only included the latter two in our list of national hubs.

This ad hoc methodology produced the set of hubs shown in Table 2. While there is some subjectivity in the classification criteria, we think it is superior to the Department of Transportation hub definitions, which are based on total passenger flow alone.

HOW STABLE ARE SCHEDULED AIR TRANSPORT MARKETS?

Kenneth Button

ABSTRACT

There has been a global trend to remove many of the traditional regulations governing the way that airlines services are provided. This has been a response to the demonstrable inefficiencies that accompanied such things as rate controls and market capacity limits. The outcome has largely been seen as beneficial to air transport users. In recent years, however, many airlines in liberalized markets have suffered large financial losses as excess capacity have emerged. This has been in the context of generally healthy financial positions of other actors (e.g. airports and global distribution systems) further back in the value chain. The difficulty stems from a lack of fully competitive markets throughout the air transportation supply chain together with the nature of competition that has emerged in the airline market itself. In particular, the need to offer a scheduled departure imposes fixed costs for each airline service that cannot be fully recovered in competitive, and possibly contestable, condition.

INTRODUCTION

In a previous book, edited jointly with Dennis Swan over 15 years ago (Button & Swan, 1989), we called the period from the 1970s 'the

Global Competition in Transportation Markets: Analysis and Policy Making
Research in Transportation Economics, Volume 13, 27–48

ISSN: 0739-8859/doi:10.1016/S0739-8859(05)13002-9

age of regulatory reform.' There seems no reason to change this view of that period and, indeed, if anything the subsequent decade or so has seen an even more dramatic global retreat of the state from economic regulation. The question addressed here is not that of defining turning points or refining the nuances of historical cycles, but rather to assess whether the implications of more liberal markets (in the British sense of the term) have proved as beneficial as advocates hoped. In doing this, the coverage is far from comprehensive.

The paper is not concerned, for example, with some of the more dramatic collapses of companies that have taken place following liberalization – this would have been expected in the transition to more market-driven environments. It is also not about short-term crises of the type created by the September 11 attacks in the United States. Nor does it focus on the apparent lack of morality that is emerging in some liberalized contexts – after all, regulated and nationalized industries have never been examples of puritanical ethics – but rather the concern is with whether reforms have always generated the welfare gains anticipated. More particularly, this paper looks at the sustainability of deregulated markets, with air transport being the specific sector of attention.

The liberalization of the U.S. domestic air transport market was the first major reform to attract both popular and academic attention.[1] Air transport is again the subject here, although the geographical confines are wider. Airlines have also been a focus for regulatory reform in many other markets, placing importance on experiences of both depth and breadth when it comes to sector analyses. The paper, however, focuses primarily on the United States, the European Union (EU) and transatlantic markets to impose some boundaries.

It also limits itself, after providing some background information and analysis, to two particular, although interconnected, aspects of the structure for air transport services. These both relate to matters that have arisen in the context of sustainability, with some imbalance in the level of treatment. The first matter concerns what is becoming known as the value chain problem. Namely, that in the full production chain that leads to the delivery of passenger and freight transport services, there are links that extract excessive rents, whereas others, and most notably the airlines, on average fail to cover their full attributable long-run costs. The implications of this are that the final delivery costs of airlines' services are higher than if all elements in the value chain were subjected to competitive pressures.

The second aspect is the competitive process itself for scheduled services. In practice, markets are neither independent of the institutional context in which they function, nor are they independent of the intrinsic nature of the technical cost and demand functions confronting suppliers. It is only in particular situations that a market will produce a sustainable outcome, and what is becoming apparent in the airline industry is whether the underlying market structure is intrinsically unstable, and whether government interventions are adding to any problems that already exist. This topic has attracted a certain amount of academic interest over the years, but was perhaps first brought into the larger arena by Smith (1995) in an article in *Fortune*. More recently, contributors to the *Wall Street Journal* have taken up the theme, "[there] is a growing band of economists, inspired by the work of Lester Telser at the University of Chicago, [that] has begun to question antitrust's primitive intolerance of cooperative acts by competitors. An important spur to the new thinking has been the visible shambles of the US airline industry" (Jenkins, 2002).

THE LIBERALIZATION MOVEMENT AND THE STATE OF THE AIRLINE INDUSTRY

The past 25 years has seen an increased liberalization of air transport markets. A trend that began with the U.S. domestic market in the mid-1970s has spread to many other national, regional, and international markets. The evidence to date suggests that while there are some variations and short-term difficulties, this liberalization has resulted in lower and more varied fare structures and a greater diversity in the types of air services that are provided. Significantly more people are also flying and more cargo is being carried, and the average customer seems to have benefited in the short term. Additionally, airlines have responded to the new institutional environment with innovative measures to capture and retain market share (e.g. frequent flyer schemes, yield management, computer sales, hub-and-spoke networks, and CRS systems). Numerous econometric and programming studies have also highlighted increases in the productivity of airlines. Economies on the cost side (e.g. of scale and scope) have helped in this as management practices have been tightened and aircraft technology and air transportation infrastructure has improved.

There is no effort here to survey the substantial technical literature in this field or to detail the spread of regulation; an overview can be found in Button and Stough (2000) for those unfamiliar with it. What is being highlighted is that as the longer-term implications of deregulation unfold, the airline industry in many markets is proving itself unstable. The industry is simply finding it impossible to recover its full costs over the business cycle.

The current situation is that airlines are in many markets suffering serious financial losses. Estimates from the International Air Transport Association (IATA) indicate that after surveying its 280 members, global losses amounted to $12 billion in 2002, with subsequent losses amounting to over $6 billion for 2003. The losses, however, have not been spread evenly. The U.S. domestic market was particularly severely affected by the events of September 11th, its after shock on travelers' confidence and economic activity. The IATA indicates that U.S. airlines lost about $6 billion in 2002 after losses of $7.7 billion in 2001. In more recent times this has largely been the result of low revenue flows as passenger traffic has grown to its pre-2001 levels. Other markets such as the North Atlantic have also been seriously affected. The air cargo market generally has recovered faster than most passenger markets. This picture, however, should be set within the context of longer-term trends in the financial conditions of the airline industry.

As with most industries, the airlines are sensitive to trade-cycle effects, and their net revenues fluctuate considerably between booms and slumps (Fig. 1). While this is to be expected, service industries are indeed often more volatile in this sense than some other sectors, and there is evidence that airlines as a whole are earning lower rates of return than industry in general over the cycle. This comes most notably from the U.S. domestic market, but also the more recently liberalized intra-European market. The return on capital is about 2–3% below the overall industrial average.[2] Linked to this, they are also not recovering their full costs across the trade-cycle. There is also evidence of frequent market entry and exit by airlines, which is particularly apparent at the route and service level, as may be expected, as carriers seek out profitable services in changing market conditions, but airlines also have a high rate of bankruptcy and of reformation. These are not the features of what would normally be seen as a sustainable industry.

One aim of the paper is to examine the reasons why this may be so, with a particular focus on the problems of common cost recovery in a network industry such as air transport. It looks at a variety of different economic models that may be used to explain the problem and can offer insights into possible solutions. It also provides an explanation for why the airlines have adopted some of their managerial strategies other than just focusing on the traditional one of 'rent' extraction. The paper argues that the key problem is that airline markets are inherently unstable if the institutions governing the markets in which they operate are inappropriate.

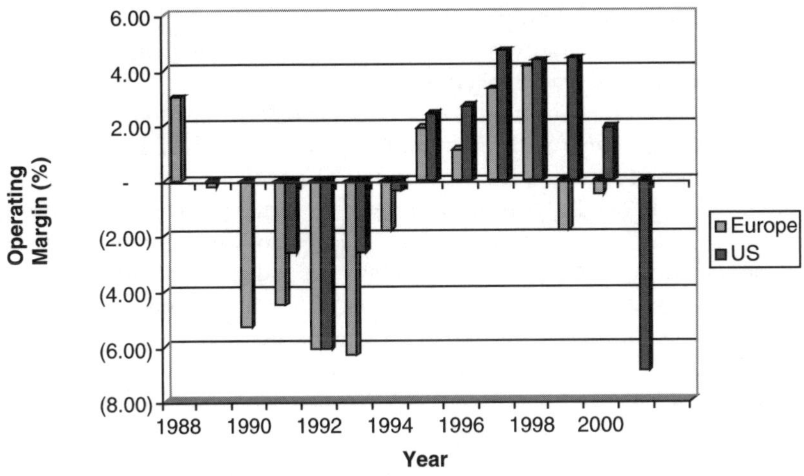

Fig. 1. Operating Margins after Interest for U.S. and European Carriers as a Percentage of Operating Revenue. *Note:* The U.S. data refers to airlines that are members of the Air Transport Association (ATA) and the European data to airlines that are members of the Association of European Airlines (excluding KLM, Finnair, Air Malta, and BMI from 1997 to 2000 and excluding Lufthansa Cargo from 1999).

THE STRUCTURE OF SCHEDULED SERVICES

The hub-and-spoke system is a long-standing structure in many transport markets, including air transport. What has changed in the air transport market since deregulation is the way that hubs are used. In the pre-1978 era in the United States, major domestic hubs already existed (Atlanta and Chicago being examples), but they served inter-lining traffic rather than online operations. This meant transferring between carriers was common with all the discontinuities this involved. At the international level, bilateral structures of air service agreements tended to focus flights through a limited number of gateways – often just one. The national flag carrier of small and medium-sized nations – as is the case for most of the EU – usually controlled about 50% of the slots under this arrangement.

Focusing on passenger airline networks, heavily regulated markets from the 1930s through to the end of the 1970s marginalized network debates. Routes were awarded to carriers on a city-pairs basis with little regard to the network economies of supply. In the U.S. domestic case, a carrier was effectively confronted with a constrained supply function when requesting new service licenses and only took network features into account in terms of instances where fleet and equipment could be used on small, inter-linked sub-sections of the carrier's overall operations.

This system limited the ability of airlines and their customers to reap the benefits that can come with interconnectivity – the ability to link services through a hub. These involve a potpourri of positive network externalities (Economides, 1996). In terms of user benefits, larger networks can offer the advantage of access to a much greater number of destinations, and often greater service frequency. Additionally, any individual user helps ensure that this range of options is available to others. What this means in terms of an individual user of an air transport network is that, by paying the full attributable cost of using the service, the user helps ensure that the service is also available for other potential users as well as benefiting from their support of the network. These benefits are not limitless but bounded by the congestion that emerges as hubs become more heavily used and air traffic controlled corridors reach capacity.

The deregulation of the U.S. domestic market led to a very rapid adoption of hub-and-spoke operations by the larger airlines as they sought to exploit potential economies of scale, scope and density that the freer market situation permitted. There also came the appreciation that network services provided economies of network presence that can stimulate demand and add to revenue flow. While this has partly been determined by the nature of demand and cost considerations, it has also often been strongly influenced by the institutional structure in which airlines operate. With this form of hubbing structure, flights are funneled in banks into a number of large hubs where substantial numbers of passengers change aircraft to complete their journeys. These banks involve the coordinated arrival of a large number of flights in a short space of time and then an equally coordinated departure of flights within a narrow time window. Larger hubs may well have up to 10 or more such banks a day. Travel time would be longer for many people but fares fell, and the frequency and the range of potential flight combinations available to any particular destination expanded considerably. In recent years, with the flexibility that freer market conditions have allowed, networks have become very dynamic with links being continually added and removed.

As domestic airlines have consolidated through mergers, and have grown with the market, so the form of the hub-and-spoke structure has evolved. The growth in international services, coupled with the move toward strategic alliances, has also influenced the pattern that has emerged in the late 1990s. For example, there has been a growth in multiple-hub operations.

This type of structure can occur when an airline seeks to serve a large market such as the entire United States. It may then locate its central transit points at two or more locations to maximize market penetration and avoid high levels of congestion at a single hub.

International alliances have led to de facto multi-hub services with a carrier in, say, Europe basing its operations around a hub but linking its services with those of a U.S. airline that has its own domestic hub-and-spoke network. Such structures grew in importance in the 1980s and 1990s as airlines sought to offer interconnecting services from origins within their base country to airports within the destination country not designated as 'gateways' under prevailing bilateral Air Service Agreements (ASAs). The restrictive ASAs that typified international aviation until recently effectively institutionalized hubs in many countries by giving monopoly rights to the flag airline. The service between national hubs would in this case be a code-shared operation involving a carrier from each country. With liberalization and the revised hub-and-spoke systems, the primary focus of competition is between the alternative networks rather than over individual links.

Not all airlines operate hub-and-spoke systems, or possibly more correctly do not operate them quite as intensely as most of the U.S. major carriers. One reason for this is that there are very dense routes where 'shuttle' styles of service meet demand considerations most effectively. These are very frequent services with the aircraft simply shuttling back-and-forth between two points. In the United States, major carriers such as United separate such services as individual operating units.

Southwest is often held up as an example of a non-hub operator in the U.S. domestic market, although it does in practice have coordinated services through a number of secondary hubs in addition to its non-hub-and-spoke operations. It, like its more recent EU emulators Ryanair and easyJet, keeps its costs down by offering minimal on-plane amenities and by selecting routes where it can turn its aircraft around quickly, thus maximizing the time its fleet can be in the air generating revenue. Route networks are selected so that, by picking up and delivering combinations of passengers at each airport, it can enjoy a high-load factor. In other cases, services may be freestanding and not be part of a linked or hub structure. This is often a function of the geography of the market that is being served. The charter services traditionally providing flights to holiday destinations in Europe or between Canada and the southern states of the United States may be seen as falling into this category.

The non-traditional scheduled carriers have enjoyed an increasing share of many markets. In Europe, for example, although the market share of no-frill carriers is still relatively small – some 5.2% of the scheduled seat kilometers in the intra-EEA market in 2000 – has grown from 0.4% in 1992. Some forecasts suggest it could be 15% of the market by 2010. In the United States, Southwest, JetBlue, AirTran and other 'low-cost' carriers are taking market share from the majors. There is also evidence that Southwest has been influential in reducing the revenues of incumbents in markets that it has entered, although with increasing gate domination at some airports, the notion that it is not a hub-and-spoke operator is becoming somewhat blurred.

THE EMERGING REALITIES: VALUE CHAIN ISSUES

The focus of air transport policy in post-liberalization countries has almost exclusively been on containing the market power of large incumbent, hub-based carriers. This has involved actions in such areas as mergers, alliances, and airport access. The underlying question is whether this is an unjustified bias in the policy-making process. In particular, while there may be competition forced into the airline market by such actions, without identical actions in complementary industries, the overall supply chain may be distorted.

For any industry operating in a commercial environment to be economically sustainable it must be able to recover its full costs over the long term. It is quite common for industries and firms within them to lose money in the short term. This may be because of business cycle effects when there are temporary reductions in demand, or because of restructuring and the adoption of more efficient technologies. The short-term losses, including any interest costs that may have been incurred to tide the industry through these times, need to be recouped during more profitable periods. Companies finance short-term losses either from reserves or by borrowing. Airlines have always followed this pattern even when regulated. In exceptional circumstances there may be a restructuring of debt through bankruptcy mechanisms; of course, nationalized industries often enjoy government support during times of financial losses.

The costs that have to be borne over this long period include the costs of capital. As with profits and losses, these capital costs are not constant but vary over time. In general, however, if an industry earns less than the average for the industry as a whole, investment will be attracted away from it. The return for industry as a whole is normally in the range of about 8–9.5%,

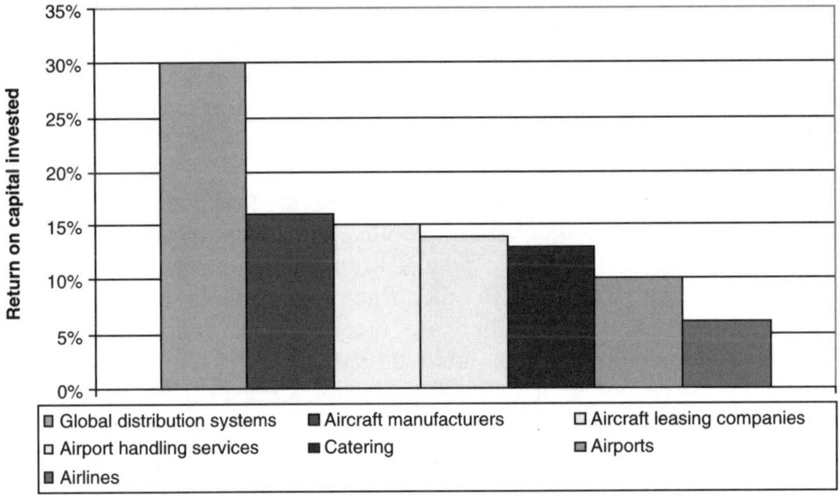

Fig. 2. Return on Capital for Industries within the Air Transport Sector.

but airlines, in contrast, are earning about 6%. As seen in Fig. 1, the operating margin for the major U.S. and EU airlines is extremely volatile, which leads to periodic cash crises that exacerbate the longer-term problems of low returns.

Calculations by the consultancy group McKinsey have also revealed (Fig. 2) that the major scheduled airlines as a group are atypical in the value chain providing air transportation services in that they are one of the few links that do not enjoy a return above the long-term cost of capital. In fact, these calculations are based on a series of assumptions, but in many ways these may be seen as extremely conservative. The returns to European airlines are likely to be somewhat lower than this, since much of the period covered still saw EU carriers at least partly protected from competition in the intra-European market. Furthermore, the returns to other elements in this value chain would seem to be underestimated. On the other side of the equation, recent data in *Airline Business* on the top 20 airports in the world (by revenues) indicates that operating margins of the top 20 airports in the world average well above the 10% or so found by McKinsey.

Of course, the poor performance of the airlines could be due to managerial efficiencies. Indeed, some of the low-cost carriers have earned higher returns, although Southwest is the only one with a genuinely durable track record. However, it seems that although some carriers may have suffered from poor management, the entire industry would have suffered from this problem. The emergence of low-cost carriers in recent years is also only going to have had a marginal effect given their relatively small market shares.

What seems more relevant is the nature of the markets in which other suppliers in the value chain operate. That many airport authorities enjoy high returns is not surprising. Although in some cities or regions there are a number of airports, in others there is only one or if there are more then they are congested. In either case, there is de facto monopoly power. This is often strengthened if international services outside of Europe are involved. The power of airports is perhaps lower in the U.S. domestic market where there are fewer capacity problems and more choices, but moving to a different hub is a major logistics exercise once an airport has been adopted.

There is also, however, potential monopoly power in other elements of the supply chain. For example, there are only two major manufacturers of airframes for large aircraft (Boeing and Airbus) and only three manufacturers of large jet aero engines. In terms of distribution systems, although this is gradually changing as airlines begin to reinternalize the process (e.g. with the development of Orbitz in the United States and Opodo in Europe), there are only four global distribution systems. In addition, there are seven aircraft-leasing companies of any size, but two of these (GECAS and ILFC) control about 45% of the global market. While the EU has attempted to introduce a degree of competition into airport handling (in the sense that carriers are not tied to the airport) these measures have not been fully successful and at many airports a monopoly remains. In terms of catering, there are two main players that control about 45% of the market and the largest seven control about 70% of the market.

In some cases, and especially where there are economies of scale in large-scale production, there are clear advantages in having a single supplier. Indeed, hub-and-spoke operations in a sense require a major, quasi-monopoly airport to be successful. In other cases, airlines may prefer to have an established link with a large supplier, which can save in transaction costs when seeking out competitive tenders, and it can also allow the airline to exercise some power over the supplier. In all cases, the issue is less the existence of the monopoly power than whether that power is exercised.

THE EMERGING REALITIES: THE PROBLEM OF AN EMPTY CORE

Theoretical Issues

Concerns over the economic stability of competitive markets developed in the last century – e.g. Edgeworth (1881) implicitly addressed the subject – yet the topic has attracted only periodic interest. There is also literature from the inter-war period that looked at overhead costs and efficient operations that fits in the lineage of core theory. The emergence of game theory in the late 1940s revived intellectual interest in market stability, though it was not until the work of Shubik (1959) that Edgeworth's work on contracts was tied to that of Nash on games. Subsequently, Scarf (1962) and Scarf and Debreu (1963) used this in developing the modern theory of the 'core.' More recently, Telser has resurrected the topic in a way relevant for the deregulated network industries such as air transport. Linked with this, but tackling the issue from a different angle, have been the developments by Baumol and others of theories concerning the contestability of markets and their potential sustainability. Although as the contestable market advocates admit there 'may well exist a more fundamental and unifying approach which encompasses both views' (Baumol, Panzar, & Willig, 1982), they find it difficult to see how coalition theory fits within their framework, as it currently exists.

It was largely theoreticians who conducted the economic analysis relevant to core conditions, and much of it is of a highly technical and mathematical nature.[3] This has not made it readily accessible. Recent applied work has also not always specifically been directed at the peculiarities of network industries but more generally at matters of anti-trust policy and mergers (Bittlingmayer, 1982, 1985; McWilliams, 1990; McWilliams & Keith, 1994). This has changed somewhat recently with empirical literature emerging that emphasizes shipping and aviation.[4]

Intuitively, the concept of the core of such an economy is related to the notion of equilibrium or the absence of further gains from trade. For an outcome to occur in the core, no further gains from trade are possible for any group or sub-group. In formal terms, the notion of the core stems from the theories of cooperative games. There are various versions of these, of which the Nash cooperative solution (Nash, 1950, 1953) has been attracting the most attention recently, although the number of those who have read the salient papers probably do not constitute enough to be depicted in a three-dimensional diagram. Nash focused on situations where coalitions of actors may be formed in a market but where side-payments are not allowed; in other words, situations where the pay-offs of forming any coalition are not transferable between the parties involved.

The literature on the core has mainly looked at cases where side-payments are possible. This means that one is not looking at a coalition where there is a Pareto improvement, but rather one where there is a potential Pareto gain in the Hicks–Kaldor tradition. It is assumed that every coalition exercises the same degree of rationality as an individual, so that the imputation (Π) allocates to each coalition at least as much as could be obtained by independent action. Imputations are vectors in Euclidean n-space summarizing the pay-offs to each player. Hence Πn in a three-player game could be (0.2, 0.7, 0.1) where player 1 receives 0.2, player 2 receives 0.7, and player 3 receives 0.1. With all players and pay-offs accounted for, the total pay-off equals the pay-off of the grand coalition of players.

To illustrate the conditions for the existence of a core, we assume that there are three players in the market (1, 2, and 3) with a three-game characteristic function form

$$v(\Pi) = 0$$
$$v(1) = 0, \quad v(2) = 0, \quad v(3) = 0$$
$$v(1,2) = 0.1, \quad v(1,3) = 0.3, \quad v(2,3) = 0.3$$
$$v(1,2,3) = v(N) = 1$$

where N is the set of players $\{1, 2, \ldots, n\}$.

The core is the set of all undominated imputations, and it is that set of imputations ($\Pi = \Pi 1, \Pi 2, \ldots, \Pi n$) that satisfies

$$\sum_{i \in S} \Pi_i \geqslant v(S) \text{ for every subset } S \text{ of } N \text{ where } N \text{ is the number of players.}$$

Fig. 3 depicts the possible outcomes of the various coalitions. The equilateral triangle represents the boundaries of the simplex in E^3, the set of imputations ($\Pi 1 + \Pi 2 + \Pi 3 = 1$) such that $\Pi i \geqslant 0$; $i = 1, 2, 3$; $\Pi 1 + \Pi 2 + \Pi 3 = 1$ where the vertices represent imputations for which one player takes all. The shaded area represents the core – a set of outcomes that is preferable to no coalition being formed.

Since all coalitions are accounted for, it would be reasonable to assume that if the core existed, then the imputation chosen would be in the core. The problem is that in many games the core is empty and no imputation satisfies the conditions of coalition rationality for any coalition. As an example, in a three-person game, if in our numerical example all coalitions of two players

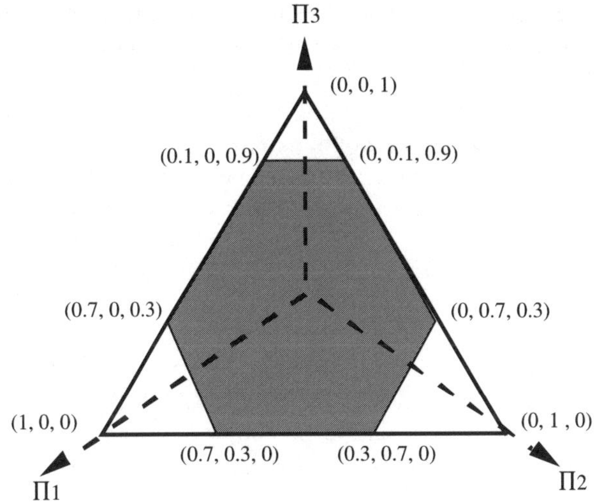

Fig. 3. Locating the Core.

receive 0.8, then the core would be empty. The number of imputations in the core is, in fact, either zero (the core is empty) or many, as in the shaded part of the figure. It is possible to obtain a unique imputation when there is a core by considering the power of each player as reflected by the additional pay-off they bring to a coalition that otherwise excluded him (Shapley, 1953). From a policy perspective, there may be a number of institutional factors such as regulations or competition policy that prevent coalitions from forming. In this case, the shaded area in the figure, even if theoretically attainable, is institutionally impossible to reach.

A simple example can illustrate the concept of the core and the possibility of an empty core. Suppose that an economy has three individuals B1, B2, and B3 who wish to fly between two points. B1 is prepared to pay $100, B2 is prepared to pay $80 and B3 is prepared to pay $75 for the trip. There is one aircraft owned by A1 that can carry up to two passengers. The plane can make the trip at a cost of $140 whether it carries zero, one or two passengers. All of the outcomes that are in the core have the property that B1 and B2 travel on the plane, while B3 fails to make a contract. The outcomes in the core differ in the rate that B1 and B2 pay for the journey. For example, the allocation in which B1 and B2 both have a flight and pay a rate of $75 is in the core. In all the outcomes in the core, airline A1 makes positive profits of at least $10. If the carrier lowered the rate further, below \$75, the third individual, B3 would also wish to fly, potentially disturbing the coalition of A1, B1, and B2. There would be an excess demand at this rate.

Not all economies, however, have a core. If another identical aircraft, owned by A2, also with a maximum capacity of two and the same costs as A1 enters the market, the model now has no core. Because the two aircraft are identical, and because in the core their owners must receive the same profit, otherwise the owner with the lower profit could negotiate with the customers of the other plane to carry them at a slightly lower tariff and steal all business, there is not enough demand to sustain both planes. One of the planes must always be idle earning zero profit, and since the two planes must earn the same profit they must, therefore, both earn zero profit. Consequently there are no outcomes in the core in which the plane owner earns zero profit. The core is then said to be empty – a situation where each and every coalition can be outbid by an alternative coalition, so that there is no stable solution.

The notion that the core may be empty provides insights into the nature of the equilibrium in more familiar economic models of oligopolistic competition. In addition, some of the limitations of the core concept can be highlighted by comparing predictions of core models with conventional economic models of monopoly. The simple examples highlight a general economic principle that is frequently overlooked in conventional economic models of oligopolies. In the presence of a minimum efficient scale (i.e. when entry must be at a minimum scale to be profitable) and free entry, often no equilibrium will exist unless restrictions are imposed on the ways in which economic actors can transact with each other. In standard models of contestable markets or monopolistic competition, equilibrium will only exist under assumptions about the ability of entrants to negotiate with potential customers.

Suppose that the minimum of the average cost for each firm occurs at a unique quantity Q and the minimum average cost corresponds to a price p. Furthermore, suppose the quantity demanded at the minimum average cost is $D(p)$. It would be fortuitous if the price corresponding to that minimum average cost corresponded to a simple multiple of the efficient scale for each firm (i.e. that $D(p) = nQ$ for some n). In the more likely case, the demand at the price corresponding to the minimum average cost $D(p)$ will lie between nQ and $(n+1)Q$ for some n. In this case, either production will be efficient (i.e. at a simple multiple of Q) and there will be excess demand, or there will be no excess demand and production will occur at a price in excess of the minimum average cost.

Consider a model in which the normal equilibrium has the property so that production occurs at the minimum average cost, such as models of Bertrand competition or contestable markets. There will typically be excess demand at a price equal to minimum average cost. A question then arises as to how those customers who do not receive service at the market price should be treated. Most of these customers will be prepared to pay a little more than the market price for service, thus providing an opportunity for profitable entry by a new entrant who offers supply at more than the market price to these customers and the same price to other existing customers.

Furthermore, in standard models of monopolistic competition the equilibrium outcome features production at a level of output insufficient to achieve full economies of production, and the market price lies above the minimum average cost. Although markets clear in this model, the question arises why a new entrant or an existing player cannot undercut the market price and expand output to the efficient scale. As long as a new entrant can attract sufficient business by cutting its price, the market price cannot be sustained above the minimum average cost. The standard monopolistic competition equilibrium is, again, meaningful only under assumptions about the manner in which entrants can contract with existing consumers.

There are other ways in which the idea of the core yields predictions in conflict with conventional economic wisdom. Specifically, the idea of the core conflicts with standard economic analysis of market power. An outcome that is in the core can never have the property that there is exploitation of market power. The core concept predicts that an efficient outcome will always occur, even in the extreme case of a pure monopoly. The intuition behind this is that within the core all opportunities for gains from trade have been exploited. As is sometimes pointed out when considering neo-classical monopoly theory, there remain opportunities for gains from trade. Gains from trade are only exhausted when the output of the monopolist is such that the price is equal to the monopolist's marginal cost. At this price, overall social welfare is maximized. As long as the level of output is such that the monopolist's price is above the marginal cost, the monopolist can sell an extra unit at a price that leaves both the customer and the monopolist better off.

To illustrate, suppose a monopolist can produce output at a constant marginal cost of \$5 and that there are three consumers who are willing to pay \$10, \$8, and \$6 for the good. If the monopolist sets a price of \$10, he sells one unit for a profit of \$5. If he sets a price of \$8, he sells two units with a profit of \$6, and if he sets a price of \$6, he sells three units for a profit of

\$3. Clearly, the profit-maximizing output is to set a price of \$8 and sell two units. This outcome, however, is not in the core. The monopolist can, after selling those units, sell an additional unit to the third customer at a price of \$6 and increase his profit to \$7. This is very much in line with ideas of price discrimination and modern yield management procedures (Brooks & Button, 1994). It is an illustration of the general principle that an outcome that is in the core can never have the property that there is a welfare loss due to the exercise of market power.

Linking into the Airline Issue

The recent legal decisions regarding a proposed merger between United Airlines and U.S. Airways and the actual take-over of TWA by American Airlines in the United States, and regarding attempted alliances such as that between American Airlines and British Airways across the Atlantic, should have resulted in coalitions within airlines debates.[5] Much of the argument that has been presented, though, has been in traditional anti-trust terms, with inevitable injections of political power plays and exploitation by vested interests. Internationally, the granting of anti-trust immunity for U.S. international carriers has been linked to political shifts in international air transportation policy – notably the acceptance of 'Open Sky' bilateral agreements.

The evidence from previous studies of mergers and alliances has been mixed and, although brought into these recent discussions, does not seem to have been very influential. One series of studies has examined the effect of major alliances and the general conclusion seems to be that they are both trade-creating and trade-diverting in their impacts; Bruecker (2003) provides a survey.

Following the game theory approach, the crucial issue from the airline perspective when considering entering alliances is whether the fully competitive market has a natural tendency toward exhibiting empty core conditions. If it does, it then leads to the public policy challenge of deciding what to do about them. On the one hand, should there be government intervention and regulation, or should the internal workings of the sector be allowed to handle them through coalitions? There has been something of a long-standing debate between lawyers and economists on the way to handle coalitions and alliances.

Evidence of the potential for there being empty cores in air transportation has only recently begun to be mustered. Some of this has implicitly been brought into our earlier general discussion of the core. We add more here. The amount of technical analysis has to date, however, been very limited. Button (1996) looked at the changing air transportation market within the EU and adopted an analytical approach akin to Sjostrom (1989) to look for conditions where there is potential for an empty core to exist. The regression analysis making use of 1990 cross-section data found that variables representing legal restrictions and market stability had signs associated with them consistent with possible empty core conditions.

Such tests, while they can be helpful, are difficult to conduct and circumstances where sensible hypotheses can be tested are few and far between. A more useful assessment of the potential for an empty core to exist is probably through simple data analysis and case studies. Within the structure, conduct, and performance framework developed by management scientists, these may offer some useful indicators of at least the extent to which some of the features of an empty core are present in airline markets. This involves looking at the types of features that Sjostrom and Telser have highlighted as accompanying empty core cases.

It is useful to consider the type of coalitions in which airlines now often participate. There has been a focus on alliances that can range in nature from full mergers, through strategic global alliances to agreements at the service level. This is indeed valid, but coalition theory extends beyond the provider of a service to embody all of those involved. This includes the user of the services. Here the aspects of coalitions embrace loyalty incentives, such as frequent flier program and airport lounge access. It can also incorporate those supplying inputs to the airlines, for example, employee participation in ownership and management, and airline involvement in airport service provision. Depending on how the boundaries of an industry are drawn, it can also extend to airport access and supplementary travel services such as hotels. Indeed, there are airlines that have some form of coalition with all of these types of activities as well as with other airlines.

Historically, what one finds in the airline industry is a clear tendency toward alliances and coalitions as the regulatory structure has been dismantled. In 2000, for example, 54% of the world's global capacity was operated under broad multi-airline alliance networks.[6] In some cases, however, and running counter to the idea that regulation has often been initiated to circumvent market instability and hence the need for coalitions, these have been in heavily controlled markets. In particular, these actions have been to circumvent capacity and other restrictions inherent in the traditional bilateral air service agreements. However, this should perhaps be discounted as a pragmatic reaction to poor regulation (intervention failure) rather than a refutation of the arguments advanced by Telser and Sjostrom. Indeed, where there has been subsequent relaxation of bilateral agreements (e.g. the U.S. Open Skies agreements with the Netherlands and Germany), this was accompanied by a considerable strengthening of strategic airline alliances (e.g. KLM/Northwest and Lufthansa/United).

Looking at some of the other patterns that have recently emerged in the air transportation sector, they show little indication that the airlines have been exploitative, but rather that instead a number of the anticipated characteristics of an empty core problem can be seen.

The deregulated air transportation markets have exhibited high levels of instability over the past two decades. In the case of the U.S. domestic market, the early years of deregulation saw 246 carriers certified between 1978 and 1997, but 168 of these have left the market or merged. Similar patterns of market entry and exit are emerging in the EU as the implications of its three Packages of reform are taking effect. Rates of return in the airline industry have never been high (an annual surplus of less than 2% since 1980), and they have been regularly subject to considerable volatility – globally, record losses were made between 1990 and 1993; record profits in the late 1990s and record losses in 2001. This is despite the coalitions that have emerged and the growth of more sophisticated pricing structures. The demand for air services is derived from that of individuals wishing to engage in some final activity. As such, the demand curve is highly sensitive to income levels.

The Viner case for a core being empty is based upon the fixed cost/increasing returns argument and has most often been applied to utilities and fixed-track transportation modes such as railways. This type of situation was essentially described in the two-airline example used earlier. The numerous studies that have been conducted on the importance of such economies in the airline industry do suggest that they are real (e.g. Brueckner & Spiller, 1994; Creel & Farell, 2001). The widespread adoption by airlines of hub-and-spoke operations is often seen as allowing such economies to be realized. Perhaps more important is the indivisibility of each airline when committed to providing a scheduled flight. Recently, technological changes have been affecting the air transportation market with the introduction of regional jets in large numbers. The use of these aircraft with 70–100 seats have, in particular, become common over medium distance routes (Dresner, Windle, & Zhou, 2002), and the wider adoption of these planes can be seen as an effort by carriers to reduce the indivisibilities associated with meeting each service obligation that creates an empty core. These may be seen as removing indivisibilities that have led to alliances. If so, then one would anticipate that the routes served by regional jets would be competitive and much less prone to being operated by coalitions.

The debate that is taking place over the relevance of the potential of empty cores not only reflects differing views amongst economists, but also between economists who feel that there are problems due to the lack of a core, and lawyers who are more focused on the conventional anti-trust paradigms. In particular, the legal position is largely that cartels are inherently against the public interest because they adversely affected output and prices. However, since there is no equilibrium price if there is no core, then no useful benchmark exists to test this. In a free market, with free entry where some players collude, for example, an alliance, optimizing capacity would only be sustainable if the alliance had costs and prices below those of potential entrance and other existing operators.

In the air transportation case, one can think of situations where low-cost carriers have entered markets previously dominated by a large carrier or an alliance, forcing the incumbents to adjust their behavior or a bandon that market. Cooperation is only tenable if the resultant coalition is more efficient than any other. The key to policy here is to ensure open entry exists and that an incumbent coalition cannot exploit its position to extract economic rent.[7] It is thus not the 'cartel' which is automatically an issue, but rather why the cartel has formed and how it is maintained. From a legal perspective this is a matter of devising appropriate criteria for structure, conduct and performance testing. McWilliams (1990) has suggested that the Sjostrom-type delineation may offer guidance in these types of situation.

There is perhaps a more fundamental issue that separates the economic from the legal analysis of coalitions. Economists in general are primarily interested in maximizing efficiency and leaving normative judgments of distribution to others. Application of Hicks/Kaldor compensation criteria suggests that as long as there is the potential for compensating losers, then whether compensation is actually provided is not an economic issue. Lawyers, however, often take the view that anti-trust policy is aimed at promoting distributional justice. Landes (1982), for instance, argues that the role of anti-trust law is to guard consumers against producer restraints on output that would appropriate their wealth.

In this latter context, anti-trust and similar policies are aimed partly at meeting a combination of efficiency and distribution goals. Hence, even if airline alliances, frequent flier programs, etc. are efficient in terms of optimizing instability and long-term capacity, these may not been seen as socially optimal as far as the distribution of benefits are concerned. If one does take this line, however, it does not rule out coalitions as de facto bad, as they may both enhance efficiency and (especially over the long term) lead to greater distributional justice. This is a matter for consideration under the rule of reason.

CONCLUSIONS

The deregulation of scheduled airline markets was based upon a combination of empirical evidence from already liberalized markets, the emergence of innovative thinking on how markets function, and mounting evidence that the administration of regulatory regimes often fails to serve the public interest. The early evidence that emerged once the short-term disruption of transition had passed is that consumers as a whole have enjoyed significant benefits from the reforms. The longer-term situation is somewhat different. Discounting the exogenous shocks of events such as the attacks on New York and Washington, the air transport market does not appear to be moving to a long-term equilibrium. The emergent picture after more than two decades of liberalization in the United States and significant, if not such extensive periods in many other markets, is that airlines are not covering all their costs.

One reason for the financial malaise of the airline industry – of course, any individual airline may be suffering because of poor management – is that it is part of a value chain where other links are extracting not only economic rent, but also not allowing airlines to earn even normal rates of return. This affects the cost structure of the airline industry and would, in aggregate, lead to under-supply. However, even if this problem were resolved, and availability of inputs into the airline industry conformed more closely with notions of competitive supply, the nature of the liberalized markets in which air services are provided may not allow full cost recovery. The nature of the production function for delivering scheduled services and the competition to extract as much revenue as possible to cover fixed costs leads to carriers accepting passengers down to the level of short-run marginal costs.

The situation has been intuitively tackled by carriers through such measures as frequent flier programs to retain customer loyalty and to contain revenue dilution, and by the formation of alliances to take some of the competitive edge from the market. In the long term, however, factors such as improved information systems and the advent of low-cost carriers have allowed travelers to find lower fares and to weigh up these against the frequent flier advantages of loyalty. Further, the competition between airlines has moved to competition between alliances and this has brought with it new levels of competition. Whether mergers and other forms of collusion that transcend that of alliances will allow the emergence of more sustainable structure is not certain, but current anti-trust policies would not seem helpful in at least allowing more experimentation in this area.

NOTES

1. In fact, the economic deregulation of the U.K. trucking industry under the 1968 Transport Act was the first major reform, though location and the less conspicuous nature of the industry meant the changes received limited attention.

2. The sustainable operating margin will depend upon the extent to which airlines seek equity capital. The suggestion that an operating margin is a judgement based upon the amount of equity financing engaged upon and the cost of equity finance. Both are highly variable across airlines, time and markets, and in the case of the cost of equity finance, the basis of calculation undertaken.

3. Telser (1994) provides a general introduction to the theory of the core. Sjostrom (1989) is also less technical and offers a practical approach to considering whether a market is likely to prove inherently unstable or not.

4. Button (1997) and Button and Nijkamp (1998) offer more general examinations of the relevance of core theory to network industries. It should be noted that the non-existence of a network service, though important for social reasons to do with policies of access or income distribution, does not imply the lack of a core.

5. One could add to this the demise of former flag carriers such as Sabena and Swissair that were linked by alliance.

6. *Airline Business* provides an annual survey of airline alliances.

7. Potential market entrance and all current participants in the market should be confronted with the same institutional conditions. This is not, for example, the case where failing non-coalition carriers are allowed bankruptcy protection of the type found under Chapter 11 in the United States.

REFERENCES

Baumol, W. J., Panzar, J. C., & Willig, R. D. (1982). *Contestable markets and the theory of industrial structure*. New York: Harcourt Brace Jovanovich.

Bittlingmayer, G. (1982). Decreasing average cost and competition. *Journal of Law and Economics*, *25*, 201–229.

Bittlingmayer, G. (1985). Did antitrust policy cause the great merger wave? *Journal of Law and Economics*, *28*, 77–118.

Brooks, M., & Button, K. J. (1994). Yield management: A phenomenon of the 1980s and 1990s? *International Journal of Transport Economics*, *21*, 177–196.

Bruecker, J. J. (2003). The benefits of codesharing and antitrust immunity for international passengers, with an application to the star alliance. *Journal of Air Transport Management*, *9*, 83–89.

Brueckner, J. J., & Spiller, P. T. (1994). Economics of traffic density in the deregulated airline industry. *Journal of Law and Economics*, *37*, 379–415.

Button, K. J. (1996). Liberalising European aviation: Is there an empty core problem? *Journal of Transport Economics and Policy*, *30*, 275–291.

Button, K. J. (1997). Regulatory reform, networks and economic efficiency. In: C. Capineri & P. Rietveld (Eds), *Policy aspects of networks*. London: Wiley.

Button, K. J., & Nijkamp, P. (1998). Economic stability in network industries. *Transport Research E, Logistics and Transport Review*, *34*, 13–24.

Button, K. J., & Stough, R. (2000). *Air transport networks: Theory and policy implications*. Cheltenham: Edward Elgar.

Button, K. J., & Swan, D. (Eds) (1989). *The age of regulatory reform*. Oxford: Oxford University Press.

Creel, M., & Farell, M. (2001). Economics of scale in the US airlines industry after deregulation: A Fourier series approximation. *Transportation Research E: Logistics and Transportation Review*, *37*, 321–336.

Dresner, M., Windle, R., & Zhou, M. (2002). Regional jet services: Supply and demand. *Journal of Air Transport Management*, *8*, 267–274.

Economides, N. (1996). The economics of networks. *International Journal of Industrial Organisation*, *14*, 673–699.

Edgeworth, F. Y. (1881). *Mathematical Physics: An essay on the application of mathematics to the moral sciences*. London: Kegan Paul.

Jenkins, H.W. (2002). How about letting airlines help themselves? Wall Street Journal, 21 August.

Landes, R. H. (1982). Wealth transfers as the origin and primary concern of antitrust: The efficiency interpretation challenged. *Hastings Law Journal*, *50*, 871–957.

McWilliams, A. (1990). Rethinking horizontal market restrictions: In defense of cooperation in empty core markets. *Quarterly Review of Economics and Business*, *30*, 3–14.

McWilliams, A., & Keith, K. (1994). The genesis of the trusts: Rationalization in empty core markets. *International Journal of Industrial Organization*, *12*, 245–267.

Nash, J. F. (1950). The bargaining problem. *Econometrica*, *18*, 155–162.

Nash, J. F. (1953). Two-person cooperative game. *Econometrica*, *21*, 405–421.

Scarf, H. E. (1962). An analysis of markets with a large number of participants. In: *Recent advances in game theory*. Princeton: Princeton University Press.

Scarf, H. E., & Debreu, G. (1963). A limit theorem on the core of an economy. *International Economic Review*, *4*, 235–246.

Shapley, L. S. (1953). A value for N-person games. In: M. Dresher, L. W. Shapley & A. W. Tucker (Eds), *Advances in game theory*. Annals of Mathematics Studies No. 52. Princeton: Princeton University Press.

Shubik, M. (1959). Edgeworth market games. In: A. W. Tucker & R. D. Line (Eds), *Contributions to the theory of games*. Annals of Mathematical Studies No. 40. Princeton: Princeton University Press.

Sjostrom, W. (1989). Collusion in ocean shipping. A test of monopoly and empty core models. *Journal of Political Economy*, *97*, 1160–1179.

Smith, T. K. (1995). Why air travel doesn't work. *Fortune*, 3 April: 26–36.y

Telser, L. G. (1994). The usefulness of core theory in economics. *Journal of Economic Perspectives*, *8*, 151–164.

THE EVOLUTION OF NETWORKS WITH CHANGES IN INDUSTRY STRUCTURE AND STRATEGY: CONNECTIVITY, HUB-AND-SPOKE AND ALLIANCES

David Gillen

ABSTRACT

This chapter examines the evolution of the structure of airline networks as dominant strategies in the industry have changed. The network strategy is an integral component of the airline strategy, and network structure will be a function of demand side externalities and uncertainty, as well as supply side network economies. Network economies have predominantly been on the demand side, but in airline networks they are viewed as being for the most part on the supply side. The paper first describes the natural move to hub-and spoke (HS) systems after deregulation and the consequences of this network configuration. Network structure choice is not just a cost issue; a network can confer revenue advantages as well. We show these advantages in a discussion of low-cost carrier strategies. Next, the paper examines the current state of the airline industry and the evolution from HS to networks of connectivity. We report on the [opposing] forces that lead to such a result. The network structure ranges from fully connected,

Global Competition in Transportation Markets: Analysis and Policy Making
Research in Transportation Economics, Volume 13, 49–73

ISSN: 0739-8859/doi:10.1016/S0739-8859(05)13003-0

to HS to alliance (fully contracted) to middle of the road strategies, which include connectivity, and "rolling hubs." The forces leading to the choice of each strategy are discussed as well as the future sustainable strategies of the airline industry.

INTRODUCTION

Transportation networks exhibit what are termed network economies, meaning the organization of production spatially confers either or both demand and supply side benefits. In most cases, the demand side is the source of benefits, but airline networks exhibit both. The choice of network will reflect the business model the carrier has chosen and will exhibit quite different costs. Airlines, for the most part, moved from a fully connected linear network to a hub and spoke (HS) network after deregulation. Certainly, this was the choice of full service carriers like American and United. Niche or regional carriers such as Alaska and USAirways pursued HS network strategies. These strategies were tied to the business model these airlines chose. Other carriers such as Southwest, which chose a different business model, selected a network that was connected but not fully. Subsequent entrants using the low-cost business model have followed the example of Southwest.[1] As the industry has come under stress following the dot-com implosion, the general world economic downturn and the events of September 11th, full service carriers have responded in different ways, one of which is to move from HS toward connectivity. Similarly, low-cost carriers have moved to more connectivity as they mature. Is this the long-run sustainable strategy of each business model?

The HS network configuration was not a consequence of deregulation but was the preferred network form of carriers once deregulation took place, because it satisfied the product values demanded by high yield business customers. Delta Airlines, for example, had organized its network into a hub at Atlanta with multiple spokes. Other carriers had evolved more linear networks with generally full connectivity and were reluctant to shift to HS for two reasons. Firstly, regulations required permission to exit markets and such exit requests would likely lead to another carrier entering to serve "public need." Second, under regulation it was not easy to achieve the demand side benefits associated with networks due to regulatory barriers to entry. Once deregulation had taken place, however, firms were free to reorganize themselves spatially. This spatial reorganization was driven by the underlying economics of the industry.

Emerging from deregulation airline strategies, including their network configurations was driven by both demand and supply-side forces. On the demand side, large firms were favored since they served large numbers of destinations, had significant geographic coverage, and had greater value in loyalty programs.[2] On the supply side there were no economies of network size, costs did not decrease as more points were added to the network, but there were significant economies of density; more passengers on a given route lowered costs per passenger.

The network effects that favored HS over linear connected networks were the compatibility of flights and the internalization of externalities between links. A carrier directly flying from A to B would achieve some direct density economies, while a carrier serving A to B traffic through H would need to consider that pricing on AH could affect traffic on HB. Furthermore, the use of a hub with feed traffic from spokes would allow more flights for a given traffic density and hence costs level. More flights reduce total trip time, with increased flexibility and serves business traffic. The supply-side network effects arose from joint production and cost complementarities between origina and destinations (OD); economies of scope.

To make an HS system work effectively also required additional features most notably, yield management and product distribution. These two elements were conditional on large information systems, and in particular computer reservation systems (CRSs). The CRS made possible sophisticated flight revenue management, administration of frequent flyer programs (FFPs), effective product distribution, revenue accounting, and load dispatch. They also drove aircraft capacity, frequency, and scheduling decisions.

Hub systems took a number of different forms. There were simple hubs where the spokes were scheduled independently from one another such as People's Express. This form of hub is not dissimilar to those of low cost carriers that create connection points such as Southwest, Westjet, and Ryanair. Complex hubs had flights coordinated to arrive in bunches or banks. It allowed more and fast connections to other flights, albeit at the cost of poor utilization of hub factor inputs (e.g. labor).[3] Complex hubs evolved in some cases to multiple hubs and directional hubs.[4] One or more of these hubs would generally be an international gateway hub.

Hub systems led to the development of feed arrangements along spokes. The feeder carriers were more cost efficient, offered greater frequency to enhance service and ensured a greater proportion of traffic were kept online with a single carrier. These domestic feeds were subsequently one of the bases for developing international alliances, in which one airline would feed another by utilizing capacity of both to increase service and pricing to internalize the double-marginalization problem.

In the mid- to late 1990s, full service airlines (FSA) like American, United, British Airways, Lufthansa, JAL and Air Canada had developed and based their business on a strategy that was highly dependent on high yield traffic, had low-cost controls and was dependent on a network configuration that was costly both in terms of inefficient use of resources and in creating its own congestion. Yet, it was this HS system that offered the service frequency and capacity that drove the demand side and in particular the demand by high yield passengers. This model was based on the proposition that as long as no subgroup of passengers could go it alone, defect from the coalition of all passenger groups, the strategy was sustainable. The opportunity to defect came in the form of low-cost carriers like Southwest.

Southwest Airlines represented a different strategy that builds the market but forces full-service carriers to lower fares when it enters a market. It has a much lower cost structure and thus is a credible threat when it enters a market.[5] The lower cost structure, discussed in detail later, is derivative from among other things a more efficient use of airport and aircraft inputs because of the design of its network and lack of a true hub. The low-cost carrier or value-based airline (VBA) creates connectivity but trades off lower levels of service, measured both in capacity and frequency, against lower fares. In all cases, the structure of the network is a key factor in the success of VBAs, even in the current economic and demand downturn.

The evolution of the FSA or legacy carrier at least in North America has been in some cases, to move away from the HS network and move in the direction of providing connections as distinct from true hubbing. American Airlines is using a "rolling hub" concept, which moves to connectivity and away from peaking. The purpose is to reduce costs through reducing factors such as aircraft and labor and to increase productivity. The first step is to "de-peak" the hub, like in electricity markets, which means not having banks as tightly integrated. This reduces the amount of own congestion created at hubs by the hubbing carrier and reduces aircraft needed. It also reduces service quality. However, as an example, American Airlines has reduced daily flights at Chicago so with the new schedules it has increased the total elapsed time of flights by an average of 10 min. Elapsed time is a competitive issue for airlines as they vie for high-yield passengers who, as a group, have abandoned the airlines and caused revenues to slump.

But that 10 min average lengthening of elapsed time appears to be a negative, American is willing to accept in exchange for the benefits. At Chicago, where the new spread-out schedule was introduced in April, American has been able to operate 330 daily flights with five fewer aircraft and four fewer gates and a manpower reduction of 4–5%.[6] This change has cleared the way for a smoother flow of aircraft departures and has saved taxi time.[7]

It is likely that American will try to keep to the schedule and be disinclined to hold aircraft to accommodate late-arriving connection passengers. While this may appear to be a service reduction, it in fact may not be, since on-time performance has improved (see footnote 7).

ECONOMICS OF NETWORKS

There has been an evolving literature on the economics of networks specifically economics of network configuration. Hendricks, Piccione, and Tan (1995) show that economies of density can explain the HS system as the optimal system in the airline networks. The key to this explanation lies in the level of density economies. However, when comparing a point-to-point network they find that an HS network is preferred when marginal costs are high and demand is low but given some fixed costs and intermediate values of variable costs a point-to-point network may be preferred. Shy (2001) shows that profit levels on a fully connected network are higher than on an HS network when variable flight costs are relatively low and passenger disutility with connections at hubs is high. What people have not explained very well is the trade-off between market size and density economies. Pels, Nijkamp, and Rietveld (2000) recently explored the optimality of airline networks using linear marginal cost functions and symmetric demand functions; $MC = 1 - \theta Q$ and $P = \alpha - Q/2$ where θ is a returns to density parameter and α is a measure of market size. What is missing in the earlier work is the presence of fixed costs, f. When f is included in the analysis, the robustness of the HS network configuration claimed by earlier authors (e.g. Hendricks et al., 1995) comes into question.

If one specifies a network of n nodes, with $n \geqslant 3$ there will be direct markets and transfer markets in an HS system. Aggregate demand will be $Q = QD + (n-2)QT$ where D is direct and T is transfer. The profits from an HS network as shown by Pels et al. is

$$\Pi_{HS} = (n-1)\left(P_D Q_D + \frac{(n-2)}{2} P_T Q_T\right) - (n-2)\left(Q_D + (n-2)Q_T - \frac{\theta}{2}(Q_D + (n-2)Q_T)^2 + f\right) \quad (1)$$

while the profits of an FC network are

$$\Pi_{FC} = \frac{n(n-1)}{2}\left(P_{FC}Q_{FC} - \left(Q_{FC} - \frac{\theta}{2}Q_{FC}^2 + f\right)\right) \tag{2}$$

One way to examine these alternative configurations is to see where they would generate the same profit levels, $\Pi_{HS} = \Pi_{FC}$ and what are the determining factors. The key parameters in this model specification are α and θ. If markets are symmetric and marginal costs are linear and with equal fixed costs, f, per link, the HS network will be more profitable.[8] We need to obtain a measure α^* which is the market size at which optimal values of prices and quantities yield equal profits, and can shown to be equal to

$$\alpha^* = \frac{\theta(2n-1) + 1 \pm \sqrt{X}}{\theta(2n - 1 + \theta)} \tag{3}$$

where, as Pels et al. (2000) show,

$$X = [1 - \theta(2n-3)](\theta - 1[2f\theta(2n - 1 + \theta) + \theta - 1]) \tag{4}$$

α^* can therefore take on two values. There are boundaries to the value of the market size parameter that affects the choice of HS or FC as the profit-maximizing network. The density parameter and fixed costs, f, play an important role. Given a value for f, there is a trade-off between market size, for a select n, and density economies. If fixed costs are low and n is small, an FC network is more profitable than an HS network. The cost savings from density economies drive the outcome but as market size increases and hence n, the affect of f outweighs the positive value of density economies. When Q and θ are large, density economies drive the result, but as market size expands n increases, which increases fixed costs. In this analysis, the conclusion is that an FC network is preferred since it is more profitable, when f is low, Q is large and density economies are low. What drives the HS network preference is reducing n, the number of connected links, and realizing density economies. As market size increases, size, α, compensates for density and HS is preferred because the fixed costs of links are assumed to drive prices (fares) too high. The claim is made that when f is low (or assumed to be zero as much of the literature assumes), an FC network may appear superior to an HS network.

In cases where f is low, fully connected networks and HS are highly dependent on density economies. As market size increases, density economies become less important. A VBA carrier with a fully or "more" connected network finds it more profitable to operate in many markets with connections and focus on market size, and not try to achieve density economies. If the fixed costs increase because more links are added to the network, density economies become more important and HS dominates.

The optimal network requires combinations of (α, θ, f). These are precisely the parameters that distinguish the FSA and VBA business models. But more importantly, the integration of these parameters will also affect the optimal network configuration. What is missing from this characterization is that f is considered fixed per link but the variable costs are very much influenced by the network configuration; in other words θ (network configuration). The key point is that f for an FSA is higher than a VBA; thus f should be subscripted, f_{HS} and f_{FC}. Furthermore, an HS network results in a higher marginal cost, which means the FSA is on a completely different cost function, not just a different point on the same cost function. The HS network results in lower productivity for FSAs. This was not an issue if the network was adding sufficient value that customers had a high willingness to pay. HS networks were focused on high yield and certainly high-cost business passengers. While f may have been low at spokes they were high at hubs. Even more important, variable costs were higher at hubs as banks of flights minimized connect time for passengers but at cost to carriers. FC networks might, in principle, have higher f than HS, but the implicit assumption is the product is the same. This is not the case for VBA. Their business model with simplicity of process, product and organization means they unbundle their airport services to assure lower f and lower variable costs; hence the remark that they are on a different and lower cost function.

The value of the HS network was based on a demand that valued frequency and connect time. The FC network provided direct time but low frequency. As these demand parameters matter less, the value proposition in the HS network diminishes. The tension in the business model of the FSA is the yield management system, which needs to be redesigned and the costs both fixed and variable of the FSA must be brought down. The significant inefficiencies associated with the HS means that there would be evolution toward a connected network where there is less reliance on density economies, a relatively greater reliance on market size through pricing.

AIRLINES: A CHANGING SET OF STRATEGIES

The trend worldwide indicates two divergent strategies. The legacy is that full-service airlines focused on developing HS networks, as this business model provided a product, which provided high value, but at a high cost. New entrants adopted a different business model with a low value but lower cost product using low-cost, point-to-point structures. The HS system places a very high value on the feed traffic brought to the hub by the spokes, especially the business traffic therein, thereby creating a complex, marketing intense business where revenue is the key. Point-to-point, low-cost, low-fare carriers have simplicity and low costs as their focus, as the low fares stimulate the latent demand, creating markets on a point-to-point basis that HS carriers cannot profitably serve. In short, the HS airlines focus on revenue, while the low-cost carriers focus on extracting every possible efficiency through simpler product, process, and organizational design. The resultant firms are radically different in appearance and underlying structure.

Table 1 provides a summary of the alternative airline strategies pursued in Canada, as elsewhere in the world. Not all are sustainable strategies and the full service carrier, such as Air Canada, is facing significant challenges to its established business model. Air Canada has responded by developing a multiple-branding strategy with the creation of Tango, Zip, Jazz, Jetz and the remaining mainline Air Canada. Few other carriers worldwide have followed such an extensive rebranding, Air New Zealand Lufthansa, Delta and United are moving somewhat in this direction with the creation of a low-price–low-cost brand, but many are watching with interest.[9] Other carriers such as American are redesigning its HS model in an attempt to lower costs. Following the lead of the low-cost carriers, they are moving away from high-cost close connection hubs to what are termed ‘rolling hubs’ where the connections are not as tight, the service lower, the costs lower but also lower fares. Others are reducing capacity, cutting jobs and moving into Chapter 11 (USAirways & United) and hoping for bailouts from the U.S. government. The move to Chapter 11 creates a dilemma for other legacy carriers since any carrier re-emerging from Chapter 11 will have a distinct cost advantage. This could lead to carriers moving to Chapter 11 so as not to be disadvantaged.

The strategy pursued by FSAs was one of relatively high-risk strategy since it is conditional on not having any group of passengers, any subcoalition, defecting to another carrier that followed a different business model. Moreover, the high risks arose from high-yield business travel being highly pro-cyclical; if economies were growing, profits were positive. The FSA strategy was also a high-cost strategy because it provided high service levels, frequency and capacity, to meet the needs of the highest yield customers-business travelers. In addition the strategy was undertaken in a HS network, which meant both reduced productivity for capital (aircraft) and labor (pilots, cabin crew, airport personnel) and increased costs due to self-induced congestion from closely spaced banks of aircraft.[10]

Table 1. Description of Strategies in the Canadian Airline Industry[a].

Strategy	High Cost, Full Service		Low Cost, No Frills		
Example	Air Canada American, United, British Airways, JAL	Roots Air	Canada 3000, Royal Airlines (pre-merger)	Air Transat	West Jet Ryanair Southwest Jet Blue
Network type	Hub-and-spoke, scheduled service	Point-to-point, scheduled service	Point-to-point, charter/ scheduled	Point-to-point, charter	Point-to-point, scheduled
Characteristics	High fixed costs, high labor costs, inflexible job tasks, full service, multiple classes, high frequencies	Moderate fixed costs, moderate labor costs, moderate job tasks, flexibility, full service, multiple classes, low frequencies	Low fixed costs, moderate labor costs, moderate job tasks flexibility, low-end full service, single and multiple classes, low frequencies	Low fixed costs, low labor costs, flexible job tasks, low-end service, single class (few wider seats), low frequencies	Low costs, lower labor costs, flexible job tasks, no frills service, single class, increasing frequencies

[a]Adapted from "Air Canada, Turning on the after-burner in Profitable Skies", April 7, 2000, Mark Korol, National Bank Financial.

The risk exposure of the FSA was therefore in part exogenous since a significant proportion of revenue was dependent on a buoyant economy. Exogenous risk also arose from the competitive entry of another carrier, but one that followed a different business model; a VBA would be such a threat. The reasoning is that entry by another FSA would mean costs would be similar, and thus, the floor to which fares could fall would be some measure of unit costs. On the other hand a carrier following a different business model focusing on low costs provided a lower floor for fares to fall to, and was a credible threat. As an example, VBAs in North America generally have CASMs (cost per available seat mile) that are 50–60% of that of FSAs.[11] Dresner and Windle (1996) have also found that when FSA compete in a market, fares fall but recover to about 90% of their pre-entry price within 6 months. If a VBA enters a market, fares fall further and return to only 70% of their pre-entry level after a longer time period. If the VBA is Southwest, fares decrease by near 58% and move back only slightly to their pre-entry levels.

Point to Point Networks and VBAs

VBAs have a fundamentally different business model and achieve their low costs through a disciplined business plan that focuses on simplicity of product, process, and organization. The choice of a point-to-point network is fundamental to achieving their high productivity, and also confers revenue advantages given the design of their product. The network design is an integral part of the business model not simply a means of distributing the product in a low-cost way.

VBAs have created a unique value proposition through product and process design that enables them to eliminate, or "unbundle" certain service features in exchange for a lower fare. These service feature trade-offs are typically: less frequency, no meals, no free, or any, alcoholic beverages, more passengers per flight attendant, no lounge, no interlining or code-sharing, electronic tickets, no pre-assigned seating, and less leg room. However, for many consumers, the lower fare more than offsets the reduction in service features.

Low fares on an everyday basis have opened up the market to a whole new segment of travelers. "Everyday" is emphasized since low fares have been available in the past, but full service carriers would only offer them during sale periods and with many restrictions. Therefore, the actual number of seats available at the lowest fare was very small, minimizing the potential traffic stimulation. Low-fare carriers make many seats available at "everyday low fares" with very few restrictions, thereby expanding the market. As a result of the product design and resulting price, VBAs attract a large amount of visiting friends and relatives (VFR), leisure, and price-conscious business travelers.

While relatively new to Canada, U.K., and Australia, VBAs have been a fixture in the United States for years. They have recently been experiencing dramatic growth in Europe due to deregulation in the EU. These include Ryanair, easyJet and GO (GO recently merged with easyjet in 2002), BUZZ (which was purchased by Ryanair in 2003) and bimybaby. More recent start-ups in Australia – Virgin Blue, Brazil – Gol, Germany – Germania, Hapag – Lloyd and Malaysia – Air Asia, seem to indicate that the low-cost, low-fare model is spreading worldwide, largely as a result of market deregulation. Furthermore, the model proven by Southwest and adopted by others in recent years has provided the method to take advantage of the opportunity.

Low-cost carriers tend to exhibit common product and process design characteristics that enable them to operate at a much lower cost per unit of output.[12] Another less obvious, but important cost saving can be found in the organization design and culture of the company. The simplicity of product design extends to the process and organizational design as well. It is worth noting at this point that the innovator of product, process, and organizational redesign is generally accepted to be Southwest Airlines. Many low-cost start-ups now attempt to replicate that model as closely as possible; however, the hardest area to replicate has proved to be the organization design and culture.[13]

Extending to the "look and feel" of the aircraft, there is a noticeable strategy for low-cost airlines. Successful VBAs focus on a homogeneous fleet type (mostly the Boeing 737 but this is changing; e.g. Jet Blue uses an A320 fleet). The advantages of a "common fleet" are numerous. Purchasing power is one – with the obvious exception of the aircraft itself, heavy maintenance, parts, supplies; even safety cards are purchased in one model for the entire fleet. Training costs are reduced – with only one type of fleet, not only do employees focus on one aircraft and become specialists, but economies of density can be achieved in training. The economies of density are achieved since it is much easier to design courses with suitable class sizes, especially with recurrent training, making better use of the trainer and facilities' time. While this may be true with all start-up airlines once each aircraft type reaches a certain number, VBAs reach that point sooner, and have a much more scalable training system. Only one training program needs to be developed, recurrent training is simple and can be applied to all employees of a specific group. There is no expensive retraining of employees when they reach a certain seniority level or need to work at a different base. Further savings are achieved through higher productivity through specialization in all departments of the airline.[14]

The choice of airports is typically another source of savings. Low-cost carriers tend to focus on secondary airports that have excess capacity and are willing to forego some airside revenues in exchange for non-airside revenues that are developed as a result of the traffic stimulated from low-cost airlines. In simpler terms, secondary airports charge less for landing and terminal fees, and make up the difference with commercial activity created by the additional passengers. Further, secondary airports are less congested, allowing for faster turn times and more efficient use of staff and the aircraft. Table 2 provides the utilization and average taxi times of major U.S. airlines, demonstrating that Southwest is a market leader in this area.[15] The cost savings are not exclusive to secondary airports however, as the way in which the airports are used can create savings. Even primary airports may have areas that allow for disembarking passengers without bridge facilities,

Table 2. Aircraft Utilization and Operating Cost of 737-300 and 737-700 fleets (3rd Q, 2001).

Airline	Departures	Block Hours	Flight Hours	Average Stage Length (miles)	Average Taxi Time in Minutes[a]	Cost per Available Seat Mile (U.S. cents)
Frontier	4.5	11.2	9.8	933	19	5.6
Southwest	7.6	10.5	8.9	472	13	4.0
ATA	3.9	10.4	8.8	1,032	25	4.3
United	5.0	9.3	7.5	639	22	8.1
Continental	3.4	8.6	7.1	895	26	6.2
America West	4.5	8.3	6.7	602	21	6.2
USAirways	5.1	8.3	6.3	466	24	8.9
Delta	4.6	7.8	6.1	546	22	7.1

Source: Aviation Daily, March 27, 2002.
[a]Calculated using the difference between block times, flight times and dividing by the number of departures.

or will allow for reduced rates to use terminals that are not up to the service standard used by the full service carriers. Primary airports also have considerable differences in efficiency, and therefore cost, of the operation. One area in which efficiencies are achieved is through operational design-making taxiways and runways more logical in relation to the bridges and the terminal. The use of staff and choices made in the expenditure of capital are also important.

Does the airport seem intent on gold plating, or is the design simple and effective to serve the primary purpose – processing passengers? These are all important considerations for low-cost carriers. Lounges and elaborate waiting areas are not part of the product design for low-cost carriers, which allow them to negotiate lower facility charges. Finally, the demand created by VBAs as a result of their low fares allow them to fly larger equipments into smaller airports than their full service carrier counterparts, and still maintain profitability. Smaller airports typically charge less in landing and terminal fees, due to their more rudimentary terminal facilities.[16] While not a savings in itself, this creates more of a discrepancy when the unit costs are compared between the same equipment of a low-cost vs. a full service carrier.

Another important source of cost savings in product design is seat density. By only providing one class of service, more seats are placed on any given aircraft type when compared to a two-class, full service carrier. The per-flight costs are quite similar, yet VBAs reduce the cost *per seat*, allowing them to charge less per seat to make a profit.

In-flight service is another source of product differentiation. Low-cost carriers fly with fewer flight attendants per passenger, as they have no need to cater to first class passengers, provide nominal, if any, food and drink services, and tend to provide little in the way of in-flight entertainment. The obvious savings in product and staffing costs are further increased in less tangible ways. Equipping an aircraft for meal service requires more galley space (and thus fewer seats), more expensive equipment and increases the flying weight of the aircraft not only with the product but the equipment as well. It has been estimated that each 1,000 pounds of excess weight on a 737 costs between $10 and 15 per flight. Further, no expensive delays are caused waiting for additional hot meals if a miscalculation is made in estimating the number of meals required for a given flight. With regards to in-flight entertainment, millions are saved in capital and maintenance costs on expensive video monitors and wiring for headsets.

Another product design, not providing cargo service, is a cost advantage for low-cost carriers. By making the decision not to take cargo in any substantial capacity (beyond mail and small packages), VBAs are able to certify their aircraft for lower maximum take-off weights (MTOW). Since most landing and navigation charges are variable based on MTOW, the savings per flight of a reduced MTOW can be substantial. It should be noted that VBAs will tend to take small packages, including mail, up to the point where cargo starts to interfere with the core business of moving passengers. In this regard, the passenger experience (and thus product) is not unnecessarily affected by delays caused by cargo movements. Expensive handling equipment does not need to be purchased and the staffing requirements for handling cargo are also unnecessary. All of these contributes to a simpler process.

Essentially, VBAs have attempted to reduce the complexity and resulting cost of the product by "unbundling" those services that are not absolutely necessary. This "unbundling" extends to airport facilities as well, as VBAs struggle to avoid the costs of expensive primary airport facilities that were designed with full service carriers in mind. While the savings in product design are the most obvious to the passenger, it is the process changes that have produced greater savings for the airline.

In the process changes, the focus is on simplification of design. There are several key areas in process design for a low-cost carrier, that is, the way in which the product is delivered to the consumer, that result in significant savings over a full service carrier.

One of the primary forms of process design savings is in the planning of point-to-point city pair flights, focusing on the local origin and destination market rather than developing hub systems. In practice, this means that flights are scheduled without connections and stops in other cities. This could also be considered product design, as the passenger notices the benefit of traveling directly to their desired destination rather than through a hub. Rather than having a bank of flights arrive at airports at the same time, low-cost carriers spread out the staffing, ground handling, maintenance, food services, bridge and gate requirements at each airport to achieve savings.

The theory is simple – the busiest time of the day drives the number of gate agents, ticket agents, security screeners, ground handlers (and their equipment), maintenance technicians, bridges, and gates required for a given shift. The HS systems create their own congestion and force staffing levels up to meet the demand, yet the interval between "banks" of flights creates idle time for the staff and capital equipment. Spreading the flights throughout the day not only minimizes staffing costs, but also reduces the amount of capital outlay for ground handling equipment, gates, check-in terminals and bridges. Having said this, the vast majority of low-cost carriers are not *pure* point-to-point carriers, primarily due to geographic constraints.[17]

Another benefit that is directly linked to the design of point-to-point networks is that the daily utilization of the aircraft increases. Utilization is defined as the number of hours flown in a day – the productive time for a very expensive asset. Point-to-point networks reduce congestion and minimize delays by eliminating the problem of waiting for connecting passengers and their luggage. Further, optimizing the daily use of aircraft is easier with a point-to-point network, since scheduled departure times from other cities do not have to be timed to meet up with many other flights in a hub city. This route design, plus the use of secondary and smaller airports, allows for quicker turn times and returns the aircraft to the sky. Turn times for low-cost carriers are typically less than 30 min. Table 2 clearly indicates that Southwest is a leader in aircraft utilization, especially given their relatively short stage length.[18]

Noncore services are often contracted out with low-cost airlines. In a sense, this furthers the concept of "unbundling" to process design. Typical processes that are contracted out include ground handling and heavy maintenance, as is the case with WestJet.[19] This can allow low-cost carriers to choose from the best of the providers in a certain area without have to invest the infrastructure and operational expertise in that function. It also allows the carrier to react quickly to changes in a pro-cyclical airline environment, as they are less constrained in staffing and capital decisions. Ryanair is a prime example of this concept – everything from call centers in foreign countries to airport staff are contracted out. The direct staffing levels at Ryanair are kept to a bare minimum – at last count, less than 50 employees per aircraft; Westjet has about 70, Air Canada 163 and the former Canadian used to have 170.

Low-cost carriers continue to re-define the cheapest distribution process as well. By pushing to a direct sales model, rather than through travel agents, using "e-tickets" rather than expensive printed tickets, and garnering a greater percentage of sales on the Internet, VBAs avoid travel agency commissions and ticket production costs.[20] Southwest Airlines was an innovator in "e-ticketing," and was also one of the first to initiate bookings on the Internet. Typically, low-cost carriers avoid expensive printed tickets and ticket jackets, and simply opt for a confirmation code or number, saving postage and handling costs. The Internet has proved to be a tremendous source of savings for airlines. Interestingly, low-cost carriers have a natural advantage in pushing their passengers to use the Internet, since the business traffic that is the mainstay of full service carriers tend to place more value on their time and will simply make their arrangements through travel agencies.[21]

Check-in procedures and seat assignments are simpler with VBAs as well. Low-cost airlines, especially the more recent examples, have had the opportunity to use newer, quicker, more advanced software to speed the passenger check-in process. Boarding passes are made to be either very cheap or re-usable, and the processing time of the boarding pass itself is designed to be fast and efficient, reducing the amount of staff required to board a flight. Most VBAs do not seat assign passengers prior to their arrival at the airport, while some do not at all. The incentive created with this process design encourages passengers to arrive early and receive a better seat, thereby improving the ability of the airline to depart on time, as there are few passengers to process at the last minute.

One result of all of the previous product and process design considerations is that the accounting and marketing functions of an airline become easier to manage. This may seem to be an odd result. However, given that many business decisions are based on information provided by accounting and the revenue management side of marketing, it is intuitive that a simpler business model will require less analysis for decision-making.

Perhaps the most important effect of all of the product and process design changes, combined with effective organization design, is the dramatic productivity difference between low-cost and full service carriers. The measurement of revenue passenger miles created per employee is considered to be a fairly representative measure of productivity that allows one to compare different airlines with different fleets. This broad measure essentially measures how many employees are required to move passengers around on the airline and its network. Almost every aspect of design contributes to the productivity increase, from the common fleet down to the fact that the product is simpler to sell on the phone or over the Internet. The revolutionary aspect behind this concept is that not only are employees more productive, they also tend to be happier through increased job satisfaction, and contrary to popular opinion, earn as much in total compensation as their counterparts at full service carriers.

All of the above reductions in cost allow low-cost carriers to offer very low fares. The focus on short haul, point-to-point markets, not only achieves lower costs but has revenue advantages as well. Further differences from full service carriers, including one-way pricing, limited fare classes, no refunds without a penalty, and fewer "fences" contributes to the simplification of revenue management. The basic objective of revenue management is to achieve the highest possible average fare, or price per mile, for each ticket sold. Low-cost carriers are well designed to maximize this metric.

Short-haul markets receive more revenue per mile than long haul markets. To a certain extent, this is a function of cost – costs per mile are greater on short haul flights for the same equipment type. However, there is also a significant difference in the demand function for short haul flights. The majority of the travelers that low-cost, low-fare carriers attract rely on disposable income to make their discretionary travel decisions. The degree to which disposable income is used up for travel obviously impacts the amount of disposable income left for other activities such as going to movies or concerts or buying products such as big screen televisions. Consumers therefore make a choice based on their preferences – if the price of travel is less *in relation* to other activities, more travel will be consumed. Not only do the pricing and time constraints favor short haul travel, there is also a greater community of interest between cities that are geographically closer. All these factors combine to create an environment, where low fares in short haul markets have the ability to stimulate much more demand than low fares in long haul markets.

VBAs worldwide have created point-to-point markets, where point-to-point service could not be warranted previously due to lower traffic volumes with higher fares. VBAs not only stimulate traffic in the direct market of an airport, but studies have shown that passengers on VBAs are derived from a much larger "catchment area" than full service carrier passengers. "Catchment area" is defined as the geographic region surrounding an airport from which passengers are derived. While a full service carrier relies on an HS network to create "catchment," low-cost carriers create the incentive for customers to become the "spokes" of their network by offering very low fares.

Flying point-to-point creates two revenue advantages. First, the passenger is willing to pay more for direct service, and second, revenue is not diluted on connecting flights. As identified above, the cost drivers behind longer flights push the cost per mile down as flights increase in length, and passengers are unwilling to pay the same per mile on longer flights as they are on shorter flights. Therefore, having connections and stops in a schedule reduces the amount per mile that can be charged for those passengers traveling on to further destinations. As an example, while WestJet may sell a Calgary–Vancouver fare, and collect the revenue entirely for that journey, Air Canada will have several passengers on the same flight connecting to flights as far away as Haneda, Japan. The pro-rated portion of the fare received for the Calgary–Vancouver sector by Air Canada will be much less than the revenue received by WestJet for the same flight sector. Pricing strategies are also less complicated with a point-to-point network, as the implications of price changes need not be assessed for connecting passengers.

FSAs support a highly complex system of connections in their HS network. Their business model is predicated on broad service in product and in geography bringing customers to an array of destinations with flexibility and available capacity to accommodate different routes, no-shows and flight changes. This business model labors under cost penalties and lower productivity of HS operations including long aircraft turns, connection slack, congestion, and personnel and baggage online connections. These features take time, resources and labor, all of which are expensive and are of fixed cost. The HS system is also conditional on infrastructure, airport and airway, information provision through CRSs and highly sophisticated yield management system – systems that are not easily changed to reflect a changing demand reality.

In summary, the design of low-cost carriers facilitates some revenue advantages in addition to the many cost advantages, but it is the cost advantages that far outweigh any revenue benefits achieved. However, what is clearly evident is that the choice of network is not independent of the firm strategy. The linear point-to-point network of VBAs allows it to achieve both cost and revenue advantages. The sources of cost advantage are identified in Figs. 1 and 2. The first figure shows the key underlying cost drivers and where a VBA like Ryanair has an advantage over FSA in crew and cabin personnel costs, airport charges and distribution costs. The first two are directly linked to network design. An HS network is service intensive and high cost. Even distribution cost savings are related indirectly to

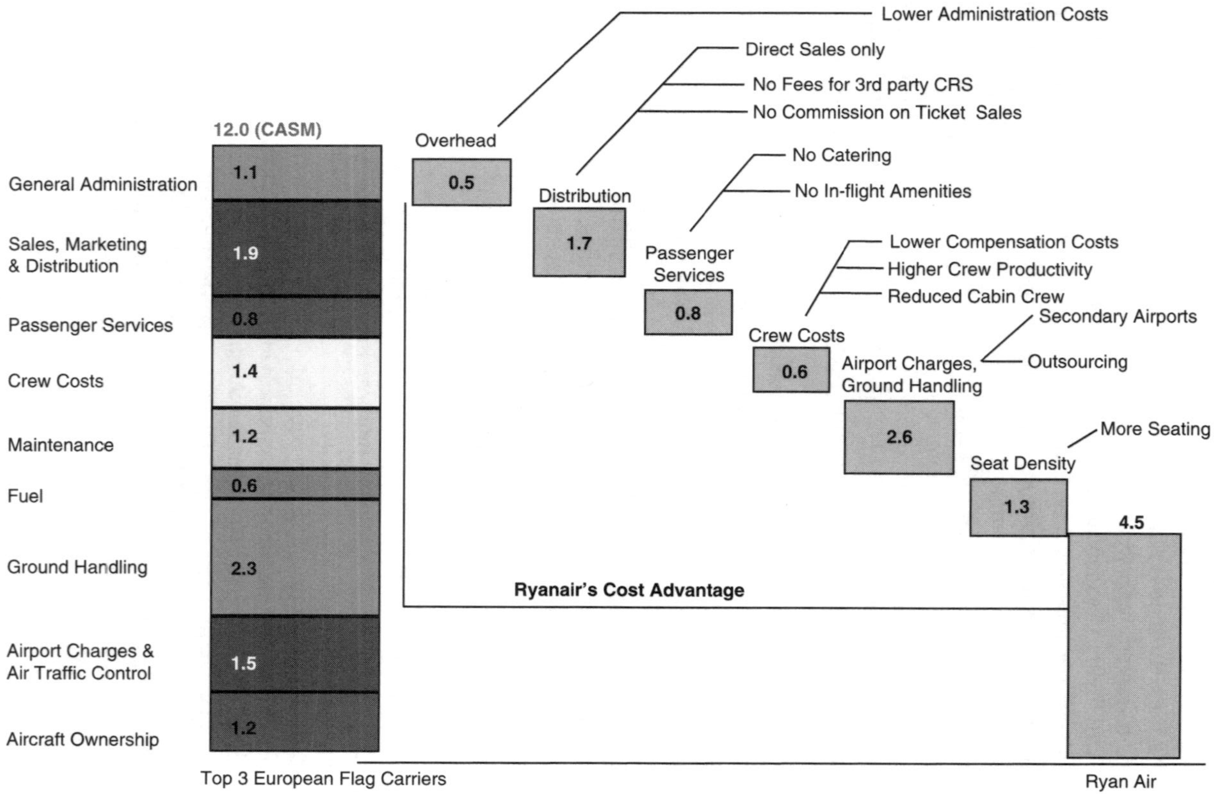

Fig. 1. Source of difference in CASM for low cost and legacy carriers. *Source:* Hyped for hopes: Europe's Low-Cost Airlines (McKinsey Quarterly, No. 4, 2002).

Unit Costs in US$ ASK adjusted for 800 km Stagelength (2001)			
	3Major EU Flag Carriers	Ryanair	easyjet
Aircraft Ownership	1.2	0.7	1.0
Airport/ATC	3.8	1.2	1.0
Distribution	1.9	0.5	0.2
Crew	1.4	0.9	0.8
Total	8.3	3.3	3.0

Fig. 2. Comparison of cost drivers for VBAS and FSAs.

network design because VBAs have simple products and use passengers' time as an input to reduce airline connect costs. In Fig. 2, the major source of cost difference and advantage for VBAs are highlighted. As in the earlier case, key cost drivers are aircraft and cabin crew and airport charges as well as distribution costs. It seems a major advantage of VBAs is the simplicity of their product design. Customers can understand and access this product through Internet or telephone. Thus distribution costs are lower. FSA with more options and complex routings means passengers rely on third parties, in-house travel planners or travel agents. As FSAs try to reduce their costs, they reduce commissions and agents now have less incentive to provide full information or loyalty to a particular carrier. The complexity of their product is tied to their HS system, in conjunction these two forces lead to higher distribution costs.

If one looks at the differences in the United States between VBAs like Southwest and FSAs, there is a 2:1 cost difference. This difference is similar to what is found in Canada between Westjet and Air Canada, as well as in Europe. The difference in cost is due in large measure to business model choices. These carriers buy the fuel and capital in the same market, although there may be some difference between carriers due to hedging for example, these are not structural or permanent changes. The vast majority of the cost difference relates to product and process complexity. This complexity is directly tied to the design of their network structure.

NETWORKS, FIRM STRATEGY AND ALLIANCES

There is a rich literature providing an economic analysis of alliances (Pels et al., 2000; Brueckner, 1997; Oum, Park, & Zhang, 1996; Park, 1997; Park & Zhang, 2000). This literature shows that social efficiency is improved with alliances due to both economies of density and the internalization of the double-mark-up in network markets provided alliances are complementary and there is some measure of alliance competition. In the first case, the summing of marginal costs on each segment internalizes the cost complementarities akin to an HS network in the second case, airline pricing in the absence of alliances ignores the affect of one airline's prices on downstream traffic. This creates an externality – known in the literature as the double marginalization problem. In an alliance, the carriers have an incentive to price to maximize joint profits, a point made very clearly in Anming Zhang's paper in this volume.[22]

There are also some claims that fares on inter-hub markets may increase due to a decrease in competition. The dispute in the literature is whether the higher fares on inter-hub links reflect market power or higher service – more frequency and more destinations. The consensus seems to be even if some rents accrue to alliance partners, the positive effects of alliances outweigh the negative aspects provided the participants are not domestic market leaders.[23]

Alliances have many of the characteristics and incentives of HS systems; cost factors, demand factors and entry deterrence. Like an HS system in an alliance the number of direct connections is reduced, thus the densities on the remaining markets are higher. Economies of density will result in lower per passenger costs. The fixed cost of each hub or gateway can also be spread over more passengers. Thus, one underlying driver of alliances is the incentive to exploit economies of density by decreasing both connections and capacity. There may be further benefits of economies of size, as well, on trunk routes. A second driver arises from the indivisibilities of equipment and schedules. These indivisibilities result in capacity economies that can only be exploited using yield management. As carriers compete for high yield passengers with frequency and destination choice, the indivisibilities mean extra seats become available. The number of seats can and does vary from flight to flight and hence can destabilize a [duopoly] market. Therefore, airlines have an incentive to align to manage the number of seats and frequency of flights. The third underlying driver is alliances leave little room for entry in a market and any entry that might occur would be either at the origin or destination. Finally, Brueckner and Zhang (2001) have shown that frequency in an HS market is higher than in a fully connected network. The strategy is to focus on high-yield traffic and trade-off more frequency for a direct flight. This strategy reduces the overall trip time and providing high value for which business customers are willing to pay more. An alliance may reduce an airlines capacity but this is replaced by a partners capacity so scheduling convenience is maintained but with no increase, and perhaps a decrease, in over market capacity.

Given the strong parallels between HS and alliance strategies, the question is what can one expect to evolve given the shift in domestic markets? Not unlike the domestic markets, firms must make three choices; the choice of network structure, the choice of capacity and the choice of prices (or quantities). One can show that if capacity costs are low and given similar marginal costs, airlines may choose a linear network over an HS. However, given the advantages of alliances the underlying drivers may offset the cost differences between FSA on HS systems and VBA on linear systems.

The underlying drivers of VBA, as we have seen, are the ability to be on a lower cost function with simplicity of product, process and organization. This is achieved through unbundling both airline and airport services. In markets in which alliances are formed it is not clear that such unbundling, beyond that offered by charter carriers, is possible. Greater aircraft use, turnaround times and use of labor and capital are all affected by route length.

The argument made earlier was that FSA in their choice of HS have created a complex high cost product. The complexity contributes to the higher costs. The HS was designed to meet business needs as have alliances but alliances while meeting the needs of high yield traffic have not provided fundamental cost advantages (other than density which is demand side driven). Alliances have to a modest degree enabled airlines to reduce costs at hubs through improved productivity, and to rationalize their networks by using partner access. However, the alliances have not provided the demand side benefits that previously increased alliance partner profits. That is, the motivation for alliances was driven by both cost and demand side benefits, failure of one side may reduce their value sufficiently as to rationalize but perhaps not abandon them.[24]

One operating model that some carriers are pursuing (Air Canada and Lufthansa) is to have parallel operations between HS and linear networks. This has been termed by some as "re-branding". To be effective, the "brand" that is competing in low-cost carrier markets must be truly low cost and not low fare. Furthermore, it must be the case that FSA subsidiaries do not cannibalize their mainline passengers. Such a parallel strategy means hubs will be less valuable and alliances will be more vulnerable.

SUMMARY

The evolution of networks in today's environment will be based on the choice of business model that airlines make. As we have seen, the HS system is an endogenous choice for FSA while the linear FC network provides the same scope for VBAs. The threat to the HS network is the threat to FSAs. The HS network will only disappear if the FSA cannot implement a lower cost structure business model and at the same time provide the service and coverage that higher yield passengers demand. The higher yield passengers have not disappeared, the market has only become somewhat smaller and certainly more fare sensitive, on average.

The ability of FSA to co-exist with VBA and hence HS networks with linear networks is to redesign their products and provide incentives for passengers to allow a reduction in product, process and organizational complexity. For example, no-shows are a large cost for FSA and they have to design their systems in such a way as to accommodate the no-shows. This includes over-booking and the introduction of demand variability. This uncertain demand arises because airlines have induced it with service to their high yield passengers. Putting in place a set of incentives to reduce no-shows would lower costs because the complexity would be reduced or eliminated. One should have complexity only when it adds value.

The HS is complex, the business processes are complex and these create costs. In a linear system it is still possible to have connectivity, more frequency and long haul flights and therefore an HS network need not substitute with a linear network. An HS lowers productivity and increases variable costs, and fixed costs, but these are not characteristics inherent in the HS design. They are inherent in the way FSA use the HS network to deliver and add value to their product. This is because the processes are complex even though the complexity is needed for a smaller, more demanding, higher yield, set of customers. The HS systems will depend on tailoring and redesigning the processes. For example, we have seen that American Airlines has increased the time between connections at some hubs; essentially substituting passenger time for expensive airline resources. This has had a significant effect on productivity and variable costs. The redesigning of business processes moves the FSA between cost functions and not simply down their existing cost function.

There are a number of ways redesigning takes place including scheduling constraints to move from HS to "rolling hubs". There may be some loss of traffic to local markets but the operational rule should be to link customer requirement and willingness to pay with capability and cost. It is possible that customers can be made better off by the price–service value proposition. A lower fare for lower service versus higher fares for higher service. However, carriers must also use incentives. Redesigning is a supply side-solution while incentives are a demand-side solution. In this new model FSA will experience a substantial change in their product attributes and support requirements but the "new" HS system will still be the dominant network structure. I refer to this as a connectivity hub, which is evolving between both VBAs and FSAs. The FSAs are increasing connect times while VBAs are building and growing their markets sufficiently to reduce connect time. The network structure drives pricing, fleet, and service strategies.

Alliances will continue as in the past. Even in the presence of fully liberalized markets, mergers will not replace alliances. One need only look to Europe to see this outcome. The United States is the only market in which nationalism does not drive the need to have a visible "national" carrier. Further problems arise when two carriers attempt to integrate work forces, organizations and work rules. One need only look to the costs of such mergers in domestic markets to appreciate these costs. On the other hand, a major issue in almost all markets, particularly North America is the excess capacity of FSAs. It is unlikely any government, particularly the United States, would allow a carrier to fail; mergers may be the only palatable way of rationalizing capacity. Alliances and HS systems are synonymous but the redesigns required in domestic markets are under less pressure in international markets. The reason is flights are longer and spokes extend into home and alliance markets. The biggest problem may be market size, which requires a focus on capacity. However, redesigning processes and creating incentive schemes in domestic markets will spillover into alliance markets; otherwise the alliance would have to be much looser than currently. A looser alliance reduces the net gains from such a relationship.

Will alliances tighten or loosen? Given the changes in the market with substantially lower demand and excess capacity, we should see the core carriers in alliances dominate and serve global hubs. Smaller alliance partners will have looser affiliations with the alliance. This argument is based on the idea that density is less important than market size and that capacity needs to be rationalized.

NOTES

1. Easyjet, because it follows a strategic focus on business travel, has multiple frequencies with fewer destinations while Ryanair has many destinations with fewer frequencies.
2. Frequent flyer points in a large carrier network are more valuable than points used in a small network.
3. The average U.S. hub has 38% of passengers originating at the hub, 61% connecting traffic and 1% from other airlines.

4. Directional hubs are less efficient since they are dependent on a traffic flow from one area to another; e.g. American's hub at St. Louis is directional and feeds traffic predominantly from east to west.

5. This is why airlines such as American, Air Canada, Northwest, and Lufthansa acted so aggressively when a low-cost carrier entered markets.

6. American has also reduced its turnaround at spoke cities from 2.5 h previously to approximately 42 min.

7. American has reported that as a result of smoother traffic flows, it has been operating at Dallas/Fort Worth International Airport with nine fewer mainline aircraft and two fewer regional aircraft. At Chicago, the improved efficiency has allowed American to take five aircraft off the schedule, three large jets and two American Eagle aircraft. Don Casey, managing director for scheduling for American, estimates savings of $100 million a year from reduced costs for fuel, facilities, and personnel, part of the $2 billion in permanent costs American has trimmed from its expense sheet. The new flight schedule has brought unexpected cost relief at the hubs but also at the many "spoke" cities served from these major airports. *Aviation Week and Space Technology*, September 2, 2002 and February 18, 2003.

8. The underlined portion is of crucial significance, as we will see below.

9. Qantas has created a low-cost, long-haul carrier, Australian Airlines, which may succeed due to the lack of a well-developed charter segment.

10. Airlines were able to reduce their costs to some degree by purchasing ground services from third parties. Unfortunately they could not do this with other processes of the business.

11. In Canada, Westjet, a low-cost carrier has a CASM of 12¢ per seat mile while Air Canada's are approximately 18¢.

12. Product design refers to the "look and feel" of a product, and is the most visible difference between low-cost and full-service carriers to the airline passenger.

13. It should also be noted that the VBA model is not generic. Different low-cost carriers do different things and like all businesses we see continual redefinition of the model.

14. On the other hand, full-service carriers will typically have a range of aircraft types for performing different missions.

15. One only has to consider the significant taxi times at Pearson Airport in Toronto to see why Hamilton is such an advantage for Westjet.

16. Westjet pays $1,200 to land in Hamilton while Pearson charges $2,000 for the same aircraft to land. Similar differences exist for San Francisco and Oakland.

17. Southwest Airlines noted in its 2001 Annual Report that over 70% of their traffic was point-to-point, while Ryanair, based in Dublin, schedules exclusively point-to-point flights and will not sell connecting tickets. Westjet has increasing amounts of connecting traffic at Hamilton and Calgary, but this is connecting not Hubbing. The passenger pays the cost in longer wait time not a higher fare.

18. Westjet and other VBAs have been increasing their stage length. JetBlue flies on over 1,000 mile stage lengths.

19. Daily as well as A and B checks are done in house while C and D checks are contacted out.

20. WestJet has estimated that a typical ticket booked through their call center costs roughly $12, while the same booking through the Internet costs around 50 cents.

21. Recent moves by full service carriers in the United States and Canada to eliminate base commissions should prove to be interesting developments in the distribution chains of all airlines. WestJet has stated that Internet booking account for approximately 40% of their sales, while Ryanair claimed an Internet sales percentage of 91% in March 2002.

22. This results in an increase in passengers. However, in the case of vertical or complementary alliances (Park, 1997) this will lead to lower prices with competition between alliances, but in a parallel alliance it will lead to an increase in prices.

23. For example, there is evidence that STAR alliance has increased fares significantly on the Canada–German market.

24. Having said this it is not clear that international traffic is going to be served by VBA in domestic markets; although interlining is much easier today than previously.

REFERENCES

Brueckner, J. K. (1997). The economics of international codesharing: An analysis of airline alliances. Office of Research Working Paper 97-0115, The University of Illinois-Urbana-Champaign.

Brueckner, J. K., & Zhang, Y. (2001). A model of scheduling in airline networks: How a Hub-and-spoke system affect flight frequency. *Fares and Welfare Journal of Transport Economics and Policy*, *35*(2), 195–222.

Dresner, M., & Windle, R. (1996). The impact of low cost carriers on airport and route competition. *Journal of Transport Economics & Policy*, *30*(3), 309–328.

Hendricks, K., Piccione, M., & Tan, G. (1995). The economics of hubs: The case of monopoly. *Rand Journal of Economics*, *28*, 291–303.

Oum, T., Park, J-H., & Zhang, A. (1996). The effect of airline codesharing agreements on firm conduct and international airfares. *Journal of Transport Economics and Policy*, *30*(2), 187–202.

Park, J-H. (1997). The effects of airline alliances on markets and economic welfare. *Transportation Research E*, *33*, 181–195.

Park, J. H., & Zhang, A. (2000). An empirical analysis of global airline alliances: Cases in the North Atlantic markets. *Review of Industrial Organization*, *16*, 367–384.

Pels, E., Nijkamp, P., & Rietveld, P. (2000). A note on the optimality of airline networks. *Economic Letters*, *69*, 429–434.

Shy, Oz. (2001). *The economics of network industries*. London: Cambridge University Press.

COMPETITION MODELS OF STRATEGIC ALLIANCES

Anming Zhang

ABSTRACT

This paper examines competition models for three types of strategic alliances (vertical, horizontal, and hybrid alliances). It is found that even if a vertical alliance creates a negative direct effect on profit, it might be pursued because it is a dominant strategy. In general, total welfare improves as a result of vertical alliances. On the other hand, a horizontal alliance reduces competition not only in the market where prior competition between the partners takes place, but also in other markets of the alliance network. The implications of hybrid alliances for international airline alliances, whose members have both complementary and overlapping routes, are discussed.

1. INTRODUCTION

Strategic alliances represent an important form of cooperation between two or more business entities and are widely and increasingly used by business firms. They are particularly prevalent in network-oriented industries such as airline, telecommunications, shipping, and logistics industries. Of these industries, the airline industry has the largest number of alliances. As of

Global Competition in Transportation Markets: Analysis and Policy Making
Research in Transportation Economics, Volume 13, 75–100

ISSN: 0739-8859/doi:10.1016/S0739-8859(05)13004-2

1999, the total number of airline alliances was 513, an increase of nearly 83% over 1994 (*Airline Business*, July 2000). Furthermore, the two largest global alliance groups, namely, Star Alliance and OneWorld, accounted for 40% of the world's revenue passenger kilometers, while the top five alliances accounted for 64%. It is predicted that because of convenience, an increasing proportion of travellers will choose to travel with these global alliance network carriers resulting in a further increase in their market share (Oum, Park, & Zhang, 2000).

Like alliances in telecommunications, logistics and other transportation industries, an airline alliance is a multi-product network, with each of its products corresponding to travel (either by people or by cargoes) in a particular city-pair market. The nature and extent of cooperation between alliance members will thus affect the nature of interaction among competing alliances in various markets. In our context, strategic alliance is considered to be a medium- to long-term partnership of two or more firms with the goal to improve partners' competitive advantages collectively vis-à-vis their competitors.[1] In other words, strategic alliances are different from ordinary alliances in that the partners make a serious commitment to cooperate, not only on operational matters but also on key strategic matters. These include new product development, market access and linkages, quality improvement, and pricing and capacity coordination.

Despite the importance of strategic alliances in network-based markets, there is only a relatively small body of literature devoted to competition between alliance networks. This is in sharp contrast to the vast literature on single-firm, single-product oligopoly and on mergers.[2] In this paper, we examine the rivalry between different alliances in a model where each alliance member maximizes its own profit and, to a certain extent, its partner's profit, in the context of both within-market and cross-market interactions. We further examine whether strategic alliances should be viewed as causes for anticompetitive concerns. Strategic alliances allow firms to expand their networks, realize economies of scale and scope, and improve product quality and customer service. For example, airline alliances enable partners to better coordinate their flight schedules to minimize travelers' waiting time between flights while providing sufficient time for connections. Joint baggage handling further eliminates the need to retrieve and re-check baggage at connecting places, thereby reducing the risk associated with interline handling in which no single carrier has the sole responsibility for the baggage. Despite these potential benefits to firms and consumers, antitrust authorities often view strategic alliances with suspicion. In the airline example, most international airline routes have only a few competitors. As a consequence, an alliance between any two significant competitors may considerably reduce the degree of competition on certain routes. An important consideration is whether government policy will be to encourage or restrict its carriers from increasing the size of their networks through alliances with foreign carriers, while taking account of any potential anticompetitive effects.

A strategic alliance is like a lesser form of a merger. Similar to a network merger, a strategic alliance in general has both horizontal and vertical effects. Our analysis shows that a *vertical alliance*, in which two firms link up their complementary products, confers a strategic advantage by allowing the partners to commit credibly to greater outputs, owing to network complementarities, and the strategic effect tends to augment its positive direct effects arising from the elimination of the double-marginalization problem in vertical integration.[3] Even if a vertical alliance creates a negative direct effect, it might be pursued because it is a dominant strategy. We find that in general, total welfare improves as a result of vertical alliances. On the other hand, a horizontal alliance reduces competition not only in the market where prior competition between the partners takes place, but also in other markets of the alliance network. Finally, for hybrid alliances (a mix of vertical and horizontal alliances), they are likely to have both pro- and anti-competitive effects. Our analysis also suggests that the incentive to form a hybrid alliance may be reduced when we move from a monopoly-pair setting to a duopoly-pair setting.

Management literature on strategic alliances has focused mainly on non-strategic motives of alliances: learning, entering new markets, creating new technology, sharing risk and cost, and gaining access to financial and technological resources (e.g., Hamel, Doz, & Prahalad, 1989; Harrigan, 1988; Hennart, 1991; Hennart & Reddy, 1997; Kogut, 1988; Kogut & Singh, 1988; Mody, 1993; Mowery & Teece, 1993; Pisano, 1990).[4] Another area of enquiry in management literature has been the effects of strategic alliances on partners' performance, using proxy variables such as stability, duration of alliances, survival ratio, perceived satisfaction of managers, and stock returns (e.g., Blodgett, 1992; Dussauge & Garrette, 1995; Geringer & Herbert, 1989; Hamel, 1991; Kogut, 1989; Osborn & Baughn, 1990; Parkhe, 1993; Veugelers, 1993; Chan et al., 1997; Das, Sen, & Sengupta, 1998).

Our paper is more closely related to recent theoretical work on international airline alliances. Park (1997), Brueckner (2001), and Park, Zhang, and Zhang (2001) examined the effect of international airline alliances on firms' strategies, traffic level, and price and welfare on the alliance routes. In particular, Brueckner (2001) and Park et al. (2001)

pointed out that an alliance would reduce competition in the gateway (or inter-hub) market that was previously served by the alliance partners, but cooperative pricing of trips by the partners would increase traffic in the connecting market since portions of a connecting trip are complements.[5] We extend this literature by analyzing generic strategic alliances and their impacts on social welfare. In addition, we explicitly consider rivalry between alliance networks where each firm decides not only on whether to join an alliance, but also on the extent of cooperation if it were to join. In the process, we develop a theory of strategic alliances – in the form of vertical, horizontal, or hybrid alliances – as opposed to other forms of inter-firm organization such as full integration of firms, or independent firms.

In Section 2, we set up a basic model of vertical alliance. In Section 3, the strategic effects of competition between vertical alliances are derived, and the impact of these alliances on price and social welfare is investigated. Sections 4 and 5 look at the models of horizontal alliances and hybrid alliances, respectively. Finally, Section 6 contains conclusions and further discusses the implications for international airline alliances, whose members have both complementary and overlapping routes.

2. MODEL OF VERTICAL ALLIANCE

We consider a four-product, four-firm model that is likely the simplest structure in which strategic alliance rivalry can be addressed. The inverse demand function is written as:

$$p^i = p^i\,(q^1, q^2, q^3, q^4) = p^i(Q) \tag{1}$$

where Q is the product vector (q^1, q^2, q^3, q^4) with

$$p_2^1 > 0, \quad p_1^2 > 0, \quad p_4^3 > 0, \quad p_3^4 > 0 \tag{2}$$

The first two inequalities of condition (2) show demand complementarity between 1 and 2, whereas the other two inequalities show demand complementarity between 3 and 4. There are four firms, A, B, C, and D, that produce 1, 2, 3, and 4, respectively. Using $c^i(q^i)$ to denote total cost, each firm's profit may be written as

$$\begin{aligned}
\pi^{A} &= \pi^{A}(Q) = q^1 p^1(Q) - c^1(q^1)\\
\pi^{B} &= \pi^{B}(Q) = q^2 p^2(Q) - c^2(q^2)\\
\pi^{C} &= \pi^{C}(Q) = q^3 p^3(Q) - c^3(q^3)\\
\pi^{D} &= \pi^{D}(Q) = q^4 p^4(Q) - c^4(q^4)
\end{aligned} \tag{3}$$

We consider two potential alliance pairs: firms A and B form one alliance pair, firms C and D the other. We explore the effect of strategic alliances in a model where each alliance member maximizes its own profit and, to a certain extent, its partner's profit. As noted in Oum et al. (2000), a significant proportion of strategic alliances in the airline industry is equity alliance in which one partner buys a share of stock in its partners. Although alliances of this type do not change production possibilities at any partner firm, they do alter the incentives of the alliance partners. Each firm now maximizes not only its own profit but also, to a certain extent, its partner's profit.[6] In effect, Oum et al. (2000, Chapter 9) found that an equity alliance is more stable and yields greater firm values, measured in stock returns, than other types of strategic alliances.

Thus, the alliance between firms A and B may be expressed as

$$\underset{q^1}{\text{Max}}(\pi^{\text{A}} + \alpha^{\text{A}}\delta^{\text{A}}) + \alpha^{\text{A}}\pi^{\text{B}} \equiv \pi^{\text{AB}}(Q; \alpha^{\text{A}})$$

$$\underset{q^2}{\text{Max}}(\pi^{\text{B}} + \alpha^{\text{B}}\delta^{\text{B}}) + \alpha^{\text{B}}\pi^{\text{A}} \equiv \pi^{\text{BA}}(Q; \alpha^{\text{B}}) \tag{4}$$

In addition to its own profit, a firm also incorporates a fraction of its alliance partner's profit in making its optimal quantity decision. The fraction is captured by parameter α^i, with $0 \leqslant \alpha^i \leqslant 1$ ($i = \text{A}, \text{B}$). For example, when $\alpha^{\text{A}} = 0$, firm A maximizes its own profit, ignoring the (positive) effect of its decision on firm B. As α^{A} increases, A pays an increasing attention to the effect of its decision on B, with $\alpha^{\text{A}} = 1$ corresponding to (full) joint-profit maximization by A. Further, note that $\alpha^{\text{A}}\delta^{\text{A}}$ and $\alpha^{\text{B}}\delta^{\text{B}}$ in (4) denote alliance "synergies" to firm A and firm B, respectively, which vary proportionally with α^i. α^i, δ^{A}, and δ^{B} are the synergies under (full) joint-profit maximization, and they can be either positive or negative. Some of the positive effects have been mentioned earlier. (Note, however, that the positive synergies defined here exclude the benefits from cooperative output decisions.) An additional positive effect is that an alliance may allow partners to have flexible organizational structures so that each can conduct complex business transactions without sacrificing autonomy (Mody, 1993). Nevertheless, strategic alliances can yield negative effects as well. For example, the partners may hesitate to share important information and behave opportunistically, since they could become would-be rivals (Hamel et al., 1989; Parkhe, 1993). The partners may also have potentially incompatible structures in financial status, asset composition, labor contracts, and management style.[7] For a given alliance, δ^{A} and δ^{B} are positive if positive effects overall dominates the negative effects.

Similarly, the alliance between firms C and D is expressed as

$$\underset{q^3}{\mathrm{Max}}(\pi^{\mathrm{C}} + \beta^{\mathrm{C}}\delta^{\mathrm{C}}) + \beta^{\mathrm{C}}\pi^{\mathrm{D}} \equiv \pi^{\mathrm{CD}}(Q;\beta^{\mathrm{C}})$$
$$\underset{q^4}{\mathrm{Max}}(\pi^{\mathrm{D}} + \beta^{\mathrm{D}}\delta^{\mathrm{D}}) + \beta^{\mathrm{D}}\pi^{\mathrm{C}} \equiv \pi^{\mathrm{DC}}(Q;\beta^{\mathrm{D}}) \quad (5)$$

where the extent of alliance is measured by β^j, $0 \leqslant \beta^j \leqslant 1$ $(j = \mathrm{C}, \mathrm{D})$, and the overall synergies are measured by $\beta^{\mathrm{C}}\delta^{\mathrm{C}}$ and $\beta^{\mathrm{D}}\delta^{\mathrm{D}}$. In (4) and (5), synergies δ's are assumed, for simplicity, to be independent of output levels.

An immediate consequence of the above specification is that

$$\pi_2^{\mathrm{A}} > 0, \quad \pi_1^{\mathrm{B}} > 0, \quad \pi_4^{\mathrm{C}} > 0, \quad \pi_3^{\mathrm{D}} > 0 \quad (6)$$

Inequalities (6) are equivalent to condition (2), since, e.g., $\pi_2^{\mathrm{A}} = q^1 p_2^1 > 0$ by (2). Inequality $\pi_2^{\mathrm{A}} > 0$ says that an increase in B's output will increase A's profit. Furthermore, the products belonging to the same alliance network are assumed to exhibit "network complementarities" in the following sense,

$$\pi_{12}^{\mathrm{A}} > 0, \quad \pi_{21}^{\mathrm{B}} > 0, \quad \pi_{34}^{\mathrm{C}} > 0, \quad \pi_{43}^{\mathrm{D}} > 0 \quad (7)$$

That is, a firm's marginal profit increases in its partner's output; as a consequence, the outputs are "strategic complements" (Bulow, Geanakoplos, & Klemperer, 1985; Fudenberg & Tirole, 1984). The fact that the partners' products are complements is conducive to network complementarities. Since, for example, $\pi_{12}^{\mathrm{A}} = p_2^1 + q^1 p_{12}^1$ and $p_2^1 > 0$ by (2) or (6), it follows that $\pi_{12}^{\mathrm{A}} > 0$ if $p_{12}^1 \geqslant 0$. Thus, given (2) or (6), condition (7) will hold if demand functions are linear. Nonetheless, (7) can hold for more general demand specifications, particularly in network-based markets. Oum et al. (1995, 1996) contained explicit derivations of network complementarities in, respectively, a hub-spoke airline network and a hub-spoke airport network. We refer to strategic alliances that satisfy (6) and (7) as "vertical alliances."

In what follows, we shall, for simplicity, impose some symmetry by considering $\alpha^{\mathrm{A}} = \alpha^{\mathrm{B}} = \alpha$ and $\beta^{\mathrm{A}} = \beta^{\mathrm{B}} = \beta$. The parameter α or β may be interpreted as the "degree of alliance" (or "degree of cooperation") between the partners. For example, when $\alpha = 0$, firm A and firm B act independently. As α increases they intensify their cooperation, with $\alpha = 1$ corresponding to "full alliance" (full integration, or a merger) in which the partners act like a single decision-making unit in their joint-profit maximization.

Furthermore, we treat the alliance decision as strictly prior to the output decision. This suggests the following two-stage alliance game. In stage 1, each alliance pair simultaneously selects its alliance structure (α, β). In stage 2, given the alliance decision, each firm chooses its own output level. As discussed in the introduction, a strategic alliance brings the participants into what is essentially a medium- to long-term contractual relationship. In practical terms, such alliance involves a strategic commitment by top management to link up a substantial part of the partners' respective networks, as well as collaborating on some key business areas. As a result, alliance development requires nontrivial sunk investments, especially the time and effort committed by senior managers. For these reasons, the alliance structure, once decided upon, cannot easily be altered in a major way. The strategic alliance relationship is a *strategic* decision, which might reasonably be regarded as given at the time when individual firms establish their own quantities in the product–market game.

In the product–market interactions, we have specified that products 1 and 2 are complements (so are products 3 and 4). Across the alliances, we assume that the two alliance pairs are rivals:

$$\pi_j^{A} + \pi_j^{B} < 0, \quad i = 1, 2; \qquad \pi_i^{C} + \pi_i^{D} < 0, \quad j = 3, 4 \tag{8}$$

That is, the outputs of the two alliances are substitutes. Following the standard practice in models of quantity competition (see Dixit, 1986; Shapiro, 1989), we further assume that an alliance's marginal revenue declines when the output of the rival alliance rises:

$$\pi_{1j}^{AB} < 0, \quad \pi_{2j}^{BA} < 0, \quad i = 1, 2; \qquad \pi_{3i}^{CD} < 0, \quad \pi_{4i}^{DC} < 0, \quad j = 3, 4 \tag{9}$$

Condition (9) implies that the outputs of the two alliances are "strategic substitutes" (Bulow et al., 1985; Fudenberg & Tirole, 1984).

3. EFFECTS OF VERTICAL ALLIANCES

3.1. Monopoly-Pair Case

Before exploring the strategic issues involved in the choice of alliances, we consider a single pair of firms who are unconcerned with entry, which serves as a useful base for comparison. It can be easily shown that the monopoly pair will form a full vertical alliance if the integration overall does not yield negative synergy for the alliance. It is useful to take a closer look at the condition, which can be written, using the first-order conditions of quantity choices, as

$$\frac{\mathrm{d}(\phi^{A} + \phi^{B})}{\mathrm{d}\alpha} = (1 - \alpha)\left(\pi_1^{B}\frac{\mathrm{d}q^1}{\mathrm{d}\alpha} + \pi_2^{A}\frac{\mathrm{d}q^2}{\mathrm{d}\alpha}\right) + (\delta^{A} + \delta^{B}) \tag{10}$$

where $\phi^{\rm A}$ and $\phi^{\rm B}$ denote the profits when q^1, q^2 take the optimal levels for any given (see (16) below), and $\pi_1^{\rm B}$ and $\pi_2^{\rm A}$ are positive by (6). It can be easily shown that $({\rm d}q^i/{\rm d}\alpha)>0$. Hence, the first term on the right-hand side of (10) is the familiar "externality effect": the alliance allows the partners to internalize the positive effect of an increase in one firm's output on its partner's profit, owing to demand complementarities. This internalization of demand externalities is similar to the mitigation of the "double marginalization" problem in vertical integration (e.g., Tirole, 1988).[8] Therefore, if the integration overall does not yield negative synergies, then a monopoly pair will form a full vertical alliance. On the other hand, if $\delta^{\rm A}+\delta^{\rm B}$ is negative and sufficiently large in magnitude, then a partial alliance may result (i.e., optimal α is between 0 and 1).

3.2. Strategic Effects

We now examine the subgame perfect equilibrium of the two-stage alliance game. To solve for the equilibrium, we start with the second-stage competition. In this stage, firms simultaneously choose their outputs to maximize profits, taking the alliance structure (α, β) as given. The Cournot equilibrium is characterized by the first-order conditions derived from (4) and (5) (subscripts denoting partial derivatives),

$$\pi_1^{\rm AB}(Q;\alpha) = \pi_1^{\rm A} + \alpha\pi_1^{\rm B} = 0 \tag{11}$$

$$\pi_2^{\rm BA}(Q;\alpha) = \pi_2^{\rm B} + \alpha\pi_2^{\rm A} = 0 \tag{12}$$

$$\pi_3^{\rm CD}(Q;\beta) = \pi_3^{\rm C} + \beta\pi_3^{\rm D} = 0 \tag{13}$$

$$\pi_4^{\rm DC}(Q;\beta) = \pi_4^{\rm D} + \beta\pi_4^{\rm C} = 0 \tag{14}$$

and second-order conditions $\pi_{ii}^K<0$, for $K = $ AB, BA, CD, DC. Regularity conditions are imposed so that the equilibrium exists and is stable.

The comparative static effects of alliance variables, α and β, on the equilibrium outputs, denoted $q^i(\alpha,\beta)$, are derived in Proposition 1 (denoting $q_\alpha^i \equiv \partial q^i(\alpha,\beta)/\partial\alpha$, etc.):[9]

Proposition 1. Under vertical alliances,

$$q_\alpha^i>0,\ \ q_\beta^j>0,\ \ q_\alpha^j<0,\ \ q_\beta^i<0,\quad i=1,2;\ j=3,\ 4 \tag{15}$$

Thus, an increase in the degree of a vertical alliance will increase own outputs, while simultaneously decreasing rival's outputs, for each member firm.

Proposition 1 gives a strong result: increasing the degree of vertical alliance will increase the alliance's own output, while simultaneously decreasing its rival's output, in both product markets. The intuition associated with this result is as follows. With the "network complementarity" effect, an increase in the degree of alliance raises the marginal profit of each alliance partner. This allows the partners to commit, credibly, to greater outputs in both markets. Since the outputs of competing alliances are strategic substitutes, such a commitment would induce a contraction in the rival's outputs. In effect, forming an alliance is a top-dog strategy in the sense of Fudenberg and Tirole (1984), which allows the alliance firms to be more aggressive in product–market competition.

These comparative static effects have important implications for alliances. The strategic interaction among firms in choosing the extent of their alliances takes place in the first stage. Taking the second-stage equilibrium outputs into account, firm i's profit, denoted as Φ^i, can be written as

$$\phi^i(\alpha, \beta) \equiv \pi^i(Q(\alpha, \beta)) + \alpha\delta^i, \quad i = \mathrm{A, B, C, D} \tag{16}$$

The alliance equilibrium arises when each alliance pair chooses its profit-maximizing alliance structure, taking the alliance structure of the other as given at the equilibrium value. The effect of α on the alliance's profit can be expressed as

$$\begin{aligned}\frac{\partial(\phi^{\mathrm{A}} + \phi^{\mathrm{B}})}{\partial\alpha} &= (1-\alpha)(\pi_1^{\mathrm{B}} q_\alpha^1 + \pi_2^{\mathrm{A}} q_\alpha^2) + (\delta^{\mathrm{A}} + \delta^{\mathrm{B}}) + [(\pi_3^{\mathrm{A}} + \pi_3^{\mathrm{B}})q_\alpha^3 \\ &\quad + (\pi_4^{\mathrm{A}} + \pi_4^{\mathrm{B}})q_\alpha^4]\end{aligned} \tag{17}$$

The first two terms on the right-hand side of (17) may be considered as the *direct effect* of the alliance variable α. This effect has already been identified in the monopoly-pair case (Eq. (10)): in addition to the alliance synergy, a vertical alliance allows the partners to internalize the positive effect of an increase in partners' output. However, the other effect, referred to as the *strategic effect*, is unique to alliances in a duopoly-pair setting. The strategic effect is represented by the third, bracketed term in (17), which is positive by Proposition 1. Here, a vertical alliance by firms A and B, given the alliance status of firms C and D, makes both A and B be able to supply more output. Such an expansion is credible, so the best response by C and D is to supply less output (i.e., $q_\alpha^j < 0, j = 3, 4$, by Proposition 1). Since the two pairs offer substitutable products, an output fall by C and D will raise the profits of A and B.

Thus, a sufficient condition for choosing a full vertical alliance is that such a full alliance overall does not yield negative synergies. In this case, full vertical alliance is each firm pair's dominant strategy. Further, since conditions (6) and (7) tend to reinforce each other, the strategic interaction usually augments the direct effect and hence full alliance. On the other hand, a positive direct effect is not a necessary condition for firms to choose full alliance. In effect, a full alliance will improve the alliance's profit if the first two terms of (17) are zero but the third term is positive. Moreover, it is possible to construct numerical examples in which even if a full alliance results in a negative direct effect, owing to substantial negative synergies, it remains the pair's dominant strategy. In these cases, the firms do not maximize direct benefits (or, minimize direct costs) in their choice of alliance structure, and the formation of a full vertical alliance is purely for strategic purposes.

While Eq. (17) captures the effect of α on the alliance's profit, the following equation captures the effect of α on the profit of the rival alliance firms:

$$\frac{\partial(\phi^{C} + \phi^{D})}{\partial\alpha} = [(\pi_1^{C} + \pi_1^{D})q_\alpha^1 + (\pi_2^{C} + \pi_2^{D})q_\alpha^2] + [(1-\beta)\pi_3^{D}q_\alpha^3 + (1-\beta)\pi_4^{C}q_\alpha^4] \tag{18}$$

Thus, an increase in α by the AB pair has two effects: first, it increases own outputs q^1 and q^2 by Proposition 1, which in turn will reduce the CD pair's profit since the two pairs produce substitutes. Second, it reduces the rival pair's outputs q^3 and q^4, which in turn will reduce the CD pair's profit as firms C and D produce complements. Similar results can be obtained for the profit effects of a change in β. One immediate implication of such analysis is that the noncooperative rivalry in vertical alliances can result in a Prisoners' Dilemma for alliance partners: that is, a full vertical alliance may yield a lower payoff for partners than a partial vertical alliance. Essentially, each firm pair will ignore the negative impact of its alliance on the rival firms, and full vertical alliances tend to increase outputs beyond the levels produced under partial alliances because of the network externality, thereby lowering prices. When both pairs engage in full alliances, the strategic gains tend to offset each other, and both will be worse off if the resulting prices are too low and/or synergies from integration are small or negative.[10]

3.3. Effects on Price and Social Welfare

From Proposition 1, it is not clear whether total output in each market increases or decreases when one vertical alliance intensifies its cooperation given the alliance structure of the other alliance. This is because the partners of the active alliance will increase their outputs while the rival firms will reduce their outputs. To answer the question, we consider that products 1 and 3 (similarly, products 2 and 4) are perfect substitutes. The vector of changes in total output in each local market can then be denoted as:

$$\frac{\partial \tilde{q}}{\partial \alpha} \equiv \frac{\partial}{\partial \alpha}\begin{bmatrix} q^1 + q^3 \\ q^2 + q^4 \end{bmatrix} = \frac{\partial}{\partial \alpha}\begin{bmatrix} q^1 \\ q^2 \end{bmatrix} + \frac{\partial}{\partial \alpha}\begin{bmatrix} q^3 \\ q^4 \end{bmatrix} \tag{19}$$

We have the following result:

Proposition 2. Assuming homogeneous products, then a vertical alliance increases total output, and hence reduces price, in at least one market. In the case of symmetric firms where all four firms face symmetric demands, both within and across the markets, and have the same costs, the alliance increases total output and reduces price in both markets.

To examine the effect on total surplus of the network, we assume the usual partial equilibrium framework in which the total surplus W can be written as

$$W = u(q^1, q^2, q^3, q^4) - \sum_{i=1}^{4} c^i(q^i) + \alpha(\delta^{\mathrm{A}} + \delta^{\mathrm{B}}) + \beta(\delta^{\mathrm{C}} + \delta^{\mathrm{D}}) \tag{20}$$

where u is the utility function. Differentiating (20) with respect to α and using $\partial u/\partial q^i = p^i$, we obtain

$$\frac{\partial W}{\partial \alpha} = \sum_{i-1}^{4} [p^i - (c^i)']q^i_\alpha + (\delta^{\mathrm{A}} + \delta^{\mathrm{B}}) \tag{21}$$

The signs of the mark-up terms (in the brackets) are positive by the first order conditions, while the q^i_α terms are either positive or negative depending on i (given in Proposition 1). The sign of $\partial W/\partial \alpha$, then, depends in general on the magnitude of mark-ups of the firms under consideration. In the case of symmetric firms we have $(\partial W/\partial \alpha) > 0$ by Proposition 2, i.e., social welfare improves as a result of the rivalry in vertical alliances.

4. HORIZONTAL ALLIANCE

Given the vertical alliance of firms A and B (i.e., given $\alpha > 0$), this section investigates the effect of horizontal alliance on output, price and welfare.

We focus on the situation where firm A and firm C contemplate whether to form a strategic alliance with each other and if so, the extent of cooperation if they were to form an alliance. For simplicity, we assume away product 4 (and firm D).[11] We further assume that firms A and C offer substitutable products

$$p_3^1 < 0, \quad p_1^3 < 0 \tag{22}$$

Similarly, the outputs of A and C are "strategic substitutes:"

$$\pi_{13}^{\mathrm{A}} < 0, \quad \pi_{31}^{\mathrm{C}} < 0 \tag{23}$$

Since $\pi_{13}^{\mathrm{A}} = p_3^1 + q^1 p_{13}^1$ and $p_3^1 < 0$ by (22), we have $\pi_{13}^{A} < 0$ if $p_{13}^1 \leqslant 0$. Thus, the fact that the outputs of A and C are substitutes is conducive to (23). Conditions (22) and (23) imply that, were there no alliances between A and C, the two firms would be direct competitors in the same market. Because of this, the alliance between them is referred to as a "horizontal alliance." Furthermore, given the AB alliance, C competes with A only. As a consequence, firms C and B are unrelated, and their profits are given by $\pi^{\mathrm{C}} = \pi^{\mathrm{C}}(q^1, q^3)$ and $\pi^{\mathrm{B}} = \pi^{\mathrm{B}}(q^1, q^2)$. Assume further that $\pi_{23}^{\mathrm{A}} = 0$, that is, firm A's marginal profit with respect to product 2 (produced by B) is unaffected by the output produced by C and vice versa.

The problem of a possible horizontal alliance between A and C may be expressed as

$$\begin{aligned}
&\operatorname*{Max}_{q^1} \quad \pi^{\mathrm{A}} + \alpha\pi^{\mathrm{B}} + \theta\pi^{\mathrm{C}} \equiv \pi^{\mathrm{ABC}}(q^1, q^2, q^3; \theta) \\
&\operatorname*{Max}_{q^2} \quad \pi^{\mathrm{B}} + \alpha\pi^{\mathrm{A}} + \theta\pi^{\mathrm{C}} \equiv \pi^{\mathrm{BA}}(q^1, q^2, q^3; \theta) \\
&\operatorname*{Max}_{q^3} \quad \pi^{\mathrm{C}} + \theta\pi^{\mathrm{A}} \equiv \pi^{\mathrm{CA}}(q^1, q^2, q^3; \theta)
\end{aligned} \tag{24}$$

where $0 \leqslant \theta \leqslant 1$ represents the extent of cooperation between A and C, and synergy effects of the alliances are for simplicity assumed away.

Given parameter θ, we write down the first-order conditions of (24) as

$$\pi_1^{\mathrm{ABC}}(q^1, q^2, q^3; \theta) = \pi_1^{\mathrm{A}} + \alpha\pi_1^{\mathrm{B}} + \theta\pi_1^{\mathrm{C}} = 0 \tag{25}$$

$$\pi_2^{\mathrm{BA}}(q^1, q^2, q^3; \theta) = \pi_2^{\mathrm{B}} + \alpha\pi_2^{\mathrm{A}} + \theta\pi_2^{\mathrm{C}} = 0 \tag{26}$$

$$\pi_3^{\mathrm{CA}}(q^1, q^2, q^3; \theta) = \pi_3^{\mathrm{C}} + \theta\pi_3^{\mathrm{A}} = 0 \tag{27}$$

As usual, assume that the second-order conditions hold, and that an equilibrium for the above problem exists and is stable. Substituting the equilibrium, denoted $q^i(\theta)$, into (25)–(27), and differentiating the resulting first-order conditions with respect to θ yields

$$\pi_{11}^{ABC} q_\theta^1 + \pi_{12}^{ABC} q_\theta^2 + \pi_{13}^{ABC} q_\theta^3 = -\pi_1^C \tag{28}$$

$$\pi_{21}^{BA} q_\theta^1 + \pi_{22}^{BA} q_\theta^2 = 0 \tag{29}$$

$$\pi_{31}^{CA} q_\theta^1 + \pi_{33}^{CA} q_\theta^3 = -\pi_3^A \tag{30}$$

where in (29), we have $\pi_{23}^{BA} = 0$ and $\pi_2^C = 0$ by using the conditions specified above. We first report the following result:

Lemma 1. Under horizontal alliance (i) q_θ^1 and q_θ^2 have the same sign, and (ii) q_θ^1 and q_θ^3 cannot both be positive.

Lemma 1 thus narrows down the sign combination of $(q_\theta^1, q_\theta^2, q_\theta^3)$ to only three possibilities: (i) $q_\theta^1 < 0$, $q_\theta^2 < 0$, $q_\theta^3 > 0$; (ii) $q_\theta^1 > 0$, $q_\theta^2 > 0$, $q_\theta^3 < 0$; and (iii) $q_\theta^1 < 0$, $q_\theta^2 < 0$, $q_\theta^3 < 0$. Provided that the horizontal alliance is feasible, we can show that the first two cases can be ruled out. The proof is done by considering, for each case, the effect of the horizontal alliance on the partners' profits. For example, consider the first case. Firm C's profit can be written as

$$\pi^C(\theta) = \pi^C(q^1(\theta), q^3(\theta))$$

From the first-order condition (27), it follows that $\pi_3^C = -\theta \pi_3^A$ and

$$\frac{d\pi^C}{d\theta} = \pi_1^C q_\theta^1 - \theta \pi_3^A q_\theta^3 \tag{31}$$

Under case (i), we obtain $d\pi^C/d\theta > 0$, owing to $\pi_1^C = q^3 p_1^3 < 0$ and $\pi_3^A = q^1 p_3^1 < 0$ by (22). Therefore, firm C prefers a full horizontal alliance in this case.

What will happen to C's potential partner, firm A? Since A has already had a vertical alliance with firm B, the proposed horizontal alliance can affect B's profit. Yoshino and Rangan (1995) argued that the major difference between an inter-firm alliance and other forms of inter-firm collaboration is in the extent of shared control, shared contribution and shared outcome. In a strategic alliance, no single firm controls the management process, decision making, or monopolizes the outcome of the alliance. Following this view, the effect of the horizontal alliance on the joint profit of A and B should be examined. The joint AB profit is

$$\pi^A(\theta) + \pi^B(\theta) = \pi^A(q^1(\theta), q^2(\theta), q^3(\theta)) + \pi^B(q^1(\theta), q^2(\theta))$$

Using the first-order conditions (25) and (26) and rearranging, we obtain

$$\frac{\mathrm{d}(\pi^{\mathrm{A}}+\pi^{\mathrm{B}})}{\mathrm{d}\theta}=(1-\alpha)\pi_1^{\mathrm{B}}q_\theta^1-\theta\pi_1^{\mathrm{C}}q_\theta^1+(1-\alpha)\pi_2^{\mathrm{A}}q_\theta^2+\pi_3^{\mathrm{A}}q_\theta^3 \tag{32}$$

Given $\pi_1^{\mathrm{B}}>0$ and $\pi_2^{\mathrm{A}}>0$ by (6), it can be easily verified that $\mathrm{d}(\pi^{\mathrm{A}}+\pi^{\mathrm{B}})/\mathrm{d}\theta<0$ under case (i). Since the joint profit of A and B will fall, the horizontal alliance is not feasible in this case.

Similarly, we can show that $\mathrm{d}(\pi^{\mathrm{A}}+\pi^{\mathrm{B}})/\mathrm{d}\theta>0$ but $\mathrm{d}\pi^{\mathrm{C}}/\mathrm{d}\theta<0$ under case (ii), so firm C will not join the alliance. The discussion leads to:

Proposition 3. For a feasible horizontal alliance, $q_\theta^1<0$, $q_\theta^2<0$, and $q_\theta^3<0$: i.e., an increase in the degree of horizontal alliance will reduce outputs in all markets.

We note that no horizontal alliance equilibrium may exist given the set up. If we impose an additional structure, namely that horizontal alliance equilibrium arises when it maximizes the joint profit of all three firms (while still keeping the condition that no individual partners are made worse off than no horizontal alliance), then the equilibrium may exist. This additional assumption is also reasonable given the work by Yoshino and Rangan (1995).

To specify the horizontal alliance equilibrium, we add (31) and (32), obtaining

$$\frac{\mathrm{d}(\pi^{\mathrm{A}}+\pi^{\mathrm{B}}+\pi^{\mathrm{C}})}{\mathrm{d}\theta}=(1-\theta)\pi_1^{\mathrm{C}}q_\theta^1+(1-\theta)\pi_3^{\mathrm{A}}q_\theta^3+(1-\alpha)\pi_1^{\mathrm{B}}q_\theta^1+(1-\alpha)\pi_2^{\mathrm{A}}q_\theta^2 \tag{33}$$

Given Proposition 3, the first two terms on the right-hand side of (33) are positive, while the other two terms are negative. The former represents the marginal benefit of reduced output in the market for products 1 and 3 following increasing output coordination by firms A and C, while the latter represents the marginal loss of reduced output because of products 1 and 2 being complements, or the loss from reducing coordination within the vertical alliance between A and B. The optimal degree of horizontal alliance, denoted θ^*, will balance the marginal benefit from increased collusion within the horizontal alliance, with the marginal cost of reduced coordination within the vertical alliance. This analysis suggests a theory of horizontal alliance in multi-product markets: a partial horizontal alliance (i.e., $0<\theta^*<1$) arises so as to trade off gains from increased collusion within the horizontal alliance against losses from reduced coordination for complementary products when a partner also produces complementary products. Note that this theory of horizontal alliance arises in the absence of any alliance synergy effects.

It can be easily seen, from (21), that

$$\frac{\mathrm{d}W}{\mathrm{d}\theta} = \sum_{i=1}^{3} [p^i - (c^i)'] q^i_\theta < 0 \tag{34}$$

where the inequality follows from Proposition 3 and the fact that the mark-up terms are positive by the first-order conditions. The above discussion then leads to the following result:

Proposition 4. At the horizontal alliance equilibrium,

(i) each firm sells less in all markets
(ii) prices are higher in all markets
(iii) total welfare is lower

than under no horizontal alliance.

Proposition 4 shows a rather interesting result: horizontal alliance reduces competition not only in the market where prior competition between the partners takes place, but also in other markets of the alliance network. This result is somewhat striking in view of possible anticompetitive effects of horizontal alliances that can occur in "secondary" markets. Our analysis suggests that a careful examination of network interactions may be warranted when competition issues of strategic alliances are examined.

5. HYBRID ALLIANCE

We have considered two "pure" strategic alliance types, namely, vertical alliance and horizontal alliance. A vertical alliance results in the extension of a firm's network and tends to benefit consumers through lower prices. On the other hand, horizontal alliances tend to be anticompetitive and hence may be restricted by regulatory agencies. However, real world alliances – in the airline industry, among others – are likely to be a mix of vertical and horizontal alliances. Such alliances may be referred to as "hybrid alliances," and they may have both pro- and anticompetitive effects. For example, an international airline alliance is likely to reduce competition in the hub-to-hub, nonstop markets in which both partner airlines operate and usually dominate, while it is likely to increase competition in connecting markets to which the partners provide cheaper airfares (see, e.g., Brueckner, 2001). The former aspect of the alliance shows the characteristic of a horizontal alliance, whereas the latter aspect reflects that of a vertical alliance.

A complete theoretical treatment of hybrid alliances in the context of oligopolistic rivalry is hard, and in most cases it is simply intractable. Here we consider a simple structure of four multi-product firms: firm A produces goods 1 and 5, firm B 2 and 6, firm C 3 and 7, and firm D 4 and 8. Products 1 and 5, denoted as (1, 5), are independent in both demand and cost, so are the other pairs, namely, (2, 6), (3, 7), (4, 8). As specified before, products within each pair (1, 2), or (3, 4), are compliments, but products (5, 6, 7, 8) are substitutes.

As before, we consider the duopoly-pair case in which firms A and B form one alliance and firms C and D form the other. Similar to Eqs. (4) and (5), the problem of duopoly hybrid alliances is formulated as follows:

$$\begin{aligned} &\underset{q^1,q^5}{\text{Max}} \quad \pi^{AB} = (\pi^A + \alpha\delta^A) + \alpha\pi^B \\ &\underset{q^2,q^6}{\text{Max}} \quad \pi^{BA} = (\pi^B + \alpha\delta^B) + \alpha\pi^A \\ &\underset{q^3,q^7}{\text{Max}} \quad \pi^{CD} = (\pi^C + \beta\delta^C) + \beta\pi^D \\ &\underset{q^4,q^8}{\text{Max}} \quad \pi^{DC} = (\pi^D + \beta\delta^D) + \beta\pi^C \end{aligned} \tag{35}$$

where

$$\begin{aligned} \pi^A &= [q^1p^1 - c^1(q^1)] + [q^5p^5 - c^5(q^5)] \\ \pi^B &= [q^2p^2 - c^2(q^2)] + [q^6p^6 - c^6(q^6)] \\ \pi^C &= [q^3p^3 - c^3(q^3)] + [q^7p^7 - c^7(q^7)] \\ \pi^D &= [q^4p^4 - c^4(q^4)] + [q^8p^8 - c^8(q^8)] \end{aligned} \tag{36}$$

with the demand functions being specified as

$$p^i = p^i(q^1, q^2, q^3, q^4), \quad i = 1, 2, 3, 4 \tag{37}$$

and

$$p^j = p^j(q^5, q^6, q^7, q^8), \quad j = 5, 6, 7, 8 \tag{38}$$

Thus each firm's profit function can be separated into two parts, e.g.,

$$\pi^A = \pi^1(q^1, q^2, q^3, q^4) + \pi^5(q^5, q^6, q^7, q^8)$$

which allows one to decompose the analysis into two parts, one for the segment of (q^1, q^2, q^3, q^4) and the other for the segment of (q^5, q^6, q^7, q^8). For the former, all earlier specifications regarding outputs (q^1, q^2, q^3, q^4) are maintained. For the market where (q^5, q^6, q^7, q^8) are all substitutes, it can be derived that

$$\begin{aligned} &\pi_i^A < 0, \;\; i = 6, 7, 8; \quad \pi_j^B < 0, \;\; j = 5, 7, 8 \\ &\pi_k^C < 0, \;\; k = 5, 6, 8; \quad \pi_l^D < 0, \;\; l = 5, 6, 7 \end{aligned} \tag{39}$$

Furthermore, we assume these outputs are also strategic substitutes, i.e.,

$$\begin{aligned} &\pi_{5i}^A < 0, \;\; i = 6, 7, 8; \quad \pi_{6j}^B < 0, \;\; j = 5, 7, 8 \\ &\pi_{7k}^C < 0, \;\; k = 5, 6, 8; \quad \pi_{8l}^D < 0, \;\; l = 5, 6, 7 \end{aligned} \tag{40}$$

The alliance equilibrium is characterized by the eight first-order conditions, which are derived from (34), and corresponding second-order conditions. As before, regularity conditions are imposed so that the equilibrium is stable.

Although we are concerned principally with alliance rivalry, the monopoly-pair case serves as a useful base for comparison. For a single pair of firms A and B, it can be easily shown as

$$\frac{d(\phi^A + \phi^B)}{d\alpha} = (1-\alpha)(\pi_1^B q_\alpha^1 + \pi_2^A q_\alpha^2) + (\delta^A + \delta^B) + (1-\alpha)(\pi_5^B q_\alpha^5 + \pi_6^A q_\alpha^6) \tag{41}$$

The first two terms are the same as those in (10), which are the corresponding expressions for a monopoly vertical alliance, so the third term in (41) represents the only additional term. Note

$$q_\alpha^5 = (\pi_6^A \pi_{56}^{AB} - \pi_5^B \pi_{66}^{BA})/\Delta^{56}, \quad q_\alpha^6 = (\pi_5^B \pi_{65}^{BA} - \pi_6^A \pi_{55}^{AB})/\Delta^{56}$$

where $\pi_6^A = q^5 p_6^5 < 0$, $\pi_5^B = q^6 p_5^6 < 0$, and $\Delta^{56} \equiv \pi_{55}^{AB}\pi_{66}^{BA} - \pi_{56}^{AB}\pi_{65}^{BA} > 0$ by the stability condition. If we assume reasonable symmetry between the two partner firms in products 5 and 6, then the stability condition $\Delta^{56} > 0$ would imply that $\pi_{66}^{BA} - \pi_{65}^{BA} < 0$ and $\pi_{55}^{AB} - \pi_{56}^{AB} < 0$. Furthermore, the symmetry would imply $\pi_6^A = \pi_5^B$, and therefore, $q_\alpha^5 < 0$ and $q_\alpha^6 < 0$.[12]

Given $q_\alpha^5 < 0$ and $q_\alpha^6 < 0$, the third term in (41) must be positive: as expected, firms that produce substitutes would normally have an incentive to engage in collusion or, in our case, alliance. The comparison between (41) and (10) therefore suggests that for a single pair of firms, a hybrid alliance strengthens the degree of cooperation as compared to a pure vertical alliance. Essentially, the partners gain through better coordination in both their complementary and substitutable products.

With two pairs of firms, however, the incentive to engage in a strategic alliance may actually be *reduced* in a hybrid setting. To see this point, we derive the following equation:

$$\frac{\partial(\phi^{A} + \phi^{B})}{\partial\alpha} = \{(1-\alpha)(\pi_1^B q_\alpha^1 + \pi_2^A q_\alpha^2) + (\delta^A + \delta^B) + [(\pi_3^A + \pi_3^B)q_\alpha^3 + (\pi_4^A + \pi_4^B)q_\alpha^4]\} + \{(1-\alpha)(\pi_5^B q_\alpha^5 + \pi_6^A q_\alpha^6) + [(\pi_7^A + \pi_7^B)q_\alpha^7 + (\pi_8^A + \pi_8^B)q_\alpha^8]\} \quad (42)$$

The terms in the first braces of the above equation are the same as those of a vertical alliance in the (q^1, q^2, q^3, q^4) segment; see Eq. (17). As shown earlier, these terms are in general positive, thereby indicating the attractiveness of a vertical alliance. The terms in the second braces represent the effects of alliance partners' coordination in the (q^5, q^6, q^7, q^8) market. For a hybrid alliance, however, these additional effects can be less desirable. Based on (39) and (40), the terms involving q^5 and q^6 are likely to be positive while the terms involving q^7 and q^8 are likely to be negative. The intuition may be explained as follows. As members of one alliance (say, A and B) strengthen their coordination, one firm (say, A) recognizes that a reduction of its own output would increase the profit of its partner firm, B, as their outputs are substitutes, which in turn benefits firm A. As a result, the members of the alliance would reduce output. On the other hand, the reduction of outputs by A and B would push up the marginal profits of the rival firms C and D as the firms all produce substitutes; consequently, the competing CD alliance would produce more in the (q^5, q^6, q^7, q^8) market.

Note that the negative effect arising from the terms involving q^7 and q^8 will dominate the positive effects when α is sufficiently large. Thus, the additional effects from the (q^5, q^6, q^7, q^8) market may dampen the incentive for firms to engage in a strategic alliance. The partners will form a full hybrid alliance only if the positive effects in the complements market are larger than the negative effects in the substitutes market.

In effect, if the effects in the substitutes market are relatively important, alliances might not emerge as an equilibrium outcome in the alliance rivalry. Consider the following simple 2×2 payoff matrix (Table 1) with four alliance configurations. The first number in each cell is the payoff to the AB pair whereas the second number is the payoff to the CD pair. Indeed, when

$$\phi^{A}(0,0) + \phi^{B}(0,0) > \phi^{A}(1,0) + \phi^{B}(1,0),$$
$$\phi^{A}(0,1) + \phi^{B}(0,1) > \phi^{A}(1,1) + \phi^{B}(1,1) \quad (43)$$

remaining independent (no alliance) will be a dominant strategy for the AB pair. We can construct numerical examples to show this is true for both pairs; in these cases $(\alpha, \beta) = (0, 0)$ is a dominant strategy equilibrium. Our

Table 1. Payoff Matrix for Hybrid Alliances.

		Pair 2: CD	
		β	
Pair 1: AB		0	1
α	0	$\phi^A(0,0)+\phi^B(0,0)$, $\phi^C(0,0)+\phi^D(0,0)$	$\phi^A(0,1)+\phi^B(0,1)$, $\phi^C(0,1)+\phi^D(0,1)$
	1	$\phi^A(1,0)+\phi^B(1,0)$, $\phi^C(1,0)+\phi^D(1,0)$	$\phi^A(1,1)+\phi^B(1,1)$, $\phi^C(1,1)+\phi^D(1,1)$

analysis thus suggests that the incentive to form a hybrid alliance may be reduced when we move from a monopoly-pair setting to a duopoly-pair setting. On the other hand, if a hybrid alliance is formed in a duopoly-pair environment, one may take it as an indication of strong complementarities between the partner firms.

6. CONCLUDING REMARKS

We have examined competition models of vertical alliance, horizontal alliance, and hybrid alliance. A vertical alliance confers a strategic advantage by allowing the partners to credibly commit to greater outputs, owing to network complementarities, and the strategic effect tends to augment its positive direct effects arising from the elimination of the double-marginalization problem in vertical integration. Even if a vertical alliance creates a negative direct effect, it might be pursued because it is a dominant strategy in an oligopolistic setting. We found that in general, total welfare improves as a result of vertical alliances. On the other hand, a horizontal alliance reduces competition not only in the market where prior competition between the partners takes place, but also in other markets of the alliance network.

In reality, most alliances are a mix of vertical and horizontal alliances. Such hybrid alliances are likely to have both pro- and anticompetitive effects. We also found that that the incentive to form a hybrid alliance may be reduced when we move from a monopoly-pair setting to a duopoly-pair setting. Thus, if a hybrid alliance is observed a duopoly-pair environment, one may take it as an indication of strong complementarities between the partner firms. Our analysis of three types of strategic alliances suggests that a careful examination of network interactions may be warranted when competition issues of strategic alliances are examined.

Our model of the hybrid alliance presented in this paper has made a simplifying assumption that the (q^1, q^2, q^3, q^4) and (q^5, q^6, q^7, q^8) segments are independent to each other. More realistically, the two sets of markets would also be interrelated. For instance, in the context of international airline alliances, the segment of the network represented by (q^1, q^2, q^3, q^4) would likely be complementary to the other segment of the network represented by (q^5, q^6, q^7, q^8). Indeed, Brueckner (2001) considered a hybrid international airline alliance with a monopoly pair and Brueckner and Whalen (2000) analyzed duopoly-pair alliances. In their models, the alliance partners provide complementary outputs when their respective domestic spoke-to-hub routes are used in combination by passengers. In the common international hub-to-hub route, however, the partners' outputs are substitutes. The spoke-to-hub routes and the hub-to-hub route are also complementary. Brueckner (2001) and Brueckner and Whalen (2000) were able to obtain some interesting results under the assumptions of linear demand functions, linear marginal costs, and symmetric demand/cost and route structures.

For our more general model, allowing complementary interactions between these two segments of the network would render the model intractable. Nevertheless, qualitatively, we may conjecture that the complementarity between the (q^1, q^2, q^3, q^4) and (q^5, q^6, q^7, q^8) segments would make a hybrid alliance more attractive than otherwise. This is because the alliance partners would be more aggressive in the (q^5, q^6, q^7, q^8) market due to its positive effect on the (q^1, q^2, q^3, q^4) segment. As a consequence, the negative effect arising from the substitutes market would be less a concern when the two segments are complementary to each other than when the two are independent. A more rigorous analysis of this more general set-up is a subject of future research.

NOTES

1. Chan, Kensiger, Kewon and Martin (1997) reported that the average life span of strategic alliances in their sample was about 5 years.

2. See Tirole (1988) and Shapiro (1989) for surveys on oligopoly rivalry among independent firms. Mergers, on the other hand, involve a complete fusion of two or more independent firms into a single decision-making unit. See Jacquemin and Slade (1989) and Farrell and Shapiro (1990a) for comprehensive discussions on mergers.

3. The analysis on vertical alliances presented here is drawn largely from Zhang and Zhang (2005).

4. Some of these studies are concerned mainly with joint ventures. To some degree a strategic alliance is similar to a joint venture as both can represent a relatively new thrust by each participant. There are some important differences, however. First, under a joint venture, two or more parent companies establish a new legal entity to manage the joint venture firm, whereas such entities are not common under a strategic alliance. Second, probably the main motive for joint ventures is to share risks, so they are especially popular for risky projects or projects requiring large capital investment. Third, in certain industries joint ventures may be not allowed. For instance, there is no joint venture company in the provision of international air service. The most likely reason for this is that each country's flag carriers are required by law to be controlled by its nationals. Therefore, joint ventures with their management being controlled by the citizens of two or more countries are not allowed (Oum et al., 2000).

5. It is noted that these papers have used fairly restrictive assumptions on demand and cost functions. Empirical investigation of international airline alliances includes Gellman Research Associates (USDOT, 1994), Youssef and Hansen (1994), the U.S. General Accounting Office (USGAO, 1995), Oum, Zhang and Zhang (1996), Brueckner and Whalen (2000), Park and Zhang (2000), and Park et al. (2001). In particular, Brueckner and Whalen, Park and Zhang, and Park et al., provided some empirical support for the pro-competitive potential of "complementary" airline alliances. Oum et al. (2000) is a useful general reference on international airline alliances. In addition, Chen and Ross (2000) explored possible entry-deterrence effects of strategic alliances involving the sharing of production capacity.

6. Farrell and Shapiro (1990b) made the similar point in the context of joint ownership of assets in oligopoly.

7. Gallacher and Odell (1994), for instance, pointed out that some airlines had been spurred into making alliances simply by the fear of being left behind.

8. This is one reason that such alliances are here referred to as vertical alliances. Note that Brueckner (2001, 2003) and Brueckner and Whalen (2000) have a clear demonstration of the double-marginalization problem in the context of an international airline alliance.

9. Unless indicated, the proofs of Proposition 1 and other results are given in the appendix.

10. Note that Eq. (18) being negative may also have implications for entry deterrence. Suppose that firms A and B have an exogenously given opportunity to choose its alliance structure prior to the entry (and alliance) decision of firms C and D. If there is a sunk cost associated with entry into a market, then for certain ranges of entry cost, a possible entry by the rival firms will be pre-empted if and only if the incumbents form an alliance. In other words, an incumbent pair can use alliance as a device to deter potential entry and will do so if the incumbent is better off with alliance and no entry than with no alliance and entry.

11. Relaxing this assumption will add algebraic complexity but will not change the basic results of this section.

12. Note that this comparative static result may also rise without the symmetry assumption. Assume, as is not uncommon in the literature, that $\pi_{66}^{BA} < \pi_{65}^{BA}$, $\pi_{55}^{AB} < \pi_{56}^{AB}$, which will ensure the existence and uniqueness of the equilibrium as

well as its stability. Then, $q_\alpha^5 + q_\alpha^6 = -\pi_5^{\mathrm{B}}(\pi_{66}^{\mathrm{BA}} - \pi_{65}^{\mathrm{BA}}) - \pi_6^{\mathrm{A}}(\pi_{55}^{\mathrm{AB}} - \pi_{56}^{\mathrm{AB}}) < 0$, which implies that q_α^5, q_α^6 cannot both be positive. Furthermore, since $\pi_\alpha^{\mathrm{A}} = -\alpha\pi_5^{\mathrm{B}} q_\alpha^5 + \pi_6^{\mathrm{A}} q_\alpha^6 < 0$ if $q_\alpha^5 < 0$, $q_\alpha^6 > 0$, and $\pi_\alpha^{\mathrm{B}} = \pi_5^{\mathrm{B}} q_\alpha^5 - \alpha\pi_6^{\mathrm{A}} q_\alpha^6 < 0$ if $q_\alpha^5 > 0$, $q_\alpha^6 < 0$, both must have the same sign if the alliance between the two firms is feasible. It then follows that $q_\alpha^5 < 0,\ q_\alpha^6 < 0$.

ACKNOWLEDGMENT

I thank the participants of the *International Symposium on Global Competition in Transportation Market – Analysis and Policy Making*, December 2002, Kobe, Japan, for helpful comments.

REFERENCES

Airline Business. (2000). *Monthly, Reed Business Information*. UK: Wyndeham Heron.

Blodgett, L. L. (1992). Factors in the instability of international joint ventures: An event history analysis. *Strategic Management Journal, 13*, 475–481.

Brueckner, J. K. (2001). The economics of international codesharing: An analysis of airline alliances. *International Journal of Industrial Organization, 19*, 1475–1498.

Brueckner, J. K., & Whalen, W. T. (2000). The price effects of international airline alliances. *Journal of Law and Economics, XLIII*, 503–545.

Bulow, J. I., Geanakoplos, J. D., & Klemperer, P. D. (1985). Multimarket oligopoly: Strategic substitutes and complements. *Journal of Political Economy, 93*, 488–511.

Chan, S. H., Kensiger, J. W., Kewon, A. J., & Martin, J. D. (1997). Do strategic alliances create value? *Journal of Financial Economics, 46*, 199–221.

Chen, Z., & Ross, T. W. (2000). Strategic alliances, shared facilities, and entry deterrence. *Rand Journal of Economics, 31*, 326–344.

Das, S., Sen, P. K., & Sengupta, S. (1998). Impact of strategic alliances on firm valuation. *Academy of Management Journal, 41*, 27–41.

Dixit, A. (1986). Comparative statics for oligopoly. *International Economic Review, 27*, 107–122.

Dussauge, P., & Garrette, B. (1995). Determinants of success in international strategic alliances: Evidence from the global aerospace industry. *Journal of International Business Studies, 26*, 505–527.

Farrell, J., & Shapiro, C. (1990a). Horizontal mergers: An equilibrium analysis. *American Economic Review, 80*, 107–126.

Farrell, J., & Shapiro, C. (1990b). Asset ownership and market structure in oligopoly. *Rand Journal of Economics, 21*, 275–292.

Fudenberg, D., & Tirole, J. (1984). The fat-cat effect, the puppy-dog play, and the lean and hungry look. *American Economic Review (Papers and Proceedings), 74*, 361–366.

Gallacher, J., & Odell, M. (1994). Dragging along. *Airline Business, July*, 25–42.

Geringer, J. M., & Herbert, L. (1989). Control and performance of international joint ventures. *Journal of International Business Studies, 20*, 235–254.

Hamel, G. (1991). Competition for competence and interpartner learning within international strategic alliances. *Strategic Management Journal, 12*, 83–103.

Hamel, G., Doz, Y. L., & Prahalad, C. K. (1989). Collaborate with your competitors – and win. *Harvard Business Review, 67*, 133–139.

Harrigan, K. R. (1988). Joint venture and competitive strategy. *Strategic Management Journal, 9*, 141–158.

Hennart, J. F. (1991). The transaction cost theory of joint ventures: An empirical study of Japanese subsidiaries in the United States. *Management Science, 37*, 483–497.

Hennart, J. F., & Reddy, S. (1997). The choice between mergers/acquisitions and joint ventures: The case of Japanese investors in the United States. *Strategic Management Journal, 18*, 1–12.

Jacquemin, A., & Slade, M. E. (1989). Cartels, collusion, and horizontal merger. In: R. Schmalensee & R. Willig (Eds), *Handbook of industrial organization*. Amsterdam: North-Holland.

Kogut, B. (1988). Joint ventures: Theoretical and empirical perspectives. *Strategic Management Journal, 9*, 319–332.

Kogut, B. (1989). The stability of joint ventures: Reciprocity and competitive rivalry. *Journal of Industrial Economics, 38*, 183–198.

Kogut, B., & Singh, H. (1988). The effect of national culture on the choice of entry mode. *Journal of International Business Studies, 19*, 411–432.

Mody, A. (1993). Learning through alliances. *Journal of Economic Behaviour and Organization, 20*, 151–170.

Mowery, D. C., & Teece, D. J. (1993). Japan's growing capabilities in industrial technology: Implications for U.S. managers and policymakers. *California Management Review, 35*, 9–34.

Ortega, J. M., & Rheinboldt, W. C. (1970). *Iterative solution of nonlinear equations in several variables*. New York: Academic Press.

Osborn, R. N., & Baughn, C. C. (1990). Forms of interorganizational governance for multinational alliances. *Academy of Management Journal, 33*, 503–519.

Oum, T. H., Park, J. H., & Zhang, A. (1996). The effects of airline codesharing agreements on firm conduct and international air fares. *Journal of Transport Economics and Policy, 30*, 187–202.

Oum, T. H., Park, J. H., & Zhang, A. (2000). *Globalization and strategic alliances: The case of the airline industry*. Oxford: Pergamon Press.

Oum, T. H., Zhang, A., & Zhang, Y. (1995). Airline network rivalry. *Canadian Journal of Economics, 28*, 836–857.

Oum, T. H., Zhang, A., & Zhang, Y. (1996). Optimal airport pricing in a hub-and-spoke system. *Transportation Research B (Methodological), 30*, 11–18.

Park, J. H. (1997). The effects of airline alliances on markets and economic welfare. *Transportation Research E (Logistics and Transportation Review), 33*, 181–195.

Park, J. H., & Zhang, A. (2000). An empirical analysis of global airline alliances: Cases in North Atlantic markets. *Review of Industrial Organization, 16*, 367–384.

Park, J. H., Zhang, A., & Zhang, Y. (2001). A model of international airline alliances. *Transportation Research B (Methodological), 35*, 865–886.

Parkhe, A. (1993). Strategic alliance structuring: A game theoretic and transaction cost examination of interfirm cooperation. *Academy of Management Journal, 36*, 794–829.

Pisano, G. P. (1990). The R&D boundaries of the firm: An empirical analysis. *Administrative Science Quarterly*, *21*, 398–418.
Shapiro, C. (1989). Theories of oligopoly behavior. In: R. Schmalensee & R. Willig (Eds), *Handbook of industrial organization* (pp. 329–414). Amsterdam: North-Holland.
Tirole, J. (1988). *The theory of industrial organization*. Cambridge, USA: MIT Press.
USDOT (1994). *A study of international airline codesharing*, Gellman Research Associates, Inc., commissioned by the U.S. Department of Transportation.
USGAO (1995). *Airline alliances produce benefits, but effect on competition is uncertain*, The U.S. General Accounting Office, GAO/RCED-95-99.
Veugelers, R. (1993). *Global cooperation: A profile of companies in alliances*. Research report DTEW 9325, K.U. Leuven, Louvain.
Yoshino, M. Y., & Rangan, U. S. (1995). *Strategic alliances: An entrepreneurial approach to globalization*. Boston: Harvard Business School Press.
Youssef, W., & Hansen, M. (1994). Consequences of strategic alliances between international airlines: The case of Swissair and SAS. *Transportation Research A (Policy and Practice)*, *28*, 415–431.
Zhang, A., & Zhang, Y. (1996). Stability of a Nash equilibrium: The multiproduct case. *Journal of Mathematical Economics*, *26*, 441–462.
Zhang, A., & Zhang, Y. (2005). Rivalry between Strategic Alliances. *International Journal of Industrial Organization*, forthcoming.

APPENDIX

Proof of Proposition 1. We show, without the loss of generality, that $q^i_\alpha > 0$ and $q^j_\alpha < 0$. Differentiating (11)–(14) with respect to α yields

$$\Phi \frac{\partial Q(\alpha, \beta)}{\partial \alpha} + \frac{\partial \Pi}{\partial \alpha} = 0 \tag{A.1}$$

where

$$\Phi \equiv \begin{bmatrix} \pi_{11}^{\mathrm{AB}} & \pi_{12}^{\mathrm{AB}} & \pi_{13}^{\mathrm{AB}} & \pi_{14}^{\mathrm{AB}} \\ \pi_{21}^{\mathrm{BA}} & \pi_{22}^{\mathrm{BA}} & \pi_{23}^{\mathrm{BA}} & \pi_{24}^{\mathrm{BA}} \\ \pi_{31}^{\mathrm{CD}} & \pi_{32}^{\mathrm{CD}} & \pi_{33}^{\mathrm{CD}} & \pi_{34}^{\mathrm{CD}} \\ \pi_{41}^{\mathrm{DC}} & \pi_{42}^{\mathrm{DC}} & \pi_{43}^{\mathrm{DC}} & \pi_{44}^{\mathrm{DC}} \end{bmatrix}, \quad Q(\alpha, \beta) \equiv \begin{bmatrix} q^1(\alpha, \beta) \\ q^2(\alpha, \beta) \\ q^3(\alpha, \beta) \\ q^4(\alpha, \beta) \end{bmatrix}, \quad \frac{\partial \Pi}{\partial \alpha} \equiv \begin{bmatrix} \pi_1^{\mathrm{B}} \\ \pi_2^{\mathrm{A}} \\ 0 \\ 0 \end{bmatrix}$$

Note that Φ can be manipulated to become $\Phi = \mathrm{diag}[\pi_{11}^{\mathrm{AB}}, \pi_{22}^{\mathrm{BA}}, \pi_{33}^{\mathrm{CD}}, \pi_{44}^{\mathrm{DC}}][I - R]$, where I is the identity matrix and

$$R \equiv \begin{bmatrix} 0 & R_{12}^{\mathrm{AB}} & R_{13}^{\mathrm{AB}} & R_{14}^{\mathrm{AB}} \\ R_{21}^{\mathrm{BA}} & 0 & R_{23}^{\mathrm{BA}} & R_{24}^{\mathrm{BA}} \\ R_{31}^{\mathrm{CD}} & R_{32}^{\mathrm{CD}} & 0 & R_{34}^{\mathrm{CD}} \\ R_{41}^{\mathrm{DC}} & R_{42}^{\mathrm{DC}} & R_{43}^{\mathrm{DC}} & 0 \end{bmatrix}$$

with $R_{ij}^{K} \equiv -(\pi_{ii}^{K})^{-1}\pi_{ij}^{K}$ for $i \neq j$. Note that R_{ij}^{K} is the derivative of alliance K's reaction function. By (7), (9) and the second-order condition it can be shown that matrix R has the following sign pattern:

$$R = \begin{bmatrix} 0 & + & - & - \\ + & 0 & - & - \\ - & - & 0 & + \\ - & - & + & 0 \end{bmatrix} \tag{A.2}$$

Solving (A.1) for $\partial Q/\partial\alpha$, we obtain

$$\frac{\partial Q}{\partial \alpha} = -\Phi^{-1}\frac{\partial \Pi}{\partial \alpha} = -[I - R]^{-1}\mathrm{diag}[\pi_{11}^{\mathrm{AB}}, \pi_{22}^{\mathrm{BA}}, \pi_{33}^{\mathrm{CD}}, \pi_{44}^{\mathrm{DC}}]^{-1}\frac{\partial \Pi}{\partial \alpha} \tag{A.3}$$

By the second-order condition, the diagonal matrix is negative. Furthermore, the first two elements of $\partial\Pi/\partial\alpha$ are positive by (6). We now determine the sign of $[I - R]^{-1}$. The stability of the Cournot–Nash equilibrium implies that the magnitude of the eigenvalues of matrix R must be less than unity (Zhang & Zhang, 1996). Hence, by Neumann Lemma (see, e.g., Ortega & Rheinboldt, 1970), $[I - R]^{-1}$ exists and

$$[I - R]^{-1} = I + R + R^2 + \cdots + R^n + \cdots$$

Hence, by (A.2), $[I - R]^{-1}$ must have the following signs:

$$[I - R]^{-1} = \begin{bmatrix} + & + & - & - \\ + & + & - & - \\ - & - & + & + \\ - & - & + & + \end{bmatrix} \tag{A.4}$$

Then, according to (A.3), we obtain Proposition 1. Q.E.D.

Proof of Proposition 2. Partition matrix Φ in (A.1) into

$$\Phi = \begin{bmatrix} \phi_{11} & \phi_{12} \\ \phi_{21} & \phi_{22} \end{bmatrix}$$

where each of Φ_{ij} is a 2×2 matrix. Then, (A.1) may be written as

$$\phi_{11} \frac{\partial}{\partial \alpha} \begin{bmatrix} q^1 \\ q^2 \end{bmatrix} + \phi_{12} \frac{\partial}{\partial \alpha} \begin{bmatrix} q^3 \\ q^4 \end{bmatrix} = - \begin{bmatrix} \pi_1^{\mathrm{B}} \\ \pi_2^{\mathrm{A}} \end{bmatrix}$$

$$\phi_{21} \frac{\partial}{\partial \alpha} \begin{bmatrix} q^1 \\ q^2 \end{bmatrix} + \phi_{22} \frac{\partial}{\partial \alpha} \begin{bmatrix} q^3 \\ q^4 \end{bmatrix} = 0$$

Solving the second equation and substituting, we obtain

$$\frac{\partial}{\partial \alpha} \begin{bmatrix} q^3 \\ q^4 \end{bmatrix} = R_{\mathrm{AB}}^{\mathrm{CD}} \frac{\partial}{\partial \alpha} \begin{bmatrix} q^1 \\ q^2 \end{bmatrix}$$

where matrix $R_{\mathrm{AB}}^{\mathrm{CD}} \equiv -\phi_{21}\phi_{22}^{-1}$ may be interpreted as the derivative of the reaction function of the CD alliance. By the condition that the equilibrium is stable, the norm of this matrix must be less than unity (Zhang & Zhang, 1996), i.e., $\|R_{\mathrm{AB}}^{\mathrm{CD}}\| < 1$. If follows that either $|q_\alpha^3| < |q_\alpha^1|$ or $|q_\alpha^4| < |q_\alpha^2|$. Thus, by Proposition 1, $\partial\tilde{q}/\partial\alpha$ must have at least one positive element.

In the "symmetric firms" case, where all four firms face symmetric demands, both within and between the markets, and have the same costs, we shall have $\mathrm{d}q^1 = \mathrm{d}q^2$ and $\mathrm{d}q^3 = \mathrm{d}q^4$. Then, as shown above, the stability condition will imply that $\partial\tilde{q}/\partial\alpha > 0$. Q.E.D.

Proof of Lemma 1. (i) From (29), it follows that $q_\theta^2 = -\pi_{21}^{\mathrm{BA}} q_\theta^1 / \pi_{22}^{\mathrm{BA}}$. Since $\pi_{22}^{\mathrm{BA}} < 0$ by the second-order condition and $\pi_{21}^{\mathrm{BA}} = \pi_{21}^{\mathrm{B}} + \alpha\pi_{21}^{\mathrm{A}} > 0$ by (7), q_θ^2 must have the same sign as q_θ^1.
(ii) In (30), we have $\pi_{33}^{\mathrm{CA}} < 0$ by the second-order condition, and $\pi_{31}^{\mathrm{CA}} = \pi_{31}^{\mathrm{C}} + \theta\pi_{31}^{\mathrm{A}} < 0$ by (23). Thus, if both q_θ^1 and q_θ^3 are positive, the left-hand side of (30) will be negative, contradicting with the fact that $\pi_3^{\mathrm{A}} = q^1 p_3^1 < 0$ by (22). Q.E.D.

NETWORK COMPETITION IN AIR TRANSPORTATION MARKETS: BI-LEVEL APPROACH

Mikio Takebayashi and Adib Kanafani

ABSTRACT

This chapter develops a model to simulate contemporary competition between network and point-to-point (PP) carriers in air transportation markets. In particular, we focus on the transition from the strategies of network carriers to the behavior of PP carriers. The concept of multilevel competition is introduced: the first level is the competition layer of network and PP carriers and the second level is the passengers' optimization layer under network service quality given as optimal solutions at the first level. Second, the properties of the developed model are examined with illustrative examples.

1. INTRODUCTION

This chapter proposes a new approach that reveals the properties of the optimal behavior of hub-spoke-shaped (HS) network carriers against the point-to-point (PP), low fare strategy of service carriers. In particular, we focus on the transition from the strategies of network carriers to the

Global Competition in Transportation Markets: Analysis and Policy Making
Research in Transportation Economics, Volume 13, 101–119

ISSN: 0739-8859/doi:10.1016/S0739-8859(05)13005-4

behavior of PP carriers. This chapter shows a multilevel competition structure between HS and PP carriers as well as carriers and passengers formulated as Mathematical Programming with Equilibrium Constraints (MPEC; Luo et.al., 1997). Illustrative examples explain the model, and the properties of the optimal behavior of HS carriers are figured out.

In the last two decades "network competition" has been studied because anticipating what will happen in deregulated markets is very important not only for airlines but also for airport authorities. Much literature has addressed competition between network carriers. Kanafani studied actual competition in the U.S. domestic air transportation market (Kanafani & Ghobrial, 1985; Hansen & Kanafani, 1988). Hansen (1990) took a theoretical approach to understanding competition among homogeneous network carriers; he introduced and applied an "n person non-cooperative" Nash game concept to network carriers' competition. He pointed out the difficulty of obtaining a unique Nash equilibrium solution under homogeneous carriers' competition and concluded that a quasi-equilibrium solution exists because of the nonconvexity of the game.

Dobson and Lederer discussed optimal schedules and pricing schemes in network carriers' competition by developing a heuristic model for optimizing a hub-spoke network system (Dobson & Lederer, 1993). Adler also adopted a game theory framework to understand competition between network carriers (Adler, 2001). She pointed out that in the competition between homogeneous network carriers it is difficult to locate a stable and unique equilibrium solution.

Some literature has addressed the differences in productivity between HS and PP carriers and concluded that the former enjoys an advantage over the latter. Hendricks, Piccione, and Tan (1995) studied carrier's choice of service type: HS or PP services. They applied a two-stage game framework to the market, which revealed that carrier's choice of network shape depends on the activity of an "economy of density" in an Origin-Destination (OD) market. Moreover, they enhanced their model to deal with a duopolystic market between HS and PP carriers (Hendricks, et al. 1997, 1999). They concluded that HS carriers have a quite stable advantage, predicting that PP service carriers will decrease because they are less productive. They pointed out that the key to survival is an "economy of density," which HS carriers enjoy more easily than PP carriers.

However, this market is facing the opposite situation: the rise of "low-cost carriers," whose operational costs are usually lower than existing network carriers. They manage service routes without insisting on operating an HS network service. Of course, most have base airports and handle connecting services, but their strategies, for example pricing schemes, flight schedules, and so forth, are usually based on PP service. It is well known that today low-cost carriers are "invading" the existing network carriers' share to increase their market powers. Hence, in the contemporary air transportation market low-cost carriers are merely "niche" carriers.

Empirical studies have pointed out the significant impact of low-cost carriers (e.g. Vowles, 2001; Mason, 2001). These studies attained some important conclusions describing the outline of the phenomena, but they do not deal with the mechanism that results in the rise of low-cost carriers. We desire to develop a theoretical model of an air transportation market that implicitly concerns the impact of low-cost carriers as contemporary PP carriers. Moreover, since frequent service is a significant strategy of low-cost carriers, a model must simultaneously evaluate both frequency and price competition.

The present chapter proposes a model that can explicitly handle price and quantity competition concerned with the behavior of passengers' route choices. However, the developed model, which does not require a two-stage game framework, as existing studies do, was developed under a hierarchical structure assumption in air transportation markets, adopting a framework of MPEC for expressing market structure.

The body of this chapter consists of five sections. In Section 2, the basic structure of the model and its formulation is discussed in detail with the concept of a multilevel MPEC problem. Illustrative examples are examined in Section 3, and the model's features are discussed. Finally, we summarize our conclusion and list further research objects.

2. THE MODEL

The structure of the developed model is very simple. Let us consider that two types of carriers exist in an air transport market: network carriers and their competitor (henceforth, HS and LC/PP carriers, respectively). HS carriers provide HS service networks, while LC/PP carriers provide point-to-point service with low fares. The HS carriers' airfare strategy is based on their hub-spoke service network, which can provide discount fares for connecting passengers, while LC/PP carriers determine their airfares per flight.

Let X_{rs} be OD volume between origin r to destination s. OD volume X_{rs} is predetermined. Hence, current OD volume is assumed to be independent from the market's service level; for example, the OD volume has no elasticity on airfares, service flight frequency, or seat availability. In this market, HS and LC/PP carriers coexist, and their services are fully compatible. Furthermore, assume that HS carriers already exist in this market and operate a hub-spoke service network; LC/PP carriers embark on their services in this market under the existence of HS carriers. Such a market can be expressed as a "bi-level" optimization problem MPEC.

For simplicity, let us consider this as a one-on-one competitive market: one HS carrier and one LC/PP carrier. This simple model can be easily extended to a general or many-on-many competition, so let us accept this assumption in the following formulation.

2.1. HS Carrier

An HS carrier maximizes profits by controlling flight frequencies and airfares for each OD market. Let a certain commercial link $l \in \Theta^{\mathrm{HS}}\Theta^{\mathrm{HS}}$ mean a set of links operated by an HS carrier. The service route provided by an HS carrier is indicated as $k \in \Omega^{\mathrm{HS}}$, where Ω^{HS} is a set of service routes that the HS carrier provides. Following the basic assumption, the control variables of the HS carrier are airfare $p_{rs}^{k,\mathrm{HS}}$ and flight frequency f_l^{HS}. As for the cost of the HS carriers, we introduce two different costs: C_l^{HS} is the operating cost at each HS link, and $C_l^{\mathrm{HS,FIX}}$ is the HS carrier's fixed cost for continuing service at link l. The value of each cost is predetermined.

The HS carrier's behavior can be expressed as follows:

$$\max \quad Z^{\mathrm{HS}}(f_l^{\mathrm{HS}}, p_{rs}^{k,\mathrm{HS}}) = \sum_{rs}\sum_{k} p_{rs}^{k} x_{rs}^{k} \delta_{rsk}^{\mathrm{HS}} - \sum_{l \in \Theta^{\mathrm{HS}}} f_l^{\mathrm{HS}} C_l^{\mathrm{HS}} - \sum_{l \in \Theta^{\mathrm{HS}}} C_l^{\mathrm{HS,FIX}} \tag{1}$$

$$\text{s.t.} \quad x_l^{\mathrm{HS}} \leqslant V_l^{\mathrm{HS}} f_l^{\mathrm{HS}} \quad \text{for } \forall l \in \Theta^{\mathrm{HS}} \tag{2}$$

$$p_{rs}^{k,\mathrm{HS}} \geqslant 0 \quad \text{for } \forall k \in \Omega^{\mathrm{HS}} \tag{3a}$$

$$f_l^{\mathrm{HS}} \geqslant 0 \quad \text{for } \forall l \in \Omega^{\mathrm{HS}} \tag{3b}$$

$$\mathbf{f}^{\mathrm{LC}} = \arg\max\{Z^{\mathrm{LC}}\} \tag{4}$$

$$x_{rs}^{k} = \arg\max\{Z^{A}(\mathbf{f}^{\mathrm{HS}}, \mathbf{p}^{\mathrm{HS}}, \mathbf{f}^{\mathrm{LC}})\} \tag{5}$$

where x_{rs}^{k} is the flow of passengers from r to s on route k. $\delta_{rsk}^{\mathrm{HS}}$ is a dichotomous variable that takes the value one when kth route in rs OD market is operated by the HS carrier; otherwise it takes the value zero. x_l^{HS} is passenger flow of link l operated by an HS carrier. V_l^{HS} is aircraft size on link l operated by an HS carrier. The relationship between path flow x_{rs}^k and link flow x_l^{HS} is determined as

$$\sum_{rsk} x_{rs}^k \delta_{rsk}^{l\mathrm{HS}} = x_l^{\mathrm{HS}} \quad \text{for } \forall l \in \Theta^{\mathrm{HS}} \tag{6}$$

where $\delta_{rsk}^{l\mathrm{HS}}$ is a dichotomous variable that takes the value one when kth route in rs OD market uses link l operated by the HS carrier; otherwise it takes the value zero. The same balance as above appears in the LC/PP carrier's decision layer. $\mathbf{f}^{\mathrm{HS}}$, $\mathbf{f}^{\mathrm{LC}}$, and $\mathbf{p}^{\mathrm{HS}}$ are vectors of optimal flight frequency and airfare operated by the HS carrier and optimal flight frequency by the LC/PP carrier, respectively. Z^A stands for the passengers' object function and is obtained in the passengers' decision-making layer.

Constraint set (1) is the object function of the HS carrier, constraint set (2) expresses the constraints of available seats on each link, constraint set (3) means the non-negative constraints of control variables, constraint set (4) which is unique, expresses the LC/PP carrier's best response and is included in the HS carrier's constraints; this means that the HS carrier has advantages in the market. Thus in this market, the HS carrier can be regarded as the leader because it can change its strategy after learning the LC/PP carrier's best reaction and constraint set (5) shows the flow of passengers defined as the best response function to carriers' behavior.

2.2. *LC/PP Carrier*

The LC/PP carrier is the follower because it embarks on its services under the HS carrier's existence. Assume that the object of the LC/PP carrier is to expand its share, make profit, and survive in the market. LC/PP carriers offer flights that are cheap. Assume that the actual control variable of LC/PP carriers is the number of flights. Hence, let the purpose of LC/PP carriers be to maximize profit with the lowest possible fares.

A key for expressing the LC/PP carrier's behavior is the setting of low fares. To provide a low fare, two additional parameters are introduced: a minimum load factor for each commercial link of the LC/PP carrier ω^{LC} that controls transportation costs per seat; and a markup ratio in airfare η that gives profit. These parameters are predetermined in this model.

LC/PP carriers provide PP service. Passengers who use an LC/PP carrier's flight can have a connecting service, but airfares are provided as a combination of PP services; no discounts are given to connecting passengers. This means that the pricing scheme of LC carriers is based on their PP service and does not provide hub-spoke network service.

Therefore, the LC/PP carrier's behavior is formulated as the following optimization problem:

$$\max \quad Z^{\mathrm{LC}}\left(f_l^{\mathrm{LC}} | p_{rs}^{k,\mathrm{LC}} = \sum_{l \in \Theta^{\mathrm{LC}}} p_l^{\mathrm{LC}} \delta_{rsk}^{l}\right) = \sum_{rs} \sum_{k} p_{rs}^{k,\mathrm{LC}} x_{rs}^{k} \delta_{rsk}^{\mathrm{LC}} = \sum_{l \in \Theta^{\mathrm{LC}}} p_l^{\mathrm{LC}} x_l^{\mathrm{LC}} \tag{7}$$

$$\text{s.t.} \quad \omega^{\mathrm{LC}} V_l^{\mathrm{LC}} f_l^{\mathrm{LC}} \leqslant x_l^{\mathrm{LC}} \leqslant V_l^{\mathrm{LC}} f_l^{\mathrm{LC}} \quad \text{for } \forall l \in \Theta^{\mathrm{LC}} \tag{8}$$

$$x_l^{\mathrm{LC}} p_l^{\mathrm{LC}} (1+\eta) - C_l^{\mathrm{LC}} f_l^{\mathrm{LC}} - C_l^{\mathrm{LC,FIX}} = 0 \quad \text{for } \forall l \in \Theta^{\mathrm{LC}} \tag{9}$$

$$f_l^{\mathrm{LC}} \geqslant 0 \quad \text{for } \forall l \in \Theta^{\mathrm{LC}} \tag{10}$$

$$x_{rs}^{k} = \arg\max\{Z^{A}(\mathbf{f}^{\mathrm{HS}}, \mathbf{p}^{\mathrm{HS}}, \mathbf{f}^{\mathrm{LC}})\} \tag{5}$$

where f_l^{LC} is number of flights at link l operated by the LC/PP carrier, p_l^{LC} is airfares at link l operated by the LC/PP carrier, $\delta_{rsk}^{\mathrm{LC}}$ is a dichotomous variable that takes the value one when kth route in rs OD market is operated by the LC/PP carrier; otherwise it takes the value zero. x_l^{LC} is link flow at link l operated by the LC/PP carrier, V_l^{LC} is aircraft size at link l operated by the LC/PP carrier, C_l^{LC} is operational cost at link l operated by the LC/PP carrier, $C_l^{\mathrm{LC,FIX}}$ is fixed cost at link l operated by the LC/PP carrier, and Θ^{LC} is a set of links operated by the LC/PP carrier.

Eq. (7) denotes the object function of the LC/PP carrier. Since we assume that the LC/PP carrier's primary control variable is the number of flights at each commercial link, airfare p_l^{LC} does not appear in object function (7). Constraint set (8) also expresses the capacity constraint of each flight. Constraint sets (9) and (10) are the profit balance constraint and non-negative constraints on the LC/PP carrier's actual control variable, respectively. Since air passengers may also choose the LC/PP carrier's services, constraint set (5) should be involved.

So far, the competition problem was located in the carrier's decision-making layer. However, another layer of competition exists in this market: competition between carriers and consumers. In the following section, the competition in the passenger's decision-making level is discussed.

2.3. Passengers

Passengers are assumed to choose the best available routes under carriers' strategies. It is also assumed that each carrier gives detailed information about route choices to passengers. Therefore, a bi-level programming problem appears between carriers and passengers as shown in Fig. 1. Finally, this market consists of two bi-level programming problems: HS–LC/PP competition and carriers–passengers competition. Since this component is complicated, its structure is decomposed in the following section.

Passengers are assumed to maximize their utility, which consists of number of flights, travel time, and cost of fares. Moreover, passengers should be concerned with the following additional costs:

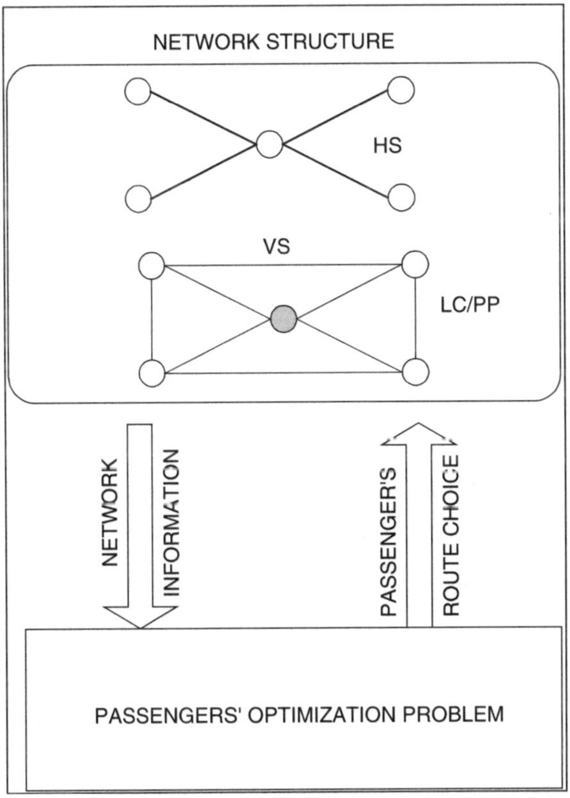

Fig. 1. Bi-Level Structure: Carriers' and Passengers' Decision-Making Layers.

(Assumptions for passengers)

(1) Average delay $D(f)$ is caused by airport congestion.
(2) Increase in link users' number x_l raises additional costs related to link congestion $\Gamma(x_l)$.

Assumption (1) expresses the additional costs due to the carrier's behavior; passengers will avoid using a congested airport. This chapter assumes that delays depend on the number of slots allocated to each carrier. This means that optimum scheduling for passengers becomes more difficult with increases of flights. Assumption (2) denotes the congestion cost per flight due to the passengers' route choices. If a passenger chooses a service route that includes a congested link, sometimes the passenger should accept an alternative flight at the same link. Hence, this congested cost reflects the passengers' route choice, expressed as the potential risk of choosing congested flights.

By introducing Assumption (2), the passenger's utility function at commercial link l, U_l, can be strictly treated as a convex function. Moreover, we introduce a random utility theory to the passengers' route choice behavior as previous research did (Adler, 2001). Thus, the optimization problem of passengers' route choice behavior results in a standard SUE/FD problem. Accordingly, passengers' flow on route k, which is formulated in constraint set (5), is obtained as

$$x^k_{rs} = \frac{\exp[\theta\{\upsilon f^k_{rs} - p^k_{rs} - \tau(t^k_{rs} + \sum_h D^m_h \delta^{l,h}_{rs,k}) - \sum_l \Gamma(x_l)\delta^l_{rs,k}\}]}{\sum_{k'} \exp[\theta\{\upsilon f^{k'}_{rs} - p^{k'}_{rs} - \tau(t^{k'}_{rs} + \sum_h \sum_m D^m_h \delta^{l,h}_{rs,k'}) - \sum_l \Gamma(x_l)\delta^l_{rs,k'}\}]} X_{rs} \tag{11}$$

where f^k_{rs} is the average frequency in route k in rs market and t^k_{rs} is the line-haul time sum of route k in rs market including average waiting time for flight l. This summed time is defined as

$$t^k_{rs} = \sum_l t_l \delta^l_{rsk} \tag{12}$$

where t_l is the line-haul time of link l and δ^l_{rsk} is a dichotomous variable that takes the value one when kth route in rs OD market includes link l; otherwise it takes zero. D^m_h are additional costs for passengers using carrier m ($m =$ HS or LC/PP) due to airport congestion. In this chapter, such additional costs are formulated as

$$D_h^m = \frac{\xi_h^1 \sum_{l \in \Theta^{\mathrm{HS}} or \Theta^{\mathrm{LC}}} f_l}{F_h^m} \tag{13}$$

where ξ is a positive parameter and F_h^m is defined as the number of allocated slots in carrier m. In this formulation, the capacity constraints, i.e. slot capacity at each airport, are relaxed to guarantee the existence of equilibrium at passengers' flow.

In passenger's flow, shown in Eq. (11), τ *and* v are positive parameters, and the congestion function Γ stands for the monetary value of congestion costs per flight on commercial link l. This cost is introduced to express the risks a passenger may experience when choosing a congested link. This value is simply defined as

$$\Gamma(x_l) = \xi_l^2 \frac{x_l}{V_l} \tag{14}$$

In Eq. (14), overflow is accepted. However, each carrier has a seat capacity constraint, so overflow is controlled at the carrier's level.

2.4. Optimization

As mentioned above, this problem consists of two MPEC problems. Usually, a network design problem regarded as MPEC is solved by sensitivity analysis (Patriksson & Rockafellar, 2002, 2003; Cho, Smith, & Friesz, 2000). In this section, we follow a sensitivity analysis framework and adjust the procedure to solve our MPEC.

As mentioned above, this problem consists of two MPEC problems. However, the behavior of both the HS and the LC/PP carriers deals with the optimal behavior of passengers. So, the competition of the HS and the LC/PP carriers can be mapped into the passengers' route choice behavior. Furthermore, since Davis shows that sensitivity analysis of network designer-users MPEC under SUE/FD is not so difficult (Davis, 1994), we concentrate on the competition between carriers.

Suppose that LC carrier's optimal behavior under the constraints of certain HS carrier behavior is $f_l^{\mathrm{HS}}\ \hat{f}_l^{\mathrm{LC}}(f_l^{\mathrm{HS}})$. Under these conditions, the basic formulation of MPEC theory is expressed below, where **f** means the vector of number of flights at each link and **g** means the function of constraints.

[MPEC-P]

$$\max_{p^{\mathrm{HS}},f_l^{\mathrm{HS}}} Z^{\mathrm{HS}}(p^{\mathrm{HS}},f^{\mathrm{HS}},\hat{f}^{\mathrm{LC}}(f^{\mathrm{HS}})) \tag{15}$$

s.t.

$$g^{\mathrm{HS}}(p^{\mathrm{HS}},f^{\mathrm{HS}},\hat{f}^{\mathrm{LC}}(f^{\mathrm{HS}}))\leqslant 0 \tag{16}$$

$$Z^{\mathrm{LC}}(p^{\mathrm{HS}},f^{\mathrm{HS}},\hat{f}^{\mathrm{LC}}(f^{\mathrm{HS}})) = \max_{f_l^{\mathrm{LC}}} Z^{\mathrm{LC}}(p^{\mathrm{HS}},f^{\mathrm{HS}},f^{\mathrm{LC}}) \tag{17}$$

s.t.

$$g^{\mathrm{LC}}(p^{\mathrm{HS}},f^{\mathrm{HS}},f^{\mathrm{LC}}(p^{\mathrm{HS}},f^{\mathrm{HS}}))\leqslant 0 \tag{18}$$

Assuming that both objective functions Z^{LC} and Z^{HS} are partially smooth, [MPEC-P] can be formulated in karush–kuhn–Tucker (KKT) style shown below.

[MPEC-D]

$$\max_{p^{\mathrm{HS}},\ f_l^{\mathrm{HS}},\ f_l^{\mathrm{LC}}} Z^{\mathrm{HS}}(p^{\mathrm{HS}},f^{\mathrm{HS}},f^{\mathrm{LC}}) \tag{19}$$

s.t.

$$g^{\mathrm{HS}}(p^{\mathrm{HS}},f^{\mathrm{HS}},f^{\mathrm{LC}})\leqslant 0 \tag{20}$$

$$\nabla_{f_l^{\mathrm{LC}}} Z^{\mathrm{LC}}(p^{\mathrm{HS}},f^{\mathrm{HS}},f^{\mathrm{LC}})^{\mathrm{T}} + \nabla_{f_l^{\mathrm{LC}}} g^{\mathrm{LC}}\varphi = 0 \quad \text{for } \forall f_l^{\mathrm{LC}} \tag{21}$$

$$\varphi g^{\mathrm{LC}}(p^{\mathrm{HS}},f^{\mathrm{HS}},f^{\mathrm{LC}}(f^{\mathrm{HS}})) = 0 \tag{18$'$}$$

where φ are Lagrangean multipliers.

The Lagrangean parameter for [MPEC-D] is obtained as

$$\begin{aligned} L^{\mathrm{HS}}(p^{\mathrm{HS}},f^{\mathrm{HS}},f^{\mathrm{LC}},\varphi,\lambda,\mu) = Z^{\mathrm{HS}} &+ \lambda(\nabla_{f_l^{\mathrm{LC}}} Z^{\mathrm{LC}}(p^{\mathrm{HS}},f^{\mathrm{HS}},f^{\mathrm{LC}})^{\mathrm{T}} \\ &+ \nabla_{f_l^{\mathrm{LC}}} g^{\mathrm{LC}}\varphi) + \mu g^{\mathrm{HS}} + \varphi g^{\mathrm{LC}} \end{aligned} \tag{22}$$

However, both HS and LC/PP carriers should satisfy constraints g^{HS} and g^{LC}. Hence, Lagrangean parameter φ can be replaced with parameter μ. Finally, Eq. (22) is redescribed as

$$L^{\mathrm{HS}}(p^{\mathrm{HS}},f^{\mathrm{HS}},f^{\mathrm{LC}},\varphi,\lambda,\mu) = Z^{\mathrm{HS}} + \lambda(\nabla_{f_l^{\mathrm{LC}}} Z^{\mathrm{LC}}(p^{\mathrm{HS}},f^{\mathrm{HS}},f^{\mathrm{LC}})^{\mathrm{T}} + \boldsymbol{\mu}^{\mathrm{T}}\nabla_{f_l^{\mathrm{LC}}}\mathbf{g}) + \boldsymbol{\mu}^{\mathrm{T}}\mathbf{g} \qquad (22')$$

where **g** stands for a set of constraints g^{HS} and g^{LC}.

For local optimization, the optimal condition of [MPEC-D] is described as

$$\frac{\partial L^{\mathrm{HS}}}{\partial f^{\mathrm{HS}}} = 0 \quad \text{and} \quad \frac{\partial L^{\mathrm{HS}}}{\partial p^{\mathrm{HS}}} = 0 \qquad (23)$$

Unfortunately, the Jacobian of μ in Eq. (22) is difficult to obtain directly. Moreover, the second-order-derivative of the Lagrangean of LC/PP carrier L^{LC}

$$\nabla_{f_l^{\mathrm{LC}} f_l^{\mathrm{HS}}} Z^{\mathrm{LC}}(p^{\mathrm{HS}},f^{\mathrm{HS}},f^{\mathrm{LC}})^{\mathrm{T}} + \nabla_{f_l^{\mathrm{HS}}}\boldsymbol{\mu}^{\mathrm{T}}\mathbf{g} + \boldsymbol{\mu}^{\mathrm{T}}\nabla_{f_l^{\mathrm{HS}}}\mathbf{g}$$

does not always satisfy the continuous differentiability of f^{HS} and p^{HS}. We should treat this problem as a nonsmooth optimization so that sensitivity analysis with a nonsmooth algorithm is adaptive (Patriksson & Rockafellar, 2002; Outrata et al., 1998; Dempe, 2000, Mifflin 1996). In this model, a bundle trust algorithm with a restart scheme (Outrata et al., 1998) is adopted to solve Eq. (22′).

3. NUMERICAL EXAMPLES

This section evaluates the model behavior and discusses the properties of competition between HS and LC carriers with numerical examples. The results that appear in the following depend on a set of parameters, but the basic properties of this model are unaffected.

In numerical computation three OD zones with three airports are considered: zone 1 has more OD traffic than the others and the HS carrier's hub airport, which the LC/PP carrier can also use. Both carriers have service networks with a predetermined shape. The basic conditions are shown in Fig. 2 and Table 1.

A list of parameters are shown in Table 2. These are artificial values adjusted to obtain realistic outcomes. Next, we discuss the model's properties under this set of value parameters.

Now, let us examine the HS carrier's pricing scheme. Since an HS carrier can independently define airfares for each OD market, it can make nonstop flight fares very low while making connecting flights expensive because of

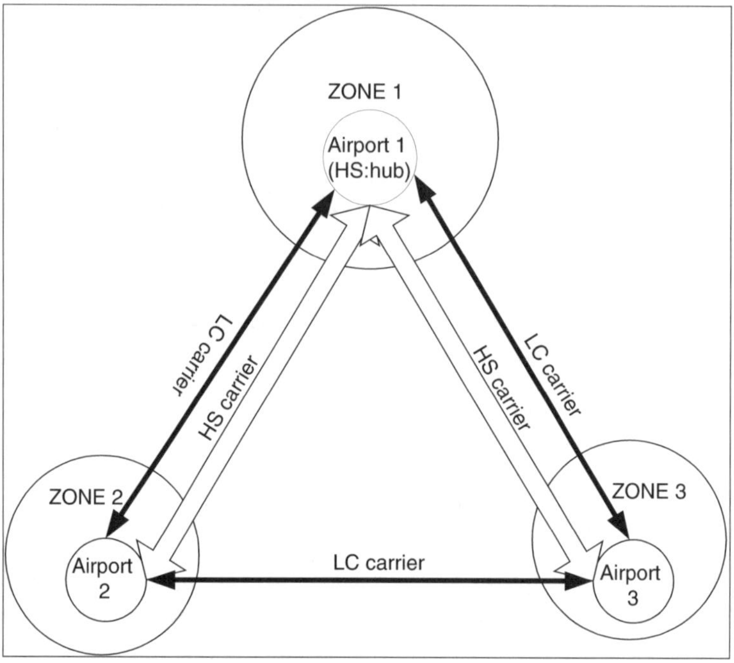

Fig. 2. Network for Computation Example.

Table 1. List of Market Conditions.

Constant Variables	Condition
OD volume	1–2 and 1–3: 100 each
(round trip)	2–3: 50 each
Aircraft size	3/flight each
Operation cost	50/flight each
Fixed cost	HS: 50 at each airport
	LC/PP: 25 at each airport
Line haul time	10/flight each
Value of *F*	HS: 100 at each airport
	LC/PP: 50 at each airport

their high profitability. However, such an unrealistic pricing scheme is avoided. In this numerical computation, the HS pricing scheme for connecting flights is defined as:

Table 2. List of Parameters.

θ	0.0125
τ	2
ξ^1	1.2
ξ^2	12.8
υ	2

(1) The airfare of a connecting flight that uses link l_1 and l_2 is the combination of nonstop flights at both links at a fixed discount rate $\boldsymbol{\rho}$.
(2) $\boldsymbol{\rho}$ is constant and applied to any given OD market that has connecting flights.

In the real world, $\boldsymbol{\rho}$ is given case by case, but for quick understanding, this simple pricing scheme is acceptable. In the computation, it is defined that $\boldsymbol{\rho} = 0.8$.

Table 3 lists the numerical results of the benchmark case and variations in which operation costs of the LC/PP carrier are reduced 10–40% more than the HS carrier's. The following outputs are listed: number of operated flights and airfares given by both carriers on main markets (1–2 and 1–3), the spoke market (2–3), the market share, and the profit of the HS carrier.

Table 3 shows a tendency previously pointed out by Mason (2001): a reduction of LC/PP carrier costs affects the profits of the HS carrier. The number of operated flights of the LC/PP carrier and the number of flights on the spoke market (2–3) are larger than the main markets (1–2 and 1–3). The load factor of the LC/PP carrier on each link reaches maximum in each case.

These features follow that active LC/PP carriers enjoy a very high average load factor on each flight. In other words, from formula (8), the lower boundary constraint on the load factor for the LC/PP carrier is not always active. Hence, these results suggest that under the HS carrier's dominance, a very high load factor is rational and inevitable for LC/PP carriers.

The LC/PP carrier's pricing scheme shows that the operational costs of the discount airfares of LC/PP carriers on each commercial link almost linearly declined. For example, when the discounting rate of operational costs is 20%, airfares are cut almost 19.5%. This linearity reflects the very high load factor, mentioned above. In addition, the 0.5% difference reflects the existence of fixed costs.

However, we find a different tendency in market shares. The reduction in the market shares of HS carriers is very moderate; when LC/PP carriers

Table 3. List of Results.

LCC's Rate of Discount	Profit of HSC[a]	HSC		LCC		HSC's Market Share (%)		HSC's Load Factor (%)
		Frequency ⟨Fare⟩[b]	⟨Fare⟩[b]	Frequency ⟨Fare⟩[b]	⟨Fare⟩[b]			
		1–2	2–3	1–2	2–3	1–2	2–3	1–2
0	100	100 ⟨100⟩	⟨162⟩	23.14 ⟨29.33⟩	24.48 ⟨29.28⟩	67.1	27.6	50.9
10%	94.05	100.59 ⟨98.34⟩	⟨157.32⟩	23.23 ⟨26.51⟩	24.37 ⟨26.90⟩	67.1	28.6	50.9
20%	88.21	98.87 ⟨98.61⟩	⟨157.84⟩	24.86 ⟨23.63⟩	25.43 ⟨23.59⟩	64.8	26.6	49.7
30%	82.37	98.39 ⟨97.6⟩	⟨156.16⟩	25.77 ⟨20.78⟩	25.97 ⟨20.74⟩	63.4	25.8	48.9

[a]Shown as proportion (HSC profit with nondiscount of LCC's operation cost is regarded as 100).
[b]Shown as proportion (the frequency (or fare) of HSC's 1–2 market with nondiscount of LCC's operation cost is regarded as 100).

reduce operational costs by 20%, HS carriers lose at most a 3.5% market share. HS carriers lose less market share in the main markets (1–2 and 1–3) than from the spoke markets (2–3). Moreover, the difference in loss ratio between the main and spoke markets widens in the discounting ratio of the operational costs of LC/PP carriers. In fact, in the benchmark case the loss ratio reaches 3.5% in the main markets and 3.8% in spoke markets; while the loss ratio is 8.5% in the main markets and 14.8% in spoke markets under 40% discount conditions.

As for profitability, HS carriers lose more profit than the discounting ratio of the LC/PP carrier's operational cost. The above discussion suggests that this large loss is due to increasing returns between the reduction of operational costs and the rise of the number of operated flights of the LC/PP carrier.

Conversely, the number of operated flights and the airfares of HS carriers did not decline so seriously, suggesting that when HS carriers enjoy the advantages from being the market leader, there is little reason to change strategy.

These results lead to the following suggestions:

(1) Although the difference in operational costs between HS and LC/PP carriers widens, the former should keep its existing strategy and as many flights and high airfares as possible. This behavior occurs because HS carriers keep operating very frequent flights to protect market share, yet retain high airfares to protect profits.
(2) A high load factor strategy is optimal for LC/PP carriers. Conversely, the HS carrier's load factor remains at around 50% in each case. This shows that HS carriers usually prefer a high flight frequency strategy to a high load factor strategy.
(3) When an LC/PP carrier behaves as a follower, it should not insist on a "high flight frequency" strategy in each market.
(4) An HS carrier can protect market share because of its advantages as leader. However, this advantage can evaporate by aggressive discount strategies of the LC/PP carrier.

Suggestion (1) reflects the advantages due to the HS carrier's market dominance. Since it can choose its strategy, the HS carrier can adopt its own optimal strategy even against strong competition. Suggestion (4) also can be accepted. However, suggestions (2) and (3) are different from the actual market behavior. LC/PP carriers, particularly Southwest clones, are usually known as "frequent service" carriers, so these results are regarded as unrealistic phenomena. One possible explanation for such unrealistic results is that the HS carrier's market dominance is supported by other factors. Hence, let us examine in detail some more numerical tests.

Sensitivity analysis on a number of allocated slots F_h^m was examined. The total number of available slots in each airport was 150, and in the benchmark test 100 slots were allocated to the HS carrier and the rest to the LC/PP carrier.

Moreover, to understand the amplification of the LC/PP carrier's low operational cost strategy, an additional condition is introduced so that the LC/PP carrier's operational costs are reduced 10% less than the HS carrier's.

Table 4 lists the results of the two scenarios: one is 90 for HS, and the other is 80. The remainder is allocated to the LC/PP carrier. The table shows that the market share of the LC/PP carrier increased in $F_h{}^{\mathrm{LC}}$. An increase of $F_h{}^{\mathrm{LC}}$ (hence, a decrease of $F_h{}^{\mathrm{HS}}$) also offers more frequent flight services and greatly lowers the airfares of the LC/PP carrier. This property comes from the definition of additional cost that relates to F_h^m in a passenger's route choice. The results suggest that the scale economy in number of operated flights is amplified by airline performance at the airport, which in this model is assumed to be the function consisting of allocated slots, F_h^m.

The HS carrier loses market share, and the profit seriously declines. Particularly, in the main market, the HS carrier reduces flights and greatly raises airfares. Conversely, the load factor at each link operated by HS carrier increases; where the LC/PP carrier gains 20 available slots, the load factor of the HS carrier reaches +10% because the reduction in airport performance for the HS carrier requires such a high load factor.

These results provide the following suggestions; if the LC/PP carrier has enough allocated slots, he can steal a large amount of shares from the HS carrier not only in spoke markets but also in the main markets. On the contrary, due to such aggressive LC/PP carrier action, the HS carrier greatly reduces the number of flights and cost of fares. In the real world, this phenomena is called the "Southwest effect," based on the entry into the market of Southwest Airlines and its big market share gains while the major network carriers reduced both airfares and flight frequencies, which reduced their profitability. Consumers benefit from such aggressive competition.

Additionally, the proposed model assumes one-on-one competition in the layers of the airline industry. However, it is very easy to modify our model to analyze one-on-many or many-on-many oligopoly competition among airlines, i.e. the model can simulate cases in which many LC/PP carriers and/or many HS carriers exist in the market (but, of course, a finite number) if we assume competition among carriers in the same level. For example, when

Table 4. Sensitivity Analysis Results for F_h^m.

LCC's Number of Allocated Gates	Profit of HSC[a]	HSC		LCC		HSC's Market Share (%)		HSC's Load Factor (%)
		Frequency ⟨Fare⟩[b]	⟨Fare⟩[b]	Frequency ⟨Fare⟩[b]	⟨Fare⟩[b]			
		1–2	2–3	1–2	2–3	1–2	2–3	1–2
0	94.05	100.59 ⟨98.34⟩	⟨157.32⟩	23.23 ⟨26.51⟩	24.37 ⟨26.90⟩	67.1	28.6	50.9
+10	58.99	81.96 ⟨84.65⟩	⟨158.86⟩	29.68 ⟨23.45⟩	27.69 ⟨23.51⟩	57.5	24	53.4
+20	36.31	68.4 ⟨73.32⟩	⟨117.32⟩	34.7 ⟨23.32⟩	30.17 ⟨23.43⟩	50	21	55.8

[a]Shown as proportion (HSC profit with nondiscount of LCC's operation cost is regarded as 100).
[b]Shown as proportion (the frequency (or fare) of HSC's 1–2 market with nondiscount of LCC's operation cost is regarded as 100).

evaluating competition between HS carriers with optimal LC/PP carriers' behavior, some additional assumption should be introduced: Cournot-type or Bertrand-type price competition theories. If the latter is adopted and elastic demand to prices is assumed, the proposed model produces the same results as their Hendricks model. If we assume a Cournot-type quantity competition, our model can be applied to the actual U.S. domestic air transportation market of network carriers' dominance.

From the above discussion, we conclude that not only the reduction of operational costs but also the improvement of performance when entering airports are two methods for LC/PP carrier to act aggressively in a market dominated by network carriers. After satisfying these conditions, the LC/PP carrier adopts "low fares with frequent services" and robs a big share from the network carriers. If LC/PP carriers do not improve their performance at the airport, which is related to the number of allocated slots in the model, a "low fare" strategy does not crucially impact the market. In other words, the advantage from performance at the airport is a significant factor for HS carriers to maintain market dominance.

4. CONCLUSION

This chapter proposed and developed a hierarchical network competition model in the air transportation market that consists of HS/PP carriers and passengers. Mathematical formulation of the hierarchical market by introducing MPEC concepts and optimal conditions were presented. Finally, how this model works is shown and its properties are discussed. The results show that HS carriers are not so sensitive to PP carrier's behavior under sufficient dominance in airport use. However, when an HS carrier loses its dominance at an airport, it becomes vulnerable to the PP carrier's behavior.

For future study, applications to actual air transportation markets are demanded. Currently, the proposed model is being modified to fit the US domestic market. Results will be presented in our next paper.

REFERENCES

Adler, N. (2001). Competition in a deregulated air transportation market. *European Journal of Operational Research*, *129*, 337–345.

Cho, H. J., Smith, T., & Friesz, T. (2000). A reduction method for local sensitivity analysis of network equilibria arc flow. *Transportation Research B*, *34*, 31–51.

Davis, G. (1994). Exact local solution of the continuous network design problem via stochastic user equilibrium assignment. *Transportation Research B*, *28*(1), 61–75.

Dempe, S. (2000). A bundle algorithm applied to bilevel programming problems with non-unique lower level solutions. *Computational Optimization and Applications*, *15*, 145–166.

Dobson, G., & Lederer, P. J. (1993). Airline Scheduling and routing in a hub-and-spoke system. *Transportation Science*, *27*(3), 281–297.

Hansen, M. (1990). Airline competition in a hub-dominated environment: An application of noncooperative game theory. *Transportation Research B*, *24B*(1), 27–43.

Hansen, M., & Kanafani, A. (1988). International airline hubbing in a competitive environment. *Transportation Planning and Technology*, *13*, 3–18.

Hendricks, K., Piccione, M., & Tan, G. (1995). The economics of hubs: The case of monopoly. *The Review of Economic Studies*, *62*(1), 83–99.

Hendricks, K., Piccione, M., & Tan, G. (1997). Entry and exit in hub-spoke networks. *RAND Journal of Economics*, *28*(2), 291–303.

Hendricks, K., Piccione, M., & Tan, G. (1999). Equilibria in networks. *Econometrica*, *67*(6), 1407–1434.

Kanafani, A., & Ghobrial, A. (1985). Airline hubbing: Some implications for airport economics. *Transportation Research A*, *19*(1), 15–27.

Luo, Z., Pang, J. S., & Ralph, D. (1997). *Mathematical programming with equilibrium constraints*. Cambridge: Cambridge University Press.

Mason, K. J. (2001). Marketing low-cost airline services to business travelers. *Journal of Air Transport Management*, *7*, 103–109.

Mifflin, R. (1996). A quasi-second-order proximal bundle algorithm. *Mathematical Programming*, *73*, 51–72.

Outrata, J., Kocvara, M., & Zowe, J. (1998). *Nonsmooth approach to optimization problems with equilibrium constraintsy*. Dordrecht: Kluwer Academic Publishers.

Patriksson, M., & Rockafellar, R. (2002). A mathematical model and descent algorithm for bilevel traffic management. *Transportation Science*, *36*(3), 271–291.

Patriksson, M., & Rockafellar, R. (2003). Sensitivity analysis of aggregated variational inequality problems, with application to traffic equilibria. *Transportation Science*, *37*(1), 56–68.

Vowles, T. M. (2001). "Southwest effect" in multi-airport regions. *Journal of Air Transport Management*, *7*, 251–258.

DEVELOPMENT AND APPLICATION OF CM-GATS ALGORITHMS IN SOLVING LARGE MULTILEVEL HIERARCHICAL NETWORK DESIGN PROBLEMS

Kerati Kijmanawat and Hitoshi Ieda

ABSTRACT

This study discusses a method that overcomes the size limitations in solving large size multilevel hierarchical hub location problem with a single allocation by using a stepwise solving approach called CM-GATS. This model uses a clustering algorithm to divide a given network into several small clusters with a manageable number of nodes then separately use hybrid heuristic algorithm based on Genetic Algorithm and Tabu Search to solve each cluster. The applicability of the proposed model in solving large multilevel hierarchical network design problems was verified by providing calculation results of CM-GATS in a large network containing up to 2,000 nodes.

Global Competition in Transportation Markets: Analysis and Policy Making
Research in Transportation Economics, Volume 13, 121–142

ISSN: 0739-8859/doi:10.1016/S0739-8859(05)13006-6

1. INTRODUCTION

Transportation networks consist of a set of nodes that share traffic of goods, people, or information. In a large transportation network where it is too expensive or unpractical to provide an exclusive link for each node, hub nodes are usually designated to act as a point of consolidation and distribution for the flow. Hierarchical networks provide an efficient way of conveying flow to destinations by reducing the number of links required to connect origin–destination (OD) pairs. An economy of scale for hub-to-hub links also provides a major incentive because hierarchical networks concentrate flow on those links, so operators can choose to improve the links or switch to a more efficient transportation mode.

There are several variations of the hierarchical network design problem. For instance, the rule can be either multiple allocation where each node is assigned to one or more hubs, or single allocation where one node is only allocated to one hub. Hubs can be designated as single level where all hubs are assigned to the same level and fully interconnected, or multilevel designation where a hierarchical relationship exists among hubs.

This chapter focuses on a multilevel single allocation uncapacitated hub location problem (MSAUHLP), which closely resembles the characteristics of freight network or postal delivery system. Solving MSAUHLP involves locating a hub system among a given set of nodes $N = \{1, \ldots, n\}$ with constant OD flows (W_{ij}). Each of the remaining nodes in the network is assigned to a single hub. The hubs are uncapacitated, which means there are no restrictions on the number of nodes assigned to a single hub in any hierarchical level. The goal in solving MSAUHLP is to find the optimum location of the hubs in each hierarchical level among the set of nodes and allocate the non-hub nodes to minimize total network costs.

In their previous work (Kijmanawat & Ieda, 2004), the authors provided a MSAUHLP formulation as a quadratic integer model as follows:

MSAUHLP:

$$\text{Min} \sum_{K=1}^{K_{\max}-1} \sum_{i_K}^{|N_K|} \sum_{h_K}^{|H_K|} X_{ih}^{K} \beta^{K} C_{ih} \sum_{j_K}^{|N_K|} W_{ij} + \sum_{K=1}^{K_{\max}-1} \sum_{j_K}^{|N_K|} \sum_{m_K}^{|H_K|} X_{jm}^{K} \delta^{K} C_{jm} \sum_{i_K}^{|N_K|} W_{ij}$$
$$+ \sum_{h_{K_{\max}}}^{|H_{K_{\max}}|} \sum_{m_{K_{\max}}}^{|H_{K_{\max}}|} X_{jj}^{K_{\max}} W_{ij} C_{ij} \alpha^{K_{\max}} + \sum_{K=1}^{K\ \max} \sum_{j} X_{jj}^{K} f_{j}^{K} \tag{1}$$

subject to

$$0 \leqslant X_{jj}^{K} \leqslant 1 \quad \text{and integers for all } j, K \tag{2}$$

$$0 \leqslant X_{ij}^{K} \leqslant 1 \quad \text{and integers for all } i, j, K \tag{3}$$

$$\sum_{h} X_{ih}^{K} = 1 \quad \text{for all } i, K \tag{4}$$

$$X_{ij}^{K} \leqslant X_{jj}^{K} \quad \text{for all } i, j, K \tag{5}$$

$$|H_K| = |N_{K-1}| \quad \text{for all } K \tag{6}$$

K represents the level of hierarchy $K \in \{1, 2, 3, \ldots, K_{\max}\}$, variable X_{ij}^{K} is equal to 1 if node i is assigned to hub j at hierarchical level K, and otherwise zero. X_{jj}^{K} is hub assignment variable. If node j is a hub at level K, then X_{jj}^{K} is equal to 1, and otherwise zero. Let W_{ij} represent the flow from node i to node j where C_{ij} is the generalized transportation cost from nodes i to j. F_{j}^{K} is the incremental fixed cost for establishing a hub at node j, and α^K, β^K and δ^K represent the factors for economy of scale, collection cost, and distribution, respectively. $|N_K|$ and $|H_K|$ are the dimensions of the set of nodes and hubs at level K.

Since O'Kelly (1987) proved that even a single level hierarchical formulation is a NP-hard problem, an algorithm called multilevel genetic-algorithm tabu search (M-GATS) was adopted, which is a hybrid heuristic model between a Genetic Algorithm (GA) and tabu search (TS), to solve the formulated problem. The study also gave extensive calculation results of M-GATS, when applied to networks of 25, 50, 100, and 200 nodes and verify the proposed model by comparing it with the available benchmark dataset. The algorithm of the M-GATS model is described in Fig. 1.

It was discussed that since MSAUHLP is an NP-hard problem, M-GATS can only efficiently solve the problem up to 200 nodes because computational time and memory requirements exponentially increase as the number of nodes increases.

This chapter presents a method to overcome the limitation of the number of nodes by developing a stepwise solving approach that uses an upward node reduction scheme. This scheme divides a given network into several small clusters with a manageable number of nodes and uses M-GATS to solve each cluster separately.

The clustering algorithm considered in this study originated from a classical Capacitated Clustering Problem (CCP), which has been extensively studied in the field of operation research. CCPs arise in such practical

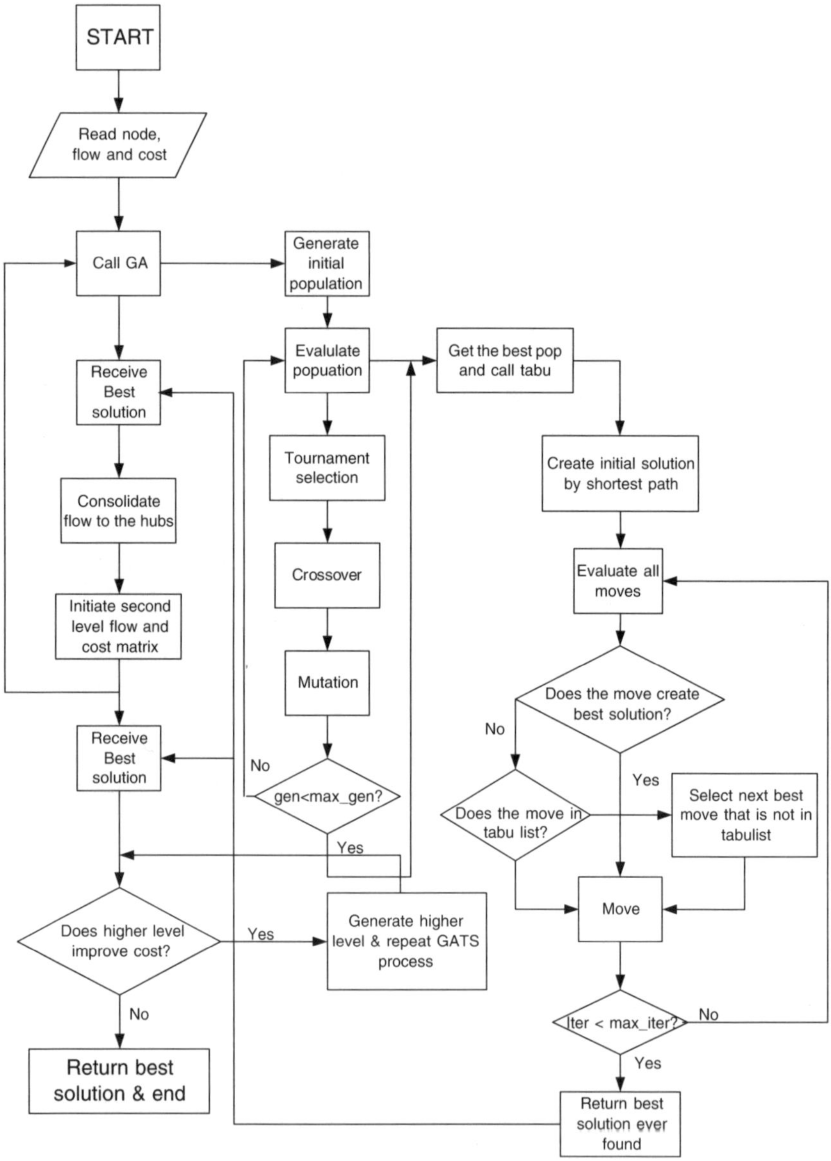

Fig. 1. Flowchart for M-GATS Heuristic.

situations as facility location problems, design sales force territories, customer service facility location problems, etc. For a comprehensive survey of CCP models and their applications, see Kaufman and Rousseeuw (1990) and Everitt (1999). Since CCP has also proven to be a NP-complete problem, recent research has focused on heuristic methods such as tabu search, genetic algorithm, simulated annealing, greedy search, and other methods that reduce computational time (Franca, Sosa, & Pureza, 1999; Osman & Christofides, 1994; Koskosidis & Powell, 1992).

This chapter is structured as follows. Section 2 provides the mathematical formulation of clustering algorithm, which is a quadratic integer program that nearly as difficult to solve as the MSAUHLP problem itself. Section 3 discusses a heuristic method that solves the clustering problem by establishing "measurement of fitness" criteria that separate nodes to into clusters. Section 4 discusses the CM-GATS algorithm. Section 5 provides a sensitivity analysis of the parameters used in the measurement of fitness. Section 6 gives an experimental calculation for a large network of 1,500 nodes and discusses the results. Section 7 provides verification of CM-GATS algorithm by comparing the calculation result of 1951 nodes with a theoretically optimal solution in a uniform boundless plane.

2. FORMULATION FOR A CLUSTERING ALGORITHM

The clustering algorithm considered in this study originated from a classical CCP. The formulation adopt a quadratic integer programming formulation proposed by O'Kelly (1987) that was originally intended to be used as an alternative approach to solve the uncapacitated p-hub location problem (UPHLP). The formulation was modified to provide criteria for clustering algorithm. The formulation can be stated as follows:

$$\text{Min} \quad \sum_i \sum_j W_{ij} \sum_g \sum_h X_{ig} X_{ih} (c_{ig} + c_{gh} + c_{ih}) \tag{7}$$

subject to

$$\sum_g X_{ig} = 1 \quad \text{for all } i \tag{8}$$

$$X_{ig} \in \{0, 1\} \quad \text{for all } i, g \tag{9}$$

$$\sum_{i} X_{ig} \leqslant n_{\max} \quad \text{for all } g \tag{10}$$

where

W_{ij} flow from node i to node j.
c_{ij} generalized transportation costs from node i to node j. c_{ij} is usually proportional to the distance from i to j and can be set to reflect the cost of travel time, transportation, and other factors.
X_{ig} nodal assignment variable. $X_{ig} = 1$ if node i is assigned to cluster g, and otherwise zero.
$n_{\max}$ maximum number of nodes in each cluster.

This formulation assigns node i to cluster g by minimizing three cost terms associated with the transportation costs of sending goods from node i to all other nodes j. These three cost terms are: c_{ig}, c_{gh}, and c_{jh}. c_{ig} is the cost of sending goods from node i to its cluster center g; c_{gh} is the cost of sending goods from one cluster center to the other; and $c_{jh\ ie}$ is the cost of sending goods from the destination center to the destination node. The graphical representation of costs associated in node clustering is described in Fig. 2.

As for constraints, constraint (8) ensures that each node is assigned to only one cluster. Constraint (9) restricts the value of the assignment matrix to either 0 or 1. Constraint (10) provides a limit on the number of nodes that can be assigned to cluster g. These constraints are directly related to the objective of introducing a clustering algorithm that divides the nodes into several clusters such that the number of nodes in each cluster is less than the capacity of the M-GATS.

The formulation assumes that the cluster center is acting as a hub for the entire cluster. However, this is not exactly true. Once clustering is completed, M-GATS will find the hub location allocation in each cluster, and the

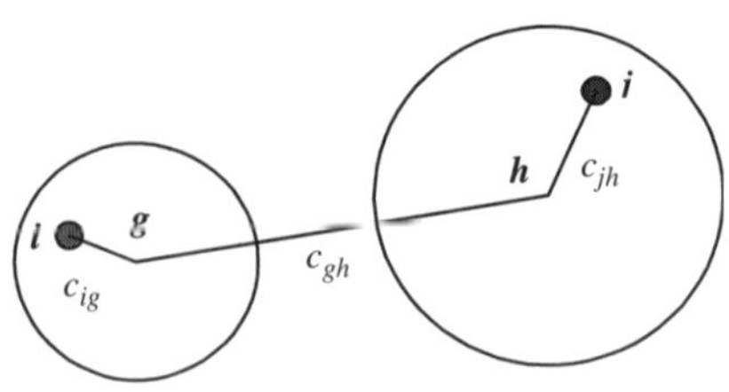

Fig. 2. Costs Associated in Node Clustering.

cluster center may not be the location of the hubs. But this assumption is necessary and logical because it provides the best approximation for the costs incurred in each cluster.

3. CLUSTERING HEURISTIC

The previous section described the formulation of a clustering algorithm. However, this problem was proven to be a NP-complete problem by Garey and Johnson (1979). Since the formulation is a quadratic integer programming, it will be as hard to solve as the MSAUHLP itself. Therefore, a constructive heuristic was developed. In this procedure, a feasible solution is constructed through successive addition of problem components according to certain criteria. This algorithm was inspired by the algorithm used by Monma and Sheng (1986) as part of a scheme to design a backbone for packet switching networks and by Klincewicz (1991) as an alternative method to solve hub location problems. In the proposed heuristic, nodes are added to each cluster individually according to fitness. Therefore, the derivation of the fitness function is crucial, since it must reflect all aspects of the objective function derived in Section 2.

3.1. Derivation for Fitness Function

The fitness function for a clustering algorithm is based on the maximization of three fitness measures: transportation cost (M_c), local traffic (M_l), and cluster balance (M_b). The physical meaning of each measure is described as follows:

1. Transportation cost measure (M_c): This measure attempts to form a new cluster at a low transportation cost within the cluster. Since costs are distance based, this measure tries to group the nodes close to one another in the same cluster. M_c can be calculated by the inverse distance from node i to cluster center g. Thus, it reflects terms c_{ig} and c_{jh} in Eq. (7) as both c_{ig} and c_{jh} represent the cost associated with sending goods from the node to its cluster center.
2. Local traffic measure (M_l): M_l attempts to form clusters of nodes which collectively send more traffic to each other than to nodes in other clusters. This means that if node i is going to be added to cluster g, the traffic

from node i should be destined for cluster g more than any other cluster. M_l can be calculated by the summation of the flow originated from node i and destined to cluster g. This measure reflects the term c_{gh} in Eq. (7). Since M_l tries to group nodes to the cluster that has the most local traffic, it reduces the amount of goods sent from the center to the center link, ultimately reducing the total cost on the center to the center link.

3. Cluster balance measure (M_b): This measure attempts to form clusters with an equal number of nodes. It is directly related to constraint (10), which limits the number of nodes that can be added to a given cluster. Actually this measure is different from the other two because it does not directly reflect the cost terms in Eq. (7), but instead the objective of establishing a clustering algorithm. A clustering algorithm allows us to divide the nodes into several clusters such that the number of nodes in each cluster is less than the capacity of M-GATS. M_b can be calculated by the inverse of the number of nodes already assigned to cluster g.

3.2. Formulation for a Clustering Heuristic

Before formulating the fitness function, the cluster center must be defined. The cluster center is defined as the "center of mass," which is the center of all nodes in the cluster weighted by nodal demand for transportation. Therefore if all nodes i in cluster g have coordinates (x_i,y_i), the cluster center $(\hat{x}, \hat{y})$ can be calculated as

$$(\hat{x}, \hat{y}) = \left(\frac{\sum_{i \in g}(O_i + D_i)x_i}{\sum_{i \in g}(O_i + D_i)}, \frac{\sum_{i \in g}(O_i + D_i)y_i}{\sum_{i \in g}(O_i + D_i)} \right) \quad (11)$$

where

O_i flows originated from node i
D_i flows destined to node i.

From the definition of the center of mass M_c, M_l, and M_b, the normalized values of M_c, M_l, and M_b can be derived as

$$M_c(i, g) = \frac{1}{\sqrt{(\hat{x} - x_i)^2 + (\hat{y} - y_i)^2}} \times \frac{1}{\hat{M}_c} \quad (12)$$

$$M_{\mathrm{l}}(i,g)=\frac{\sum_{i\in g}W_{ij}+W_{ji}}{\hat{M}_{\mathrm{l}}} \tag{13}$$

$$M_{\mathrm{b}}(g)=\frac{\hat{M}_{\mathrm{b}}}{\sum_{i}X_{ig}} \tag{14}$$

where

$$\hat{M}_{\mathrm{c}}=\operatorname*{Max}_{\text{all } g}\left(\frac{1}{\sqrt{(\hat{x}-x_i)^2+(\hat{y}-y_i)^2}}\right) \tag{15}$$

$$\hat{M}_{\mathrm{l}}=\operatorname*{Max}_{\text{all } g}\left(\sum_{i\in g}W_{ij}+W_{ji}\right) \tag{16}$$

$$\hat{M}_{\mathrm{b}}=\operatorname*{Max}_{\text{all } g}\left(\frac{1}{\sum_{i}X_{ig}}\right) \tag{17}$$

Finally, the fitness function can be derived as

$$M(i,s)=A_{\mathrm{c}}M_{\mathrm{c}}+A_{\mathrm{l}}M_{\mathrm{l}}+A_{\mathrm{b}}M_{\mathrm{b}} \tag{18}$$

where A_{c}, A_{l}, and A_{b} are the weight parameters for M_{c}, M_{l}, and M_{b}, respectively. A_{c}, A_{l}, and A_{b} are weighted such that

$$A_{\mathrm{c}}+A_{\mathrm{l}}+A_{\mathrm{b}}=1 \tag{19}$$

3.3. Algorithm for a Clustering Heuristic

The algorithm for a heuristic that clusters n nodes into p clusters can be described as follows:

Step 1. *Initialize*: Select p nodes with the largest amount of flows (O_i+D_i) as initial cluster centers.

Step 2. *Node selection*: Select the node that has the highest flow that was not selected as an initial center to be assigned to one of the clusters.

Step 3. *Compute fitness*: Compute the fitness value for adding the selected node to each of the clusters.

Step 4. *Add the node*: Add the node to the cluster that has the highest fitness value.

Step 5. *Move the center*: Recompute the coordinates of the mass of the cluster center as in Eq. (11).
Step 6. *Continue adding nodes*: Repeat steps 2–5 for all nodes in descending order by flow.

4. CM-GATS ALGORITHM

The clustering algorithm developed in Section 3 can be combined with M-GATS to enhance M-GATS's ability to solve large problems. This combination of clustering and M-GATS algorithm is called as CM-GATS, which is described below and depicted in Fig. 3.

Step 0. *Initialize*: Acquire the nodal flow, nodal coordinates, and other parameters.
Step 1. *Clustering*: From a given network, divide the nodes into small clusters by a clustering algorithm.
Step 2. *GATS*: Use GATS to solve the single level hub location-allocation in each cluster.
Step 3. *Consolidation*: Consolidate all the flows in the local nodes to the hubs found in step 2.
Step 4. *Reclustering*: If the total number of hubs in all clusters is still too large for M-GATS to handle, recluster the hubs by repeating steps 1–3.
Step 5. *M-GATS*: If the total number of hubs in all clusters is small enough to be handled by M-GATS, apply it to solve the higher level hubs in the hierarchy.

5. SENSITIVITY ANALYSIS FOR CM-GATS

In Section 3.2, the fitness function was described as

$$M(i,s) = A_c M_c + A_l M_l + A_b M_b \tag{18}$$

In the formulation, the weight for M_c, M_l, and M_b must be set by parameters A_c, A_l, and A_b. Sensitivity analysis is carried out by solving a problem from the AP dataset with 100 nodes by CM-GATS with different parameters, comparing the results with the optimal solution provided in Kijmanawat and Ieda (2004). A test problem with the following cost

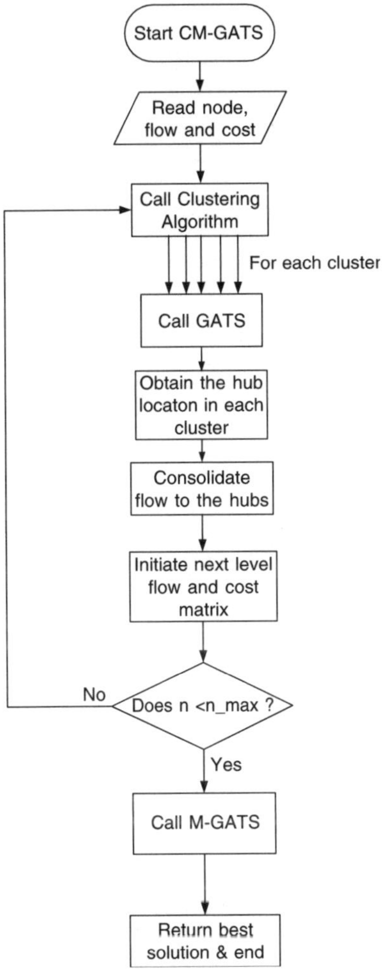

Fig 3. Flowchart for CM-GATS.

parameters: $\delta_1 = 2$, $\beta_1 = 3$, $\delta_2 = 0.75$, $\beta_2 = 0.75$, $\alpha = 0.2$, $F_j^1 = 2,000$, and $F_j^2 = 3,000$ was selected. The optimal objective value for this case is 114,789.30. This case should find that 17 nodes are selected as the first level hub, and among those 17 nodes, 7 nodes are selected as the second level hub. The original node arrangement and optimal hub location-allocation for this case are described in Figs. 4 and 5, respectively.

Fig. 4. Node Arrangement in AP 100-Node Dataset.

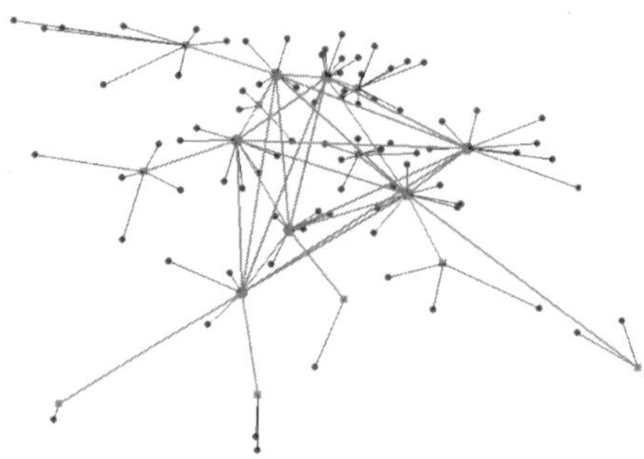

Fig. 5. Hub Location–Allocation for a 100-Node Problem in AP Dataset; Case 2 by M-GATS.

5.1. Effects of Parameters on Objective Value

This experiment studied the effects of parameters on objective value. Nodes are divided into five clusters with 18 variations of A_c, A_l, and A_b. Different cluster results and objective values were observed. The calculation results are tabulated in Table 1.

Table 1 shows changes in objective value with different combinations of A_c, A_l, and A_b. In the table, the value of one parameter is changed at a time while keeping the other two constant and equal. For example, when $A_c = 0.2$, then A_l, and A_b equal 0.6; when $A_c = 0.4$, then A_l, and A_b equal 0.3. The effects of A_c, A_l, and A_b on objective value are shown in Fig. 6(a)–(c).

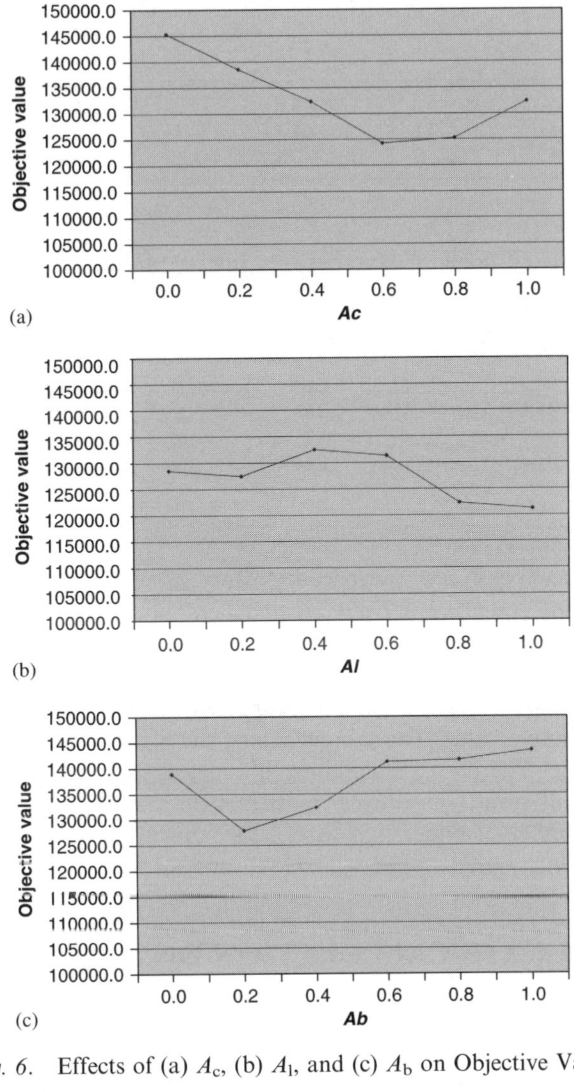

Fig. 6. Effects of (a) A_c, (b) A_l, and (c) A_b on Objective Value.

5.2. *Effects of Number of Clusters on Objective Value*

In this experiment, the effect of number of clusters on objective value was investigated. The test was performed on cases 1–3 of the AP 100-node

Table 1. Effects of A_c, A_l, and A_b on Objective Value.

	0	0.2	0.4	0.6	0.8	1
A_c	145,347.6	138,546.3	132,346.6	124,234.2	125,222.3	132,341.3
A_l	128,512.4	127,453.8	132,454.4	131,345.4	122,323.3	121,234.3
A_b	138,883.5	127,841.9	132,343.3	141,239	141,588.9	143,432.3

Table 2. Effects of Number of Clusters on Objective Value.

Test Set	No. of Clusters	M-GATS Obj. Value	CM-GATS Obj. Value	% Diff	Test Set	No. of Clusters	M-GATS Obj. Value	CM-GATS Obj. Value	% Diff
1	2	95,179.22	99,175.21	4.03	2	2	114,789.3	118,545.5	3.17
	5	95,179.22	102,874.5	7.48		5	114,789.3	127,453.8	9.94
	10	95,179.22	103,852.3	8.35		10	114,789.3	129,546.7	11.39
	20	95,179.22	107,484.3	11.45		20	114,789.3	131,924.5	12.99

dataset. For each case, the number of clusters was varied from 2, 5, 10, and 20, respectively, creating a total of 12 calculation tests. The calculation results for comparisons between M-GATS and CM-GATS are shown in Table 2.

The table shows that dividing the problem into several small clusters increases objective value. The more clusters, the higher is the discrepancy in objective value. Therefore, the number of clusters should be set to the minimum that still results in clusters with less than 200 nodes, which is the maximum number of nodes that M-GATS can solve.

6. CM-GATS CALCULATION ON A 1500-NODE PROBLEM

This section shows calculation example of CM-GATS on a 1500-node problem prepared by randomly generating 1500 node coordinates in a plane. The flow for each node was also randomly generated based on uniform distribution. The cost matrix was calculated from the nodal coordinates by a distance-based rule. Parameters were set as follows: $\delta_1 = 2$, $\beta_1 = 3$, $\delta_2 = 0.75$, $\beta_2 = 0.75$, $\delta_3 = 0.2$, $\beta_3 = 0.2$, $\alpha = 0.1$, $F_j^1 = 2{,}000$, $F_j^2 = 3{,}000$, $F_j^3 = 4{,}000$, $A_c = 0.4$, $A_l = 0.2$, and $A_b = 0.4$.

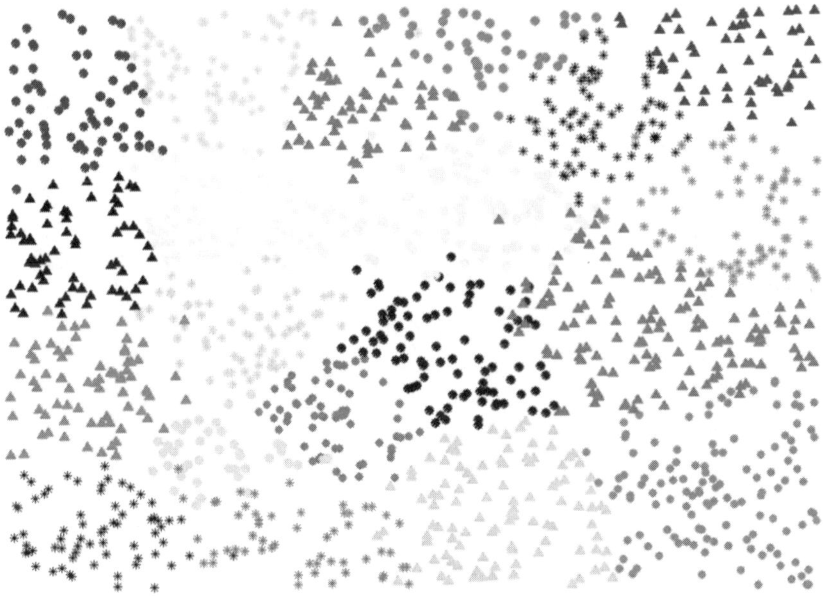

Fig. 7. Node Clusters for 1500-Node Problem.

The calculation results for the 1500-node problem are shown in Figs. 7–9. The complete (three levels) hub location–allocation in the 1500-node problem has an objective value of 287,535.23, whereas the one level hub location–allocation in the same problem has an objective value of 325,212.74. The difference in objective value is as much as 11.58%. This shows that in a large network a multilevel hierarchical system has more merit than a single level one.

7. VERIFICATION OF CM-GATS

This section provides verification of CM-GATS by comparing calculation results with a theoretically optimal solution achieved by Ieda (1997) on a uniform boundless plane. Outline of the study is described in Section 7.1. Then a large problem of 1951 nodes was created to simulate these uniform boundless plane assumptions. The calculation results and comparisons are carried out in Section 7.2.

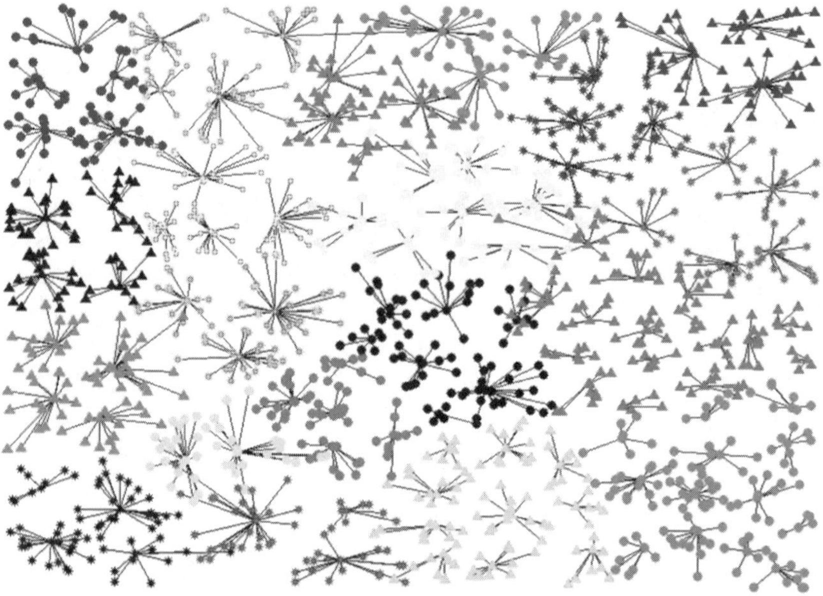

Fig. 8. Nodal Clusters with First-Level Hub for 1500-Node Problem.

7.1. Outline of Modeling and Solving Hierarchical Transport Systems in a Uniform Boundless Plane

Ieda (1997) proposed a model to address multilevel hierarchical transport systems in uniform boundless planes by using Dynamic Programming (DP) to solve the formulated problem with calculation results. This work, the first to deal with multilevel hierarchical problems, solved the problem optimally. However, the main assumption in this work is that nodes must be uniformly distributed throughout the boundless plane, as shown in Fig. 10, which makes it difficult to apply the model to solve practical problems. Nevertheless, this work offers a theoretically optimized solution and thus can be used as a benchmark to validate the proposed CM-GATS model.

7.2. Verification

In this section, the calculation results of CM-GATS is compared to those of DP as described in Section 7.1. A uniform node structure was created based

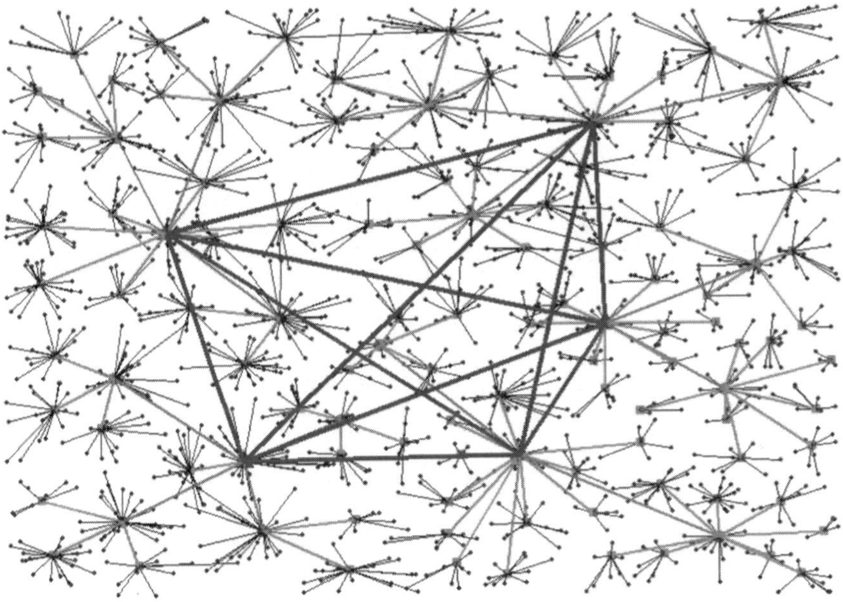

Fig. 9. Complete Hub Location–Allocation for 1500-Node Problem.

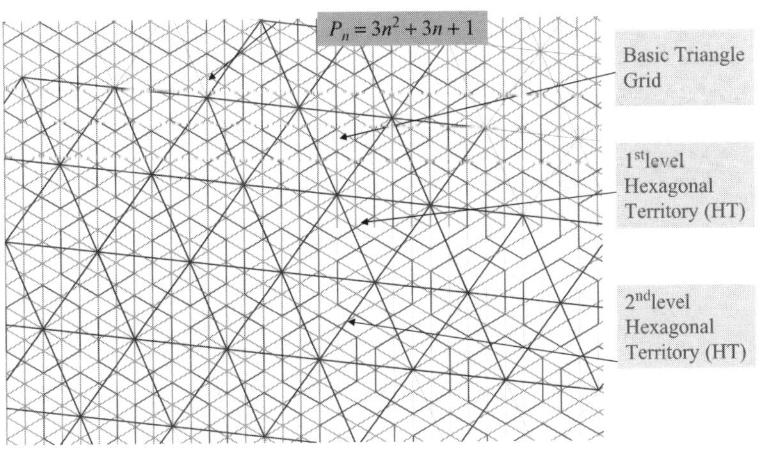

Fig. 10. Hexagonal Territory and Basic Triangle Grids.

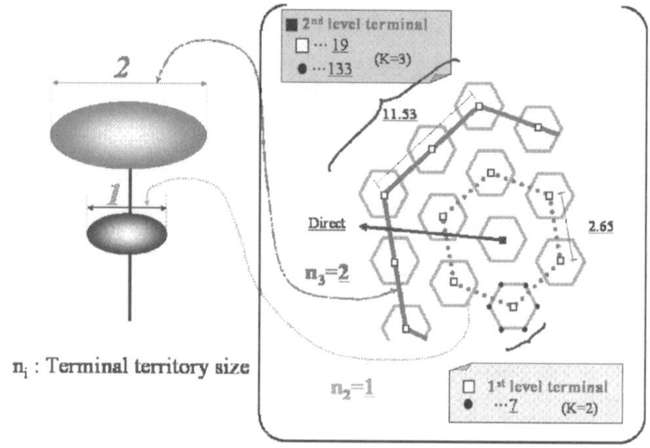

Fig. 11. Calculation Results from DP Model.

Fig. 12. Clustering Results for Hexagonal Terminal with 1951 Nodes.

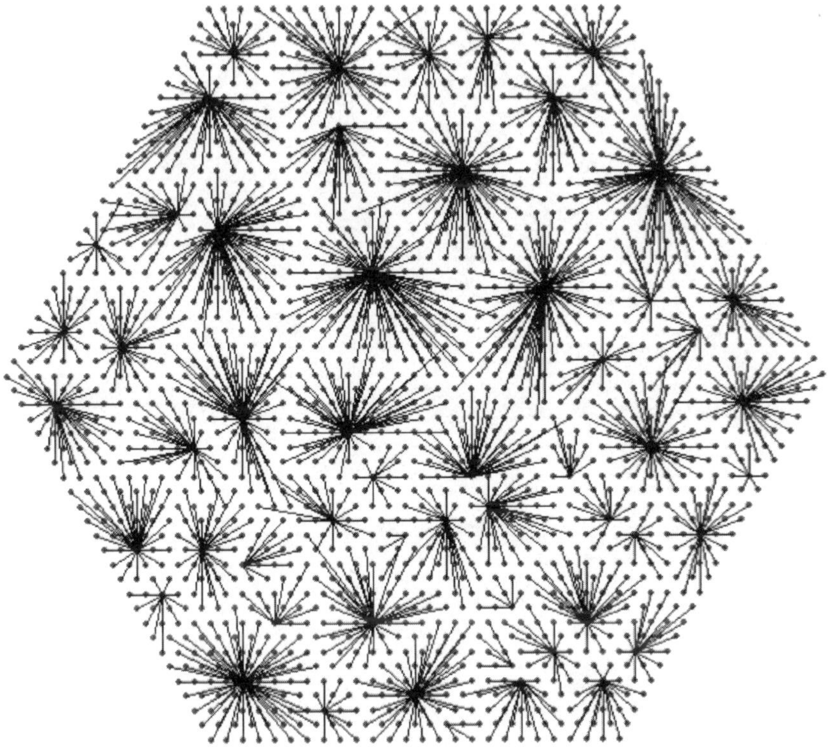

Fig. 13. First-Level Hub Location–Allocation for Hexagonal Terminal with 1951 Nodes.

on a basic triangle grid assumption in a hexagonal terminal with $n = 25$. The hexagonal terminal contained a total of 1,951 nodes. Note that the dynamic programming model was performed on a uniform boundless plane, but the model was performed on a hexagonal terminal with only 1,951 nodes. Therefore, the result around the boundary might not be comparable; nevertheless, such general characteristics of the two systems as the average terminal size should be similar.

To compare, one calculation case was chosen from the literature. The parameters in this case were set as follows:

$$D_1 = 1,\ Q_1 = 10^{-2},\ c_1 = 10^{-2},\ s_1 = 0.7,\ \text{and}\ s_2 = 0.7$$

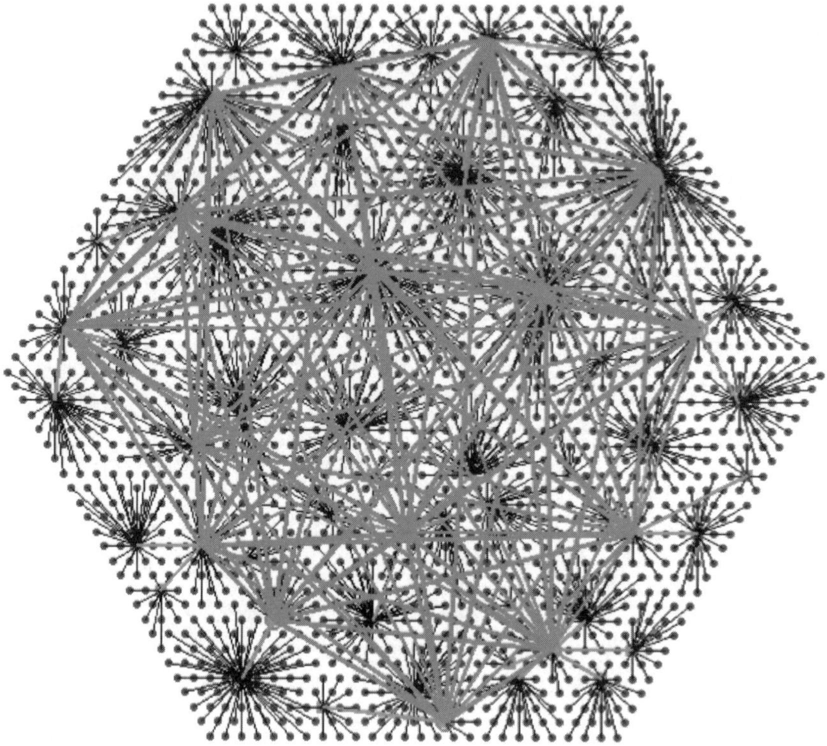

Fig. 14. Complete Hub Location–Allocation for Hexagonal Terminal with 1951 Nodes.

The calculation results for this case are given in Fig. 11. The results from CM-GATS calculations are shown in Figs. 12–14.

Comparisons for total transport cost per grid and terminal size from a uniform boundless plane model and CM-GATS model are shown in Table 3.

Table 3 shows that the number of grids per terminal for both first- and second-level terminals from the DP and CM-GATS models are almost the same, even though the transport cost per grid is very different. This happened because the problem solved by CM-GATS consisted of only 1,951 nodes, while the DP model had an infinite number of nodes. Thus, it is logical that the transportation cost per grid in CM-GATS would be considerably less than the DP model.

Table 3. Calculation Results Between Uniform Boundless Plane Model and CM-GATS Model.

	DP Model	CM-GATS Model
Transport cost per grid	0.288	0.0821
No. of grids in first-level terminal	37	35.4 (avg.)
No. of grids in first-level terminal	3	3.43 (avg.)

8. CONCLUSIONS

This chapter presents a method to overcome the limitations in the total number of nodes in a MSAUHLP that M-GATS can solve. It also proposes a stepwise solving approach by using a clustering algorithm to divide the network into several small clusters with a manageable number of nodes and used M-GATS to separately solve each cluster. The enhanced M-GATS model was subsequently called CM-GATS.

The formulation for this clustering algorithm was based on a quadratic programming problem. Since it is an NP-hard problem, a constructive heuristic was developed.

It is shown that CM-GATS can solve an extremely large network up to 2,000 nodes, which clearly eliminates the size limitations found in M-GATS model (200 nodes). This study also provides a verification of a CM-GATS model by comparing the results obtained by a uniform boundless plane model, which is a theoretically optimal solution, and those obtained by CM-GATS. The results obtained by CM-GATS were very close to the theoretical optimal one.

ACKNOWLEDGMENT

We thank M.E. O'Kelly, A.T. Ernst, and M. Krishnamoorthy for providing CAB and AP datasets in the OR-library (http://mscmga.ms.ic.ac.uk/jeb/orlib/phubinfo.html). We also thank Y. Yafeng and H. Kato for their insightful comments and discussion. We also gratefully acknowledge the productive comments of our referees and reviewers.

REFERENCES

Everitt, B. S. (1999). *Cluster analysis.* New York: Edward Arnold.

Franca, P. M., Sosa, N. M., & Pureza, V. (1999). An adaptive tabu search algorithm for the capacitated clustering problem. *International Transactions in Operational Research, 6*, 665–678.

Garey, M. R., & Johnson, D. S. (1979). *Computers and intractability: A guide to the theory of NP-Completeness*. San Francisco: Freeman.

Ieda, H. (1997). Modeling and solving of hierarchical transport systems such as hub-spokes/point-to-point and/or consolidated/direct delivery in the uniform boundless plane. *Journal of Infrastructure Planning and Management, 14*, 733–782.

Kaufman, L., & Rousseeuw, P. (1990). *Finding groups in data: An introduction to cluster analysis*. New York: Wiley.

Kijmanawat, K., & Ieda, H. (2004). Multilevel hierarchical network design: Formulation and development of M-GATS algorithm. *Journal of Infrastructure Planning and Management, 751*(IV-62), 139–150.

Klincewicz, J. G. (1991). Heuristics for the *p*-hub location problem. *European Journal of Operation Research, 53*, 25–37.

Koskosidis, Y. A., & Powell, W. B. (1992). Clustering algorithm for consolidation of customer orders into vehicle shipments. *Transportation Research B, 26*, 365–379.

Monma, C. L., & Sheng, D. D. (1986). Backbone network design and performance analysis: A methodology for packet switching networks. *IEEE Journal on Selected Areas in Communication, SAC-4*, 53–72.

O'Kelly, M. E. (1987). A quadratic integer program for the location of interacting hub facilities. *European Journal of Operation Research, 32*, 393–404.

Osman, I. H., & Christofides, N. (1994). Capacitated clustering problems by hybrid simulated annealing and tabu search. *International Transactions in Operational Research, 3*, 317–336.

AIRLINE NETWORK STRUCTURE WITH THICK MARKET EXTERNALITY

Emine Yetiskul, Kakuya Matsushima and Kiyoshi Kobayashi

ABSTRACT

In the past decade, low-cost carriers that offer point-to-point connections with frequent services have been consistently more profitable than those operating hub-and-spoke networks. Therefore, we study the advantages of each network as compared with the other. In addition to actual fares, time costs also affect the consumers' preferences, so the increase in flight frequency causes an increase in demand. Besides, the more passengers the airline carries, the more frequent services it offers. Thus, a positive feedback mechanism is incorporated into an economy of frequency. Additionally, the complementarity that arises from the demand for two-way trips is investigated.

1. INTRODUCTION

Deregulation has become a widespread trend in most of the countries over the past two decades. Even though the intensity of the reforms and the

Global Competition in Transportation Markets: Analysis and Policy Making
Research in Transportation Economics, Volume 13, 143–163

ISSN: 0739-8859/doi:10.1016/S0739-8859(05)13007-8

instruments applied have varied between sectors, deregulation has appeared in markets such as transportation, telecommunications, banking and financial services, broadcasting and energy services. Under government intervention, services had not been improved at the pace of technological advances and regulators had failed to control prices and quality of services due to information asymmetries. Thus, governments realized that competition might improve social welfare. Although the airline industry exhibits different characteristics compared with other sectors, deregulation of the airline industry is also welfare improvement.

An important case in aviation industry is the U.S.'s deregulation of its domestic passenger market in 1978. This approach enlarged on the international aviation in 1979. Many European countries followed the United States in air transport deregulation and left air carriers largely free from economic regulation since the middle of 1997 (Button & Stough, 2000). The universe of a Single European Market from 1993 also accelerated the liberalization of international air transport within Europe. Deregulation in Europe has taken two forms: bilaterally negotiated reform of Air Service Agreements and multilateral reform initiated by the European Commission (Schipper, 2001). After the U.S. initiatives for open skies agreements with many European countries, successfully concluded in mid-1990s, the U.S. Department of Transportation has shifted its focus to Asia (Oum & Park, 1997). Moreover, the United States and Canada signed their "Open Skies" agreement in 1995. Liberal bilateral Air Service Agreements have been made also with Japan, France and Korea in 1998 although they are outside the Open Skies framework (Button & Taylor, 2000). The Australian market was also regulated with Two Airline Policy in 1990 (Schipper, 2001).

The Airline Deregulation Act of 1978 proposed relaxation of the Civil Aeronautics Board's regulation of the industry. The Board's authority over routes was to end in 1981 and its authority over fares in 1983. The Board would cease operations entirely in 1985 (Bailey, Graham, & Kaplan, 1985). This laissez-faire policy caused many unpredicted changes in the airline industry, such as intensive use of hub-and-spoke (HS) networks, brand loyalty programs, travel agent commissions and strategic alliances. Perhaps the most debatable consequence of deregulation was the transformation of air networks into HS networks. An HS network is defined as "a system of routing air traffic in which a major airport serves as a central point for coordinating flights to and from other airports or spokes" (Merriam–Webster Internet Dictionary). By combining passengers with different origins and destinations, a carrier can increase the average number of passengers per flight and thereby reduce costs.

A large empirical and theoretical body of literature has analyzed these operational changes in networks.[1] Most of them have shown "economies of density" as a major reason why the carriers choose HS networks. While they have concentrated on the economy of HS networks, there are many policymakers and dissatisfied travelers who discuss the disadvantages of HS network choices, for example, transferring passengers from a hub airport between spokes means the increase in inconvenience and time incurred costs for the passengers. Additionally, hub operators began to apply predatory fare strategies to deter the entrance of smaller carriers to the market and eventually getting control of the market, which allows them to offer monopoly prices in the itineraries, beginning from or ending in their hub airports.[2]

For that reason, smaller entrants in order to compete the HS operators developed low-cost operating strategies. The US DOT (1996) defines "low-cost carriers" on the basis of

(i) estimated passenger expenses per seat mile for passenger service,
(ii) average prices in all markets served.

Adopting a low-cost strategy, new entrants offer very low fares. In markets that do not involve a dominated network hub, low-cost service results in average fare savings of $46 per passenger, while in markets that do involve dominated network hubs, fare savings are $70 per passenger (US DOT, 1996). Additionally, low fares stimulate demand. Gillen and Morrison (2003) explain the reasons of lower costs in Southwest model. By bringing extra benefits to the passengers, these carriers have increased their competitiveness and shares not only in the airports on the edge but also in the concentrated hubs of the leading carriers.

This trend began with the U.S. interstate carrier "Southwest Airlines" in 1991, so the term "southwest effect" is used to explain the success of low-cost, low-fare carriers. Imitating the success of Southwest Airlines, new carriers have emerged in all cities in and out of the United States, such as ValuJet, Reno Air, Air South, American Trans Air and Frontier in the United States and Ryanair in Europe. In the literature many papers have discussed low-cost carriers and shown econometric evidence indicating that low-cost carriers decreased fares and increased traffic on routes they operate.[3] Finally, Rhoades (2003) declares that the mega-carrier concept in North America is dead as a result of acceleration in the growth of low-cost carriers and the tendency of the traditional business travelers to change flights from high-fare–high-restriction traditional carriers toward either less travel overall or lower-cost carriers.

In this paper, we try to explain the success of low-cost airlines within a theoretical model. In doing so, we distinguish a simple network model between two basic network types, a point-to-point (PP) and an HS, and highlight the cost heterogeneity under two alternative networks because

(i) while major, large airlines have been offering services with their HS network systems, Southwest Airlines has expanded its markets with a direct connection market strategy.
(ii) USAir, the highest-cost airline, exhibits unit costs 64% above Southwest Airline's (Borenstein, 1992). Gillen and Morrison (2003) also find that the low-cost airlines are substantially lower cost than the full service carriers. The Southwest and others' low-cost strategies are partly explained by their simplicity of operation, partly by their lower input costs, especially wages, and partly by their no-frills service policies.

Additionally, this paper analyzes the relation between flight schedule and network choice. While there is a large theoretical literature about two alternative networks, only a few theoretical papers, Berechman and Shy (1998), Brueckner and Zhang (2001) and Brueckner (2004) analyze scheduling decisions. The frequency decision is very important for both airlines and passengers, because flight frequency on the one hand fixes an airline's cost with a large proportion and on the other determines the quality of services for passengers, it also stimulates market demand. Demand for trips is derived by aggregating travelers' choices. A traveler's demand is a function of actual fare and time cost. While the importance of actual fare on passengers' choices is continuing, the importance of time cost is rising as a result of the increase in the number of business activities as well as consumer willingness to increase leisure activities. These two changes cause not only a shift in travel demand but also an increase in time pressure on consumers. Hence, our model incorporates time costs.

In this paper, a potential passenger has a desired arrival time that is the same as the start time of the activity in the destination city. Moreover, each has also two inter-activity times. While one of the inter-activity times equals to the interval between the end of the activity in the origin city and the desired arrival time, the other one arises from the interval between the end of the activity in the destination city and the next activity in the origin city. A passenger's choice for travel depends whether his inter-activity time is long enough to cover the travel time and schedule delay, which is the difference between the actual and desired arrival times. While travel time is related to the network configuration, schedule delay depends on the frequency of flights. Hence, the increase in frequency causes a decrease in schedule delays, resulting in shifting the demand curve outwards. The increase in flexibility time for consumers, that is the interactivity time minus the travel time minus the schedule delay, increases the possibility that they make trips, which automatically causes an increase in market size. This phenomenon is called "thick market externality" (Matsushima & Kobayashi, 2005). As a result, the increase in the number of passengers gives an additional payoff to the airline.

In addition, we focus on the demand for two-way trips because outbound and return legs of itineraries complement each other. Under two-way trip demand, an airline tries to increase the probability that a passenger takes one of its flights on the outbound leg of his trip to guarantee the same passenger for the return leg. Hence, a shift in demand curve outward in one direction as a result of the increase in frequency causes the same shift in the other direction. This encourages transactions between passengers' choices and the airlines' frequency decisions.

This paper is organized as follows: In Section 2, we firstly introduce the model for a regulated monopoly airline firm under a PP network choice; later we carry out the similar analysis under a HS one. In Section 3, we discuss the advantages of each network structure as compared with the other in terms of fare, flight frequency and traffic volume and illustrate two numerical examples. Concluding remarks follow in Section 4. The proof of the proposition is included in the appendix.

2. NETWORK CHOICE MARKET EQUILIBRIUM

We consider a network economy of three cities labeled A, B, and C, and three possible city-pair markets, AB, BC, and AC in which passengers originate in one city and terminate in the other as illustrated in Fig. 1. In our model, we assume that only one carrier offers services for these city-pair

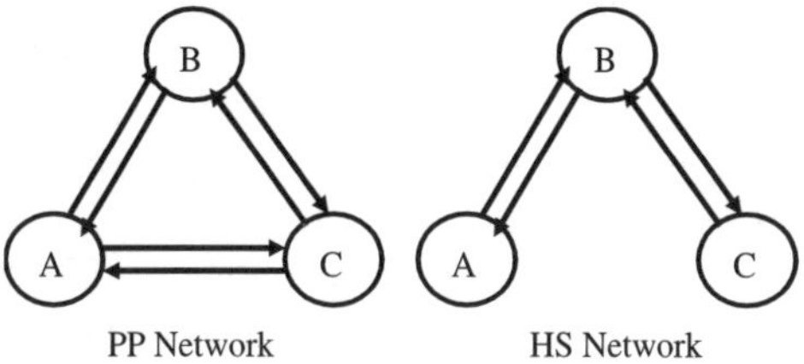

Fig. 1. A Point-to-point Netwok and a Hub- and-spoke Network.

markets because of governmental restrictions. The single company formalizes his choice according to the network economies.

In a "HS network", the airline directly carries the passengers whose city of origin or destination is that hub city and indirectly carries the passengers of connecting trips who transfer at the hub city (city B in Fig. 1). However, in a "PP network", the airline connects each pair cities. After analyzing the model in two alternative network types, the comparison of fares, flight frequency and load factors under cost heterogeneity are focused because the cost of operating larger aircraft and transferring passengers at congested hub airports is not the same as the cost of operating smaller aircraft and carrying passengers between secondary airports.

In addition, the model is characterized for the demand for two-way trips. In the literature, even though that all passengers travel in both ways in one-city pair market is assumed, the profit maximization solutions are characterized in the case that carriers set one-way trip ticket fares and consumers purchase one-way trip tickets on each way, separately. However, in the following model, optimum two-way ticket price and flight frequency according to the demand for two-way trips are found in the monopolist profit-maximization problem.

2.1. Point-to-Point Network

2.1.1. Passengers

In AB city-pair market, M denotes the number of potential passengers who want to travel from city A to city B and back. Travel demand is identical on both directions in each city-pair market as well as in all markets. We firstly modelize the behavior of passengers who are taking two-way trips, originating from and ending at city A in AB market, then extend the model to the other city-pair passengers because in a PP network, an airline offers direct services between each city-pair market. We assume that each potential passenger has a scheduled activity in the destination city so each wants to take two-way trips. For each consumer, there is a fixed amount of income, Y, and a start time for his activity in the destination city. The term θ denotes the start time of the activity, which can be also called as desired arrival time in the destination city for a consumer, and these start times are distributed continuously and uniformly in a circular time interval $[0, 2\pi]$. "Desired arrival time" and "most preferred departure time," termed firstly by Douglas and Miller (1974), are very similar.

In addition to consumer heterogeneity in the desired arrival times, we assume that consumers differ in their inter-activity times. The variable s denotes the time interval from the activity before leaving the origin city till that after arriving in the destination city for a consumer; t denotes inter-activity time that arises from the interval, beginning from the end of the activity in the destination city and ending at next activity in the origin city. While s has a uniform distribution with support $[0, \bar{s}]$, t has with $[0, \bar{t}]$. Each potential passenger has two inter-activity times that are independent (Fig. 2). Throughout the paper, it is assumed that consumers have three appointments at different scheduled times so they have to take flights on both legs of their two-way trips during their inter-activity times. Hence, each has to take a flight after the time $\theta - s$ at his origin city and return before the time $\theta + \alpha + t$, where α denotes the duration of the activity in the destination city. For simplicity, we normalize α to a fixed value or zero.

Ignoring any distance differences between cities, we assume that the duration of a nonstop travel between any city pairs is identical and it is shown as *f*. In a PP network, the actual flight time is same on all city-pairs; in an HS network for the connecting market, the duration of the trip is equal to two actual flight times (*2f*) and layover time at the hub airport. We suppose that there is no restriction in the capacity of aircraft. The airline company offers n flights on each direction in each city-pair market and the intervals (i.e. the headways) between departure times of flights are same. The flights originating from city A to city B are indexed by i, and $a(n) = (a_1, \ldots, a_n)$ denotes the set of departure times of the flights i. The headway is equal to $2\pi/n$. As it is assumed that the flights offered for each direction in each city-pair market

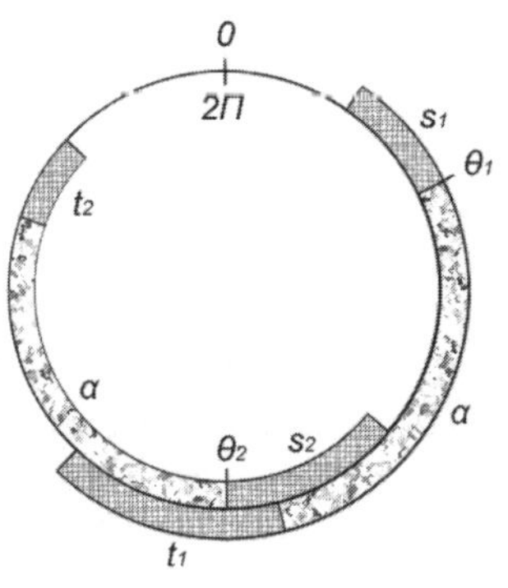

Fig. 2. Example of θ, α, s, t Distribution for Two Passengers.

are symmetric, the departure times can be written as

$$a_i = (i-1)\frac{2\pi}{n} \quad (i = 1, \ldots, n) \tag{1}$$

The indirect utility of a potential passenger on a two-way trip is

$$U(\theta, w : \mu) = Y + w - p \tag{2}$$

where w denotes the consumer-specific gross utility from taking the two-way trip, and has a uniform distribution with support $[0, \bar{w}]$. Consumers with identical θ, s, t also differ in their gross utilities. p is the price of a two-way trip ticket. $\mu = \{p, n, a(n)\}$ is the vector set of the flight characteristics in which a two-way ticket price, the number of flights in one direction and the flight departure times are denoted, respectively. Each potential passenger takes one of the flights on both legs only if the utility of taking the two-way trip exceeds his income, Y:

$$U(\theta, w : \mu) \geqslant Y \tag{3}$$

Eq. (3) means that if the condition $w - p \geqslant 0$ is satisfied, and the consumer takes the flights.

In the model, the role of the inter-activity times is in determining the market thickness, which is related to the vertical differentiation between potential passengers. It runs in defining the following conditions. As each passenger has to take one of the flights on the outbound leg after the activity ends in the origin city and on the return leg before the next activity starts in the destination city, actual flight time and the time arising from the deviation between the actual and desired arrival times have to be equal to or smaller than s. This condition is given by

$$SD + f \leqslant s \tag{4}$$

where SD denotes the deviation time. Assuming that the utility loss caused by taking a flight earlier and later than the desired arrival time is same, it is allowed that the passengers choose the flight belonging to the smaller value of time cost. The term "Schedule Delay" is used to capture both types of utility loss. As the desired arrival times of consumers are distributed continuously and uniformly and the headways of the flights are identical, schedule delay costs increase linearly, from zero for consumers located exactly at the arrival time to $2\pi/4n$ for consumers located on the boundary of the next flight. Then, the average schedule delay is found as $SD = 2\pi/4n$. Substituting SD into (4), we can rewrite the condition for the outbound leg of the trip and similar condition for the return leg is also written as

$$\frac{\pi}{2n}+f\leqslant t \tag{5}$$

As a result, we can say that if a potential passenger on the outbound/ return leg of his trip has an inter-activity time that satisfies conditions (4) and (5), there is a possibility that he takes one of the flights on that leg; finding the total number of flights yields the market density on that leg.

To derive the cost of schedule delay, we use the same method, presented in Schipper (2001) and Brueckner (2004), in which consumer heterogeneity arises from both desired times and gross utility values. In our model, consumers also differ in their inter-activity times. Taking advantage from average schedule delay in the solution is a way to simplify the analysis, which has horizontal and vertical differentiation in potential passengers. Otherwise, the model solutions become very complex to analyze.

In the model, condition (3) shows the number of passengers who find a two-way trip worthwhile. Conditions (4)/(5) shows the number of passengers for whom inter-activity time is sufficiently large to cover the actual flight time and schedule delay on the outbound/return leg of the trip. Recalling that the inter-activity times of a potential passenger on both legs are independent, we realize the possibility that while condition (4) is satisfied for a consumer but not condition (5) and vice versa. Hence, a two-way trip exists only if both conditions as well as condition (3) are satisfied, otherwise there is no trip. The probability found from conditions (4) and (5) shows the density of consumers on that leg. However, the total mass of consumers whose both inter-activity times are large enough to satisfy both conditions is found from the multiplication of the probabilities. Assuming that $\bar{s}=\bar{t}$ gives the symmetry between the probabilities, so total demand for the flights on one direction can be expressed as

$$\begin{aligned} X(\mu) &= M\int_0^{2\pi}\int_0^{\bar{s}}\int_0^{\bar{w}}\left\{\Pr\left(\frac{\pi}{2n}+f\leqslant s\right)\right\}^2\Pr(w-p\geqslant 0)\,\mathrm{d}w\,\mathrm{d}s\,\mathrm{d}\theta \\ &= M\left\{\frac{\bar{s}-(f+\pi/(2n))}{\bar{s}}\right\}^2\frac{\bar{w}-p}{\bar{w}} \end{aligned} \tag{6}$$

where $\Pr(\cdot)$ defines the probability. While the first expression of the right side of (6) shows the probability of passengers whose both inter-activity times are long enough to satisfy conditions (4) and (5), the second one is the share of passengers whose net utilities, derived from taking two-way trips by purchasing a two-way ticket, are non-negative.

2.1.2. Firm Behavior and Market Equilibrium

The monopolist maximizes his profit according to two-way trip demand and optimizes the number of flights, n and fare of a two-way ticket, p. Before analyzing firm behavior in a PP network, firstly we note that an airline's fixed cost, d_i, per flight on each connection between two cities differs belonging to network types, $i = p, h$ where f denotes the PP network and h the HS. As the fixed cost per flight consists of those related to operating a flight – including salaries of the flight crew, maintenance, landing fee, and services provided to the aircraft while it is on the ground – the size of an aircraft under a PP/HS network, landing fees and the services of hub/secondary airport, etc. can be different. Therefore, we distinguish between two alternative networks. Additionally, it is assumed that the passenger cost, c_i, associated with services such as providing food and drink on board and passenger baggage, is different under each network type; c_p and c_h denote, respectively, the variable cost per passenger on a PP and an HS network.

Assuming that the demands for two-way trips are symmetric in AB city-pair market (from city A to B and back, from city B to A and back), the profits earned by the carrier from operating the flights on each direction in AB market can be written as

$$\pi(\mu) = (p - c_p)X(\mu) - nd_p \tag{7}$$

Additionally, there is symmetry between the demands of each city-pair market in the three-city-model, so the monopolist that offers n flights with a departure time set, $a(n)$, in each direction and sets two-way trip fare, p, faces the following maximization problem

$$\max_{p,n,s(n)} \{3\pi(\mu)\} \tag{8}$$

We can characterize the solution for (8) in two stages. The optimal two-way price, p^*, and the number of flights, n^*, are found sequentially, where p is conditional on n, and then n is chosen in a second stage. The departure times of flights, $a(n)$, are captured spontaneously after n^* is found, because the time interval between neighboring flights are the same and it is assumed that the first flight is departing at a time, *0*. Throughout the paper we also suppose that p and n are continuous variables.

Substituting (6) and (7) into (8) and holding n fixed, we can write the profit as a function of p. Taking the first derivative with respect to p yields

$$M\left\{\frac{\bar{s} - (f + \pi/(2n))}{\bar{s}}\right\}^2 \left(\frac{\bar{w} - 2p + c_p}{\bar{w}}\right) = 0 \tag{9}$$

From (9), the optimum price p^* for two-way trip is found as

$$p^*(n) = 1/2(\bar{w} + c_p) \tag{10}$$

Eq. (10) shows that the monopolist sets a price that covers the marginal cost of a seat partially. After finding the optimum price, we can rewrite the total profit function, conditional on n, as

$$\Pi(n, p^*(n)) = \frac{M\pi(\bar{w} - c_p)^2}{4\bar{s}^2\bar{w}} \left\{ \bar{s} - f - \frac{\pi}{2n} \right\}^2 - nd_p \tag{11}$$

To find the optimal n^*, the derivative of (11) is equalized to zero. Then, the first order gives the following condition

$$\frac{M\pi(\bar{w} - c_p)^2}{4\bar{s}^2\bar{w}} \left\{ (\bar{s} - f)n - \frac{\pi}{2} \right\} = d_p n^3 \tag{12}$$

The left-hand-side (LHS) and the right-hand-side (RHS) of condition (12) are illustrated in Fig. 3, where the S-shaped curve and the line represent the RHS and LHS of the expression, respectively. The curve and the line intersect at two positive points, in which the second solution illustrates the optimum number of flights when the second-order condition of the airline's optimization problem is checked. The way outlined here is essentially the way presented in Brueckner (2004). A quick observation of Eq. (12) generates a number of comparative-static results. For example, an increase in one unit of variable cost, causing an increase in two-way ticket fares, results

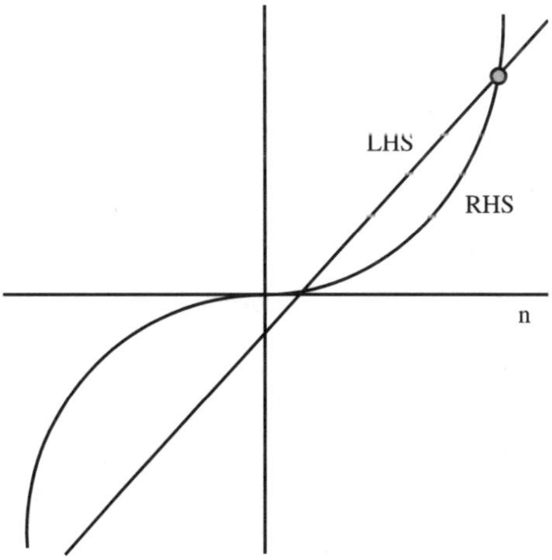

Fig. 3. The Frequency Solution Under the PP Network.

in a frequency decrease. However, when the gross valuation of consumers increases, the slope of the line in Fig. 3 increases, raising n. Additionally, the smaller the value of fixed cost, the higher the frequency.

Given the distribution of desired arrival times of passengers over time, the inter-activity times of them over $[0, \bar{s}]$ and $[0, \bar{t}]$ and the gross utilities, the monopolist first sets the ticket price for two-way trips and then schedules the flights. Recalling that under PP network, the airline serves each of the three city-pair markets AB, BC and AC with separate flights so p^* and n^* are same for all.

2.2. Hub-and-spoke Networks and Market Equilibrium

2.2.1. Passengers

In this section we consider that the monopolist chooses a "HS network" and offers services for the three city-pair markets of city A, B and C. It transfers the passengers of AC market via city B as its hub so its aircrafts are flown only on the AB and BC routes. The monopolist offers $\bar{n}$ flights on each connection in a circular time interval $[0, 2\pi]$ and the headways between the flights are same. The flights originating from city A to the hub and from the hub to city C are indexed by i, and $a(\bar{n}) = (a_1, \ldots, a_{\bar{n}})$ denotes the set of departure times of the flights. The headway is equal to $2\pi/\bar{n}$. As it is assumed that the flights offered for each direction are also symmetric, the number of flights and their departure times in both directions of each connection are identical.

The passengers' behavior in AB and BC city-pair markets of the HS network is the same as that in the markets of the PP network but the behavior of AC market passengers is different, because the utility loss derived from time cost increases when traveling via the hub. Hence, under the HS network, we distinguish between two different types of price: $\bar{p}$, the price set for the direct passengers, and $\bar{q}$, the price for connecting passengers. The indirect utility of a connecting passenger who can take a two-way trip is given by

$$\bar{U}(\theta, w : \bar{\mu}) = Y + w - \bar{q} \tag{13}$$

where $\mu = \{\bar{q}, \bar{n}, a(\bar{n})\}$ denotes the vector set of the flight characteristics on a connecting market. Each potential passenger flies only if the utility of taking the two-way trip exceeds his income, Y,

$$\bar{U}(\theta, w : \bar{\mu}) \geqslant Y \tag{14}$$

Eq. (14) shows that if the condition $w - \bar{q} \geqslant 0$ is satisfied, the consumer takes the flights.

Under the HS case, taking one of the scheduled flights on the outbound and return legs of the trip depends upon s and t, respectively, the same as that under PP case. However, under the HS case, a passenger of the connecting market bears additional cost that arises from additional flight time and layover time at the hub. As a passenger takes the trip on the outbound leg after the activity ends in the origin city and before the next activity starts in the destination city, the inter-activity time has to involve the duration of the total travel time, layover time and schedule delay for each connecting passenger. Letting the layover time and the total actual travel time be denoted by v and $2f$, this condition can be written as

$$\frac{\pi}{2\bar{n}} + 2f + v \leqslant s \tag{15}$$

Given symmetry between both legs of a two-way trip, the condition for the return leg of the trip is the same as (15). As a two-way trip generates only if all conditions are satisfied, the total demand for the connecting flights can be found from

$$\begin{aligned} \bar{X}(\bar{\mu}) &= M \int_0^{2\pi} \int_0^{\bar{s}} \int_0^{\bar{w}} \left\{ \Pr\left(\frac{\pi}{2\bar{n}} + 2f + v \leqslant s\right) \right\}^2 \Pr(w - \bar{q} \geqslant 0)\, \mathrm{d}w\, \mathrm{d}s\, \mathrm{d}\theta \\ &= M \left\{ \frac{\bar{s} - (2f + v + \pi/(2\bar{n}))}{\bar{s}} \right\}^2 \frac{\bar{w} - \bar{q}}{\bar{w}} \end{aligned} \tag{16}$$

2.2.2. Firm Behavior and Market Equilibrium

The monopolist under an HS network maximizes his profit by finding optimum number of flights, $\bar{n}$. As the firm carries the passengers of the local and connecting markets in the same aircraft under an HS network, it optimizes the flight frequency according to total passenger volume. $\bar{n}$ is same on each of the two local routes. Under an HS network, there is an asymmetry in the ticket prices of the local and connecting markets, so the monopolist optimizes the fare of a ticket, $\bar{p}$ for direct passengers and $\bar{q}$ for connecting passengers. As we clarified in Section 2.1.2, we separate the airline's costs into fixed cost, d_i per flight and variable cost, c_i per passenger, and assume cost heterogeneity under both types of networks. The revenue of the firm under an HS network comes from carrying the passengers of two direct markets and one connecting market. However, the total fixed cost is identified by the cost of flying only on the AB and BC markets, so, HS operator can utilize from economies of density if the total fixed cost does not increase under the HS network. Given the symmetry between both directions in each city-pair market, the profit earned by the carrier from operating the flights under an HS network is given by

$$\bar{\Pi}(\bar{\mu}) = 2(\bar{p} - c_h)X(\bar{\mu}) + (\bar{q} - c_h)\bar{X}(\bar{\mu}) - 2\bar{n}d_h \tag{17}$$

where $X(\bar{\mu})$ and $\bar{X}(\bar{\mu})$ are demands for the flights of the direct and connecting market, respectively. $X(\bar{\mu})$ has the same form as (6). Then, substituting (6) and (16) into (17), we can write the profit of the monopolist as a function of $\bar{p}$ and $\bar{q}$. Taking the first derivatives with respect to the price of the local and connecting market, yields the following:

$$M\left\{\frac{\bar{s} - (f + \pi/(2\bar{n}))}{\bar{s}}\right\}^2 \left(\frac{\bar{w} - 2\bar{p} - c_h}{\bar{w}}\right) = 0 \tag{18a}$$

$$M\left\{\frac{\bar{s} - (2f + v + \pi/(2\bar{n}))}{\bar{s}}\right\}^2 \left(\frac{\bar{w} - 2\bar{q} - c_h}{\bar{w}}\right) = 0 \tag{18b}$$

Solving (18a) and (18b), the optimum prices for two-way trips in the direct and connecting market are found as

$$\bar{p}^* = \bar{q}^*(n) = 1/2(\bar{w} + c_h) \tag{19}$$

From Eq. (19), we can say that the monopolist sets the same ticket prices for the passengers of both markets. Additionally, it is the same as the optimum price set in the PP case. After substituting (19) into (17), we can rewrite the total profit function of the monopolist under an HS network as

$$\bar{\Pi}(\bar{n}, \bar{p}^*(\bar{n}), \bar{q}^*(\bar{n})) = \frac{M\pi(\bar{w} - c_h)^2}{4\bar{s}^2\bar{w}} \times \left\{\left(\bar{s} - 2f - v - \frac{\pi}{2\bar{n}}\right)^2 + 2\left(\bar{s} - f - \frac{\pi}{2\bar{n}}\right)^2\right\} - 2\bar{n}d_h \tag{20}$$

Taking the first derivative of (20) with respect to $\bar{n}$ and rearranging the first order yields the condition for frequency

$$\frac{M\pi(\bar{w} - c_h)^2}{4\bar{s}^2\bar{w}}\left\{\left(\frac{3}{2}\bar{s} - 2f - \frac{v}{2}\right)\bar{n} - \frac{3\pi}{4}\right\} = d_h\bar{n}^3 \tag{21}$$

If $d_p = d_h$ is assumed, the RHS of the condition is the same as that in (12). However, the slope and the intercepts of the LHS are different. The diagram of the condition is like in Fig. 3. As before, there are two positive intersection points of the S-shaped curve and the line that are economically relevant. When the second-order condition is scrutinized, the second one, which has a higher position than the other, represents the optimum flight frequency. Focusing on heterogeneity in costs under two alternative network types, we compare frequency levels in the next section.

3. COMPARISON OF TWO ALTERNATIVE NETWORKS

In Sections 2.1 and 2.2, the optimum price and frequency solutions of the monopolist under two network structures are scrutinized without discussing which type of network offers greater fares or frequency or traffic volumes. As cost heterogeneity in network types is assumed, a change in one cost parameter alters the levels, causing a difficultly in identifying the network choice of the monopolist. However, the following results can be established:

Proposition 1. The comparison of fare levels between two alternative networks does not depend on the frequency parameters, so an observation of Eqs. (10) and (19) leads to the conclusion that "the fares in direct markets as well as connecting one under an HS network is higher (lower) than those under a PP network if the variable cost of a passenger when operating an HS network is higher (lower) than that when operating a PP network."

The comparison of frequency levels under both network types will be held on in two cases: under the assumption $d_p = d_h$ and $c_p = c_h$; and the assumption $d_p < d_h$ and $c_p < c_h$.

Proposition 2. Under the first case, comparing the first-order conditions for flight frequency in Eqs. (12) and (21) yields the solution, "flight frequency is higher in the HS network than in the PP network, $n < \bar{n}$," because the LHS of the HS network in Eq. (21) has a higher position than that of the PP network in (12) while the RHS of two conditions are the same (see the appendix). As cost assumptions of the first case are the same as in Brueckner (2004), the result, established in Proposition 2, is also same. The line of the HS network has a higher level than that of the PP network because of the passenger volumes in each connection. In other words, the marginal flight cost on each HS route is paid by the local market and the connecting market passengers. However, in the PP case, only the passengers on that route pay the flight cost. Higher frequency under an HS network on one connection does not mean that the total numbers of flights operated under an HS network is greater than under a PP one. If this expectation is true, it can be said that the HS operator saves costs and exploits economies of density.

Under the second case, if the costs of the airline are low enough when operating a PP network, the comparison of frequency levels yields $n > \bar{n}$, implying that the number of flights in a PP network is higher than in an HS network on each connection, because the positions of both sides of condition (12) change. While a decline in the variable cost makes the intercept of the line more negative, a decline in fixed cost makes the rate of change in the slopes of the curve less steep, so the intersection point in the positive quadrant moves to the right, raising n. A decrease in ticket fare causes an increase in the number of passengers who find the trip worth taking, thus adopting low variable cost per passenger under a PP network; the monopolist sets lower fares, which results in an increase in flight frequency. Lower operating flight cost also causes greater traffic volumes because the higher frequency, lowering the schedule delays, results in an increase in the probability of passengers whose inter-activity times are large enough to cover actual flight time and schedule delay. As a result, it can be said that the higher the flight frequency, the more the PP operator utilizes economies of frequency as a result of higher demand.

In the case that Proposition 2 is valid, comparison of the traffic levels between the PP and HS network leads to the same result for direct markets of HS network, which is that the traffic volumes in city-pair markets AB and BC are higher in the HS network than in the PP network. However, this transaction between higher frequency and higher demand cannot be said for the connecting market of the HS network. The level of traffic in city-pair market AC can be higher or lower than that under PP network because the increase in trip duration for connecting passengers as a result of flying two legs and waiting at hub causes a shift in the demand curve downward. Hence, it is difficult to conclude that higher frequency generates higher profit level.

To show the effect of parameter changes in the comparison of the HS and PP solutions and to reconfirm the results established above, two numerical examples are illustrated. The focus of the first example is to capture the changes in frequency and profit levels under both network types as a result of an increase in actual travel time and the ratio between two types of operating flight costs. Given a set of parameters as $M = 100$, $\bar{w} = \pi/2$, $\bar{s} = \pi/6, v = \pi/150$, $c_p = c_h = \pi/10$, $d_p = 1$, the HS and PP solutions are illustrated in Fig. 4, in which the lower and upper lines show the boundaries between the lower line and the upper line levels, respectively. While profit is determined by setting equilibrium profit $\Pi = \bar{\Pi}$, the upper line is from setting equilibrium flight frequency $n^* = \bar{n}^*$. Therefore, the lower line demarcates the regions where the profit level of the monopolist is more under the HS case and less under the PP network, respectively. However, in both

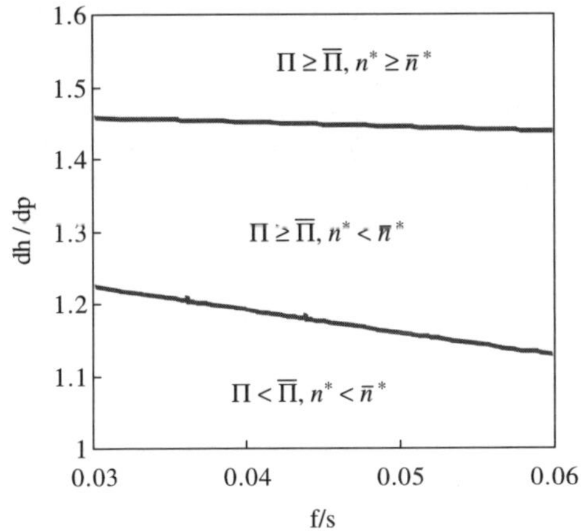

Fig. 4. The Effect of *f*.

regions, flight frequency is higher in the HS network than in the PP network. It can be seen from the diagram that the profitability of the HS network decreases when either the actual flight time or the ratio between two flight operating costs, d_h/d_p, increases. If the ratio continues to increase, implying that the carrier's operating cost is lower under the PP network than under the HS one, frequency is higher in the PP network than in the HS network. The impact of the increase in the actual travel time on the equilibrium frequency line is less than that on the equilibrium profit line.

In addition, the effect of market size, *M*, on frequency and profit levels under both network types is examined. Holding the actual travel time equal to $\pi/100$ yields the result in Fig. 5, in which the lower and upper lines again show the boundaries between profit and frequency levels, respectively. Except that the demarcation lines have different slopes, Fig. 5 is the same as Fig. 4. As can be followed from the diagram, the increase in *M* causes a decrease in the profitability of the monopolist under the HS network, leading to the conclusion that the effect on the profit of the downward shift in the demand curve of the connecting market under the HS network dominates the effect of the upward movement due to the higher frequency.

4. CONCLUSION

This chapter models the success, development and competitive advantages of new entrants maintained over large, major airlines in a theoretical setting.

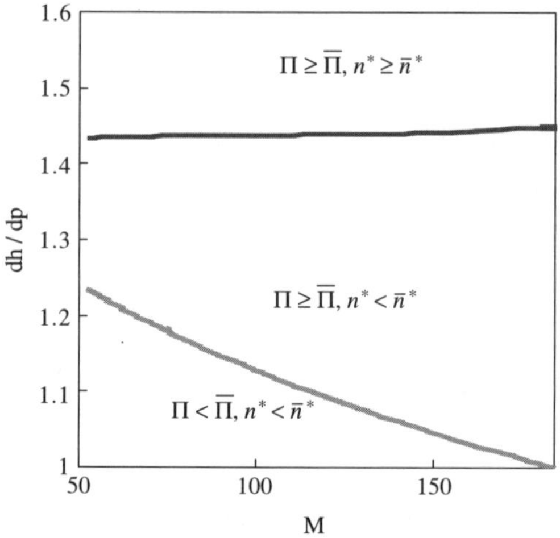

Fig. 5. The Effect of *M*.

While incumbents offer indirect connections via hub city and benefit from economies of density, new entrants develop low-cost operating strategies that make them offer very low fares. The reason for their lower costs is the simplicity in their operations, which involves offering direct frequent services in a PP network, using uncongested secondary airports, servicing with the same type of aircraft and providing no-frills services. As a result, new entrants have expanded their market shares not only on the secondary routes that leading carriers don't offer services but also on the routes that they dominate. Hence, distinguishing a simple network model between two basic network types, PP and HS, under cost heterogeneity, we examine the effects of the network structure on the fares, flight frequency and traffic volumes.

The analysis shows that offering services in an HS network leads to increase in flight frequency under the equal cost assumption. However, when the operating and variable costs are low enough, PP operator offers more services. As the ticket price covers the variable cost per passenger, the difference between variable costs under the two networks affects the levels of fares. The decrease in ticket fares, causing an increase in the number of passengers who find the trips worth taking, results in higher frequency.

Additionally, introducing inter-activity times, which are consumer-specific, we observe market thickness under both alternative networks. The inter-activity time for a consumer has to be larger than the duration of travel and schedule delay. Hence, extra travel time required for the passengers of connecting markets under the HS network decreases the probability that a potential passenger can take one of the flights during his inter-activity time, causing a decrease in market thickness. In any case, the demand curve is shifted up when frequency increases as a result of decrease in schedule delays. There is a positive interaction between the number of services in the market and the market thickness, termed as thick market externality, which implies that the higher the frequency the airline offers, the more passengers it carries.

In the model, we introduce the demand complementarity inherent to two-way trips to analyze the total demand for two-way trips and find the optimum two-way ticket price. Analyzing the factors that affect the airline's choice about network type, the model captures the advantages of each network structure as compared with the other. However, the model compares the fare, frequency and traffic levels for a monopolist, so it yields biased solutions. A further discussion on the duopoly case can be conducted. Schipper (2001) analyzes the frequency choice in air transport markets without distinguishing between two basic network types.

NOTES

1. The theoretical papers include Brueckner and Spiller (1991), Starr and Stinchcombe (1992), Hendricks, Piccione, and Tan (1995, 1997 and 1999), Oum, Zhang, and Zhang (1995), Zhang (1996) and Shy (2001). The empirical ones are Caves, Christensen, and Tretheway (1984), Brueckner, Dyer, and Spiller (1992), Brueckner and Spiller (1994) and Morrison and Winston (1995).

2. See Berry (1990, 1992), Borenstein (1992) and Berry, Carnall, and Spiller (1997).

3. The research includes Cohas, Belobaba and Simpson (1995), Dresner, Lin, and Windle (1996), US DOT (1996), Windle and Dresner (1999) and (Volwes, 2001).

REFERENCES

Bailey, E., Graham, D., & Kaplan, D. (1985). *Deregulating the airlines*. Cambridge, MA: The MIT Press.

Berry, S. (1990). Airport presence as product differentiation. *American Economic Review*, *80*, 394–399.

Berry, S. (1992). Estimation of a model in the airline industry. *Econometrica*, *60*(4), 889–917.

Berry, S., Carnall, M., & Spiller, P. T. (1997). Airline hubs: Costs, markups and the implications of customer heterogeneity. NBER Working Paper # 5561.

Berechman, J., & Shy, O. (1998). The structure of airline equilibrium networks. In: J. C. M. J. van der Bergh, P. Nijkamp & P. Rietveld (Eds), *Recent advances in spatial equilibrium modeling* (pp. 138–155). Berlin: Springer.

Borenstein, S. (1992). The evolution of U.S. Airline competition. *Journal of Economic Perspectives*, *6*, 45–73.

Brueckner, J. K. (2004). Network structure and airline scheduling. *The Journal of Industrial Economics*, *52*, 291–312.

Brueckner, J. K., Dyer, N. J., & Spiller, P. T. (1992). Fare determination in airline hub-and-spoke networks. *Rand Journal of Economics*, *23*, 309–333.

Brueckner, J. K., & Spiller, P. T. (1991). Competition and mergers in airline networks. *International Journal of Industrial Organization*, *9*, 323–342.

Brueckner, J. K., & Spiller, P. T. (1994). Economies of traffic density in the deregulated airline industry. *Journal of Law and Economics*, *37*, 379–415.

Brueckner, J. K., & Zhang, Y. (2001). A model of scheduling in airline networks: How a hub-and-spoke system affects flight frequency, fares and welfare. *Journal of Transport Economics and Policy*, *35*(2), 195–222.

Button, K., & Stough, R. (2000). *Air transport networks: Theory and policy implications*. Cheltenham: Edward Elgar.

Button, K., & Taylor, S. (2000). International air transport and economic development. *Journal of Air Transport Management*, *6*, 209–222.

Caves, D. W., Christensen, L. R., & Tretheway, M. W. (1984). Economies of density versus economies of scale: Why trunk and local service Airline costs differ. *Rand Journal of Economics*, *15*, 471–489.

Cohas, F., Belobaba, P., & Simpson, R. (1995). Competitive fare and frequency effects in airport market share modeling. *Journal of Air Transport Management*, *2*(1), 33–45.

Douglas, G., & Miller, J. C. (1974). *Economic regulation of domestic air transport: Theory and policy*. Washington, DC: Brookings Institution Press.

Dresner, M., Lin, J. C., & Windle, R. (1996). The impact of low-cost carriers on airport and route competition. *Journal of Transport Economics and Policy*, *30*(3), 309–328.

Gillen, D., & Morrison, W. (2003). Bundling, integration and the delivered price of air travel: Are low cost carriers full service competitors? *Journal of Air Transport Management*, *9*, 15–23.

Hendricks, K., Piccione, M., & Tan, G. (1995). The economics of hubs: The case of monopoly. *Review of Economic Studies*, *62*, 83–99.

Hendricks, K., Piccione, M., & Tan, G. (1997). Entry and exit in hub-spoke networks. *Rand Journal of Economics*, *28*, 291–303.

Hendricks, K., Piccione, M., & Tan, G. (1999). Equilibria in networks. *Econometrica*, *67*(6), 1407–1434.

Matsushima, K., & Kobayashi, K. (2005). Endogenous market formation with matching externality. In: K. Kobayashi, T. R. Lakshmanan & W. P. Anderson (Eds), *Structural change in transportation and communications in the knowledge economy: Implications for theory, modeling and data*. Cheltenham: Edward Elgar.

Merriam–Webster Internet Dictionary: http://www.m-w.com.

Morrison, S. A., & Winston, C. (1995). *The evolution of the airline industry*. Washington, DC: The Brookings Institute.

Oum, T. H., & Park, J. H. (1997). Airline alliances: Current status, policy issues, and future directions. *Journal of Air Transport Management*, *3*, 133–144.

Oum, T. H., Zhang, A., & Zhang, Y. (1995). Airline network rivalry. *Canadian Journal of Economics*, *28*, 836–857.

Rhoades, D. L. (2003). *Evolution of international aviation-phoenix rising*. Hants: Ashgate.

Schipper, Y. (2001). *Environmental costs and liberalization in European air transport*. Cheltenham: Edward Elgar.

Shy, O. (2001). *The economics of network industries*. New York: Cambridge University Press.

Starr, R., & Stinchcombe, M. (1992). An economic analysis of the hub and spoke system. Mimeo, University of California, San Diego.

US DOT (1996). The low cost Airline service revolution. US Department of Transportation, (available at: http://ostpxweb.dot.gov/aviation/domav/lcs.pdf).

Vowles, T. M. (2001). The "southwest effect" in multi-airport regions. *Journal of Air Transport Management*, *7*, 251–258.

Windle, R., & Dresner, M. (1999). Competitive responses to low cost carrier. *Transportation Research Part E*, *35*, 59–75.

Zhang, A. (1996). An analysis of fortress hubs in Airline networks. *Journal of Transport Economics and Policy*, *30*, 293–307.

APPENDIX

Proof of Proposition 2. Using $d_p = d_h$ and $c_p = c_h$, compare (12) and (21). The RHS and the first expression $M\pi(\bar{w} - c_h)^2/4\bar{s}^2\bar{w}$ in LHS are same in both. Denoting the second expression in the LHS of (12) and (21) as $g_p(n)$ and $g_h(n)$, the difference can be written as $g_h(n) - g_p(n) = (\bar{s}/2 - f - v/2)n - \pi/4$. The condition $n \geqslant \pi/(2\bar{s} - 4f - 2v)$ is satisfied, the frequency is higher in HS than PP. On the other hand, the condition to satisfy positive demand in AC city-pair market is $n \geqslant \pi/(2\bar{s} - 4f - 2v)$. Since both are same, $g_h(n) > g_p(n)$, implying that in HS network, the monopolist offers more flights than in the PP network.

SPATIAL BENEFIT INCIDENCE ANALYSIS OF AIRPORT CAPACITY EXPANSION: APPLICATION OF SCGE MODEL TO THE HANEDA PROJECT

Takayuki Ueda, Atsushi Koike,
Katsuhiro Yamaguchi and Kazuyuki Tsuchiya

ABSTRACT

In this chapter, we develop a SCGE model that covers the entire nation of Japan to clearly analyze the spillover impacts of the Haneda airport project. We then describe the spatial incidence of the project's benefits. Benefit incidence is summarized in a benefit incidence table (BIT), which clearly shows the canceling out of indirect benefits and distribution of benefits by region and economic sector. Major findings include (i) the expansion of Haneda can bring a large amount of benefit to all regions in Japan, particularly peripheral regions and (ii) consumption at trip destination increases in Hokkaido, Kanto, Kyushu, and Okinawa.

Global Competition in Transportation Markets: Analysis and Policy Making
Research in Transportation Economics, Volume 13, 165–196

ISSN: 0739-8859/doi:10.1016/S0739-8859(05)13008-X

1. INTRODUCTION

An improvement of a major airport affects the entire aviation network and then triggers structural changes in spatial dimensions, impacting not only the city where the airport is located but also other regions directly or indirectly. In Japan, the Haneda Airport in Tokyo is to be expanded to provide more frequent flights. Since Haneda is the biggest airport and the central node of a domestic aviation network as illustrated in Fig. 1, we

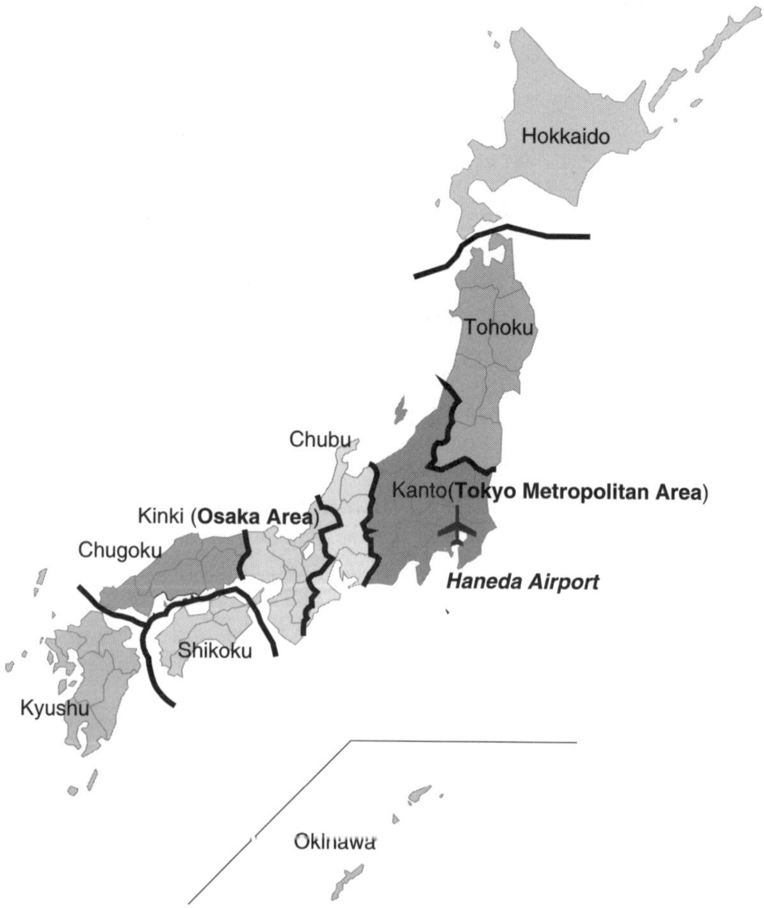

Fig. 1. Location of Haneda Airport and Nine Regions in Japan.

anticipate that it will benefit consumers not only residing in Tokyo but also in other regions. Benefits may be distributed spatially over the whole of Japan.

To comprehensively analyze the spatial impacts of the airport improvement, we must model economic behavior of all involved parties of the project and spatial interaction of the regions. We employ the techniques of a spatial computable general equilibrium (SCGE) model, which is a powerful tool for analyzing impact in various dimensions.

Many SCGE models have already been developed in Japan, especially two streams of economic modeling as illustrated in Fig. 2. One is a non-spatial computable general equilibrium (CGE) model in the tradition of Walrasian general equilibrium theory. It has become a tool for the quantitative evaluation of tax, trade, environmental policies, industrial development as well as econometric models. The other stream is a spatial

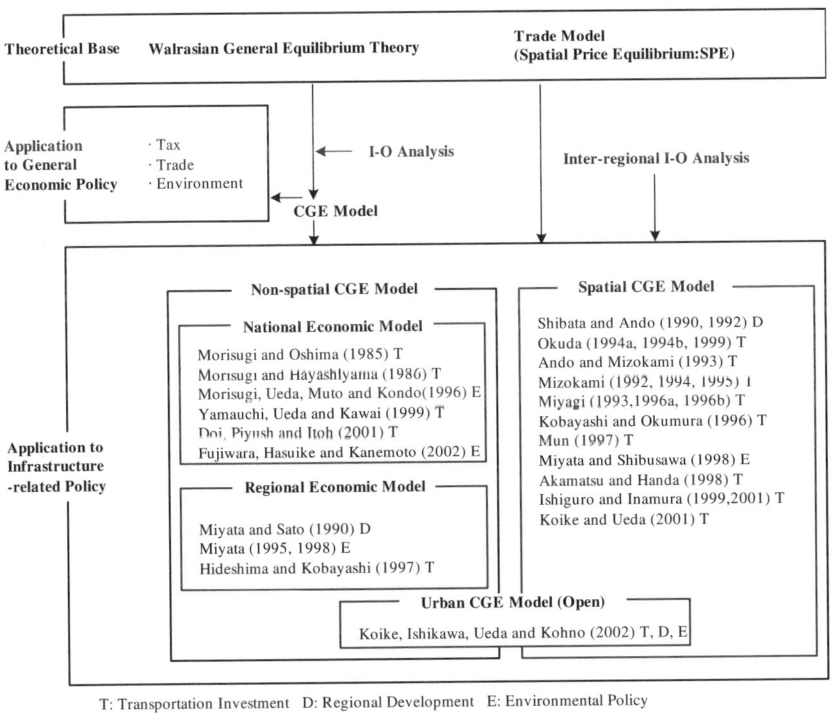

Fig. 2. Development of CGE and SCGE Models in Japan.

price equilibrium model that has dealt with the spatial flow of commodities between markets. It is therefore applicable to describe an economy with a transportation network. Since the spatial impacts of a large-scale transportation infrastructure are of great interest in Japan, many SCGE models have been developed and applied in practical works of project evaluation. However, no SCGE model exists for passenger transportation infrastructure. Therefore, we have to develop a new SCGE model which can explicitly describe passenger transportation to evaluate Haneda expansion. This SCGE model can deal with induced transportation demand in a manner consistent with microeconomic theory and estimate the impact of passenger transportation on industrial sectors.

In this study, we developed a model that covers the entire area of Japan, divided into nine regions and seven industrial sectors, to describe the impacts of inter-sectoral and spatial propagation. Unlike transportation models usually used in practical demand forecasting, this model can describe the flow of passenger trips completely consistent with microeconomic behavior theory. We can therefore analyze transportation demand induced by improvements in transportation systems and the linkage of trips with consumption or production of non-transport services and commodities.

We will describe the spatial incidence of project benefits and interregional distribution of benefits in a benefit incidence table (BIT) developed by Morisugi and Ohno (1995). BIT clearly shows the cancellation of indirect benefits and the final incidence of benefits by region and economic sector. It should be a basis for discussing the balance of beneficiaries and cost burdens.

This paper is organized as follows. In Section 2, we formulate the SCGE model. In Section 3, we apply it to the Haneda expansion project. Finally in Section 4, we conclude the paper.

2. MODEL

2.1. Outline and Assumptions

Based on Koike and Ueda (2001), we formulated a SCGE model with the following major assumptions:

(1) The model spatially covers regions labeled by $s \in \mathbf{S} = \{1, \ldots, s, \ldots, S\}$. In each region, a representative household, a transportation sector, and industrial sectors exist. The industrial sectors are labeled by $i \in \mathbf{I} = \{1, \ldots, i, \ldots, I\}$.

(2) Each industry produces commodities/services by inputting factors labeled by $k \in \mathbf{K} = \{1, \ldots, k, \ldots, K\}$ and intermediate goods traded between sectors.
(3) The household earns income by supplying factors, consumes commodities/services provided by industries, and takes private trips provided by the transportation sector.
(4) The transportation sector labeled $i = 0$ provides passenger transportation service to households and industries. The passenger transportation service includes several modes labeled $a \in \mathbf{A} = \{1, \ldots, a, \ldots, A\}$. Excess profit of the transportation sector is divided and distributed to households.
(5) Any markets in the economy are assumed to be in long-run competitive equilibrium.

2.2. Industries' Behavior

An industry produces a commodity or a service by inputting factors and intermediate goods. The production technology is illustrated in Fig. 3.

Industry i in region s produces output Q_i^s. We adopt a Leontief-type technology for the intermediate goods provided from industry i', $X_{i'i}^s$, and value-added VA_i^s. The input coefficients, $a_{i'i}^s$, $i \in \mathbf{I}$, a_{vi}^s are constant. Then output Q_i^s is formalized as the function

$$Q_i^s = \min\left[\frac{VA_i^s}{a_{vi}^s}, \frac{X_{1i}^s}{a_{1i}^s}, \ldots, \frac{X_{ii}^s}{a_{ii}^s}, \ldots, \frac{X_{Ii}^s}{a_{Ii}^s}\right] \tag{1}$$

Value-added VA_i^s is regarded as a composite factor whose ingredients are inputs f_{ki}^s and $k \in \mathbf{K}$. We assume that value added follows the constant elasticity of substitution (CES) technology:

$$VA_i^s = \phi_{1i}\left[\sum_{k \in \mathbf{K}} \delta_{ki} f_{ki}^{s(\sigma_{1i}-1)/\sigma_{1i}}\right]^{\sigma_{1i}/(\sigma_{1i}-1)} \tag{2}$$

where ϕ_{1i}, δ_{ki}, and σ_{1i} (>0) are constant parameters for industry i.

We assume that an industry behaves to minimize production costs for a unit of value-added as follows:

$$W_i^s = \min_{f_k(k \in \{1,2\}, k=3)} \left[\sum_{k \in \{1,2\}} w_k^s f_{ki}^s + w_{3i}^s f_{3i}^s\right] \tag{3a}$$

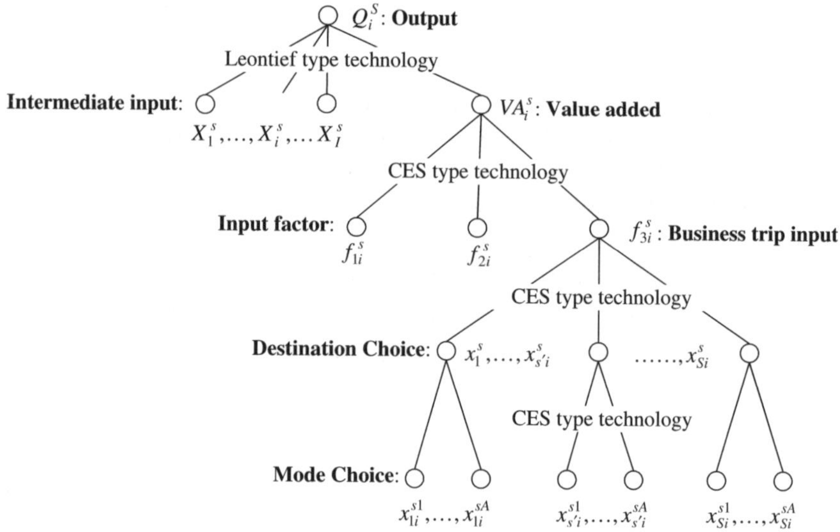

Fig. 3. Structure of Production Technology.

$$\text{s.t.} \quad (VA_i^s =)\phi_{1i}\left[\sum_{k\in\mathbf{K}} \delta_{ki} f_{ki}^{s(\sigma_{1i}-1)/\sigma_{1i}}\right]^{\sigma_{1i}/(\sigma_{1i}-1)} = 1 \tag{3b}$$

where w_k^s is the rent of input factor $k \in \{1, 2\}$ and w_{3i}^s the rent of business trips. By solving the programming in (3), we obtain the following input factor demand function and the unit cost of value-added W_i^s:

$$f_{ki}^s = \phi_{1i}^{-1}\left[\frac{\delta_{ki}^{\sigma_{1i}} w_k^{s^{1-\sigma_{1i}}}}{\sum_{k=1,2}\delta_{ki}^{\sigma_{1i}} w_k^{s^{1-\sigma_{1i}}} + \delta_{3i}^{\sigma_{1i}} w_{3i}^{s^{1-\sigma_{1i}}}}\right]\left[\frac{W_i^s}{w_k^s}\right] \quad \text{for all } k \in \{1,2\} \tag{4a}$$

$$f_{3i}^s = \phi_{1i}^{-1}\left[\frac{\delta_{3i}^{\sigma_{1i}} w_{3i}^{s^{1-\sigma_{1i}}}}{\sum_{k=1,2}\delta_{ki}^{\sigma_{1i}} w_k^{s^{1-\sigma_{1i}}} + \delta_{3i}^{\sigma_{1i}} w_{3i}^{s^{1-\sigma_{1i}}}}\right]\left[\frac{W_i^s}{w_{3i}^s}\right] \quad \text{for } k = 3 \tag{4b}$$

$$W_i^s = \left[\sum_{k=1,2}\delta_{ki}^{\sigma_{1i}} w_k^{s^{1-\sigma_{1i}}} + \delta_{3i}^{\sigma_{1i}} w_{3i}^{s^{1-\sigma_{1i}}}\right]^{1/(1-\sigma_{1i})} \tag{4c}$$

The input of business trips plays a key role in determining productivity. Business trip input f_{3i}^s in (4b) is again regarded as a composite input which is a function of trips $x_{s'i}^s$ labeled by destination s' and the attractiveness of destination $A^s_{s'i}$, $s' \in \mathbf{S}$, $a \in \mathbf{A}$. The industry chooses destinations of business trips to minimize the unit cost for business trips or the rent of business trip input w^s_{3i}:

$$w^s_{3i} = \min_{x^s_{s'i}(s' \in \mathbf{S})} \sum_{s' \in \mathbf{S}} p^s_{s'i} x^s_{s'i} \tag{5a}$$

$$\text{s.t.} \quad (f^s_{3i} =) \phi_{2i} \left[\sum_{s' \in \mathbf{S}} A^s_{s'i} x^{s(\sigma_{2i}-1)/\sigma_{2i}}_{s'i} \right]^{\sigma_{2i}/(\sigma_{2i}-1)} = 1 \tag{5b}$$

where $p^s_{s'i}$ is the generalized price of passenger transport services from s to s' and ϕ_{2i}, σ_{2i} (>0) are parameters specific to industry i.

Business trips for each destination can be disaggregated into trips by transport mode; the trips counted by destination are also regarded as a kind of composite input which is a function of trips labeled by mode $x^{sa}_{s'i}$, $a \in \mathbf{A}$. The industry chooses modes to minimize the unit cost of business trips to destination s', $p^s_{s'i}$:

$$p^s_{s'i} = \min_{x^{sa}_{s'i}(a \in \mathbf{A})} \sum_{a \in \mathbf{A}} (p^{sa}_{s'} + w^s_1 t^{sa}_{s'}) x^{sa}_{s'i} \tag{6a}$$

$$\text{s.t.} \quad (x^s_{s'i} =) \left[\sum_{a \in \mathbf{A}} (\gamma^{saB}_{s'})^{1/\sigma_{3i}} (x^{sa}_{s'i})^{(\sigma_{3i}-1)/\sigma_{3i}} \right]^{\sigma_{3i}/(\sigma_{3i}-1)} = 1 \tag{6b}$$

where $p^{sa}_{s'}$ is the price of passenger transport service from s to s' by mode a, $t^{sa}_{s'}$ is travel time from s to s' by mode a, $\gamma^{saB}_{s'}$ is the attractiveness of mode a, and σ_{3i} (>0) is a parameter specific to industry i.

After solving the programming in (6), we obtain a demand function for business trips labeled by transport mode and price labeled by destination:

$$x^{sa}_{s'i} = \left[\frac{\gamma^{saB}_{s'} (p^{sa}_{s'} + w^s_1 t^{sa}_{s'})^{1-\sigma_{3i}}}{\sum_{a \in \mathbf{A}} \gamma^{saB}_{s'} (p^{sa}_{s'} + w^s_1 t^{sa}_{s'})^{1-\sigma_{3i}}} \right] \left[\frac{p^s_{s'i}}{p^{sa}_{s'} + w^s_1 t^{sa}_{s'}} \right] \tag{7a}$$

$$p^s_{s'i} = \left[\sum_{a \in \mathbf{A}} \gamma^{saB}_{s'} (p^{sa}_{s'} + w^s_1 t^{sa}_{s'})^{1-\sigma_{3i}} \right]^{1/(1-\sigma_{3i})} \tag{7b}$$

We also obtain a demand function for the trip labeled by destination and the rent of the business trip as factor price from (5):

$$x^s_{s'i} = \phi_{2i}^{-1} \left[\frac{A^{s\sigma_{2i}}_{s'i} p^{s1-\sigma_{2i}}_{s'i}}{\sum_{s' \in \mathbf{S}} A^{s\sigma_{2i}}_{s'i} p^{s1-\sigma_{2i}}_{s'i}} \right] \left[\frac{w^s_{3i}}{p^s_{s'i}} \right] \tag{8a}$$

$$w_{3i}^s = \left[\sum_{s' \in \mathbf{s}} A_{s'i}^{s\sigma_{2i}} p_{s'i}^{s1-\sigma_{2i}} \right]^{1/(1-\sigma_{2i})} \tag{8b}$$

Since the destination choice for business trips is formulated in a CES-type structure, the share of trips to each destination resembles a Gravity Model:

$$\frac{p_{s'i}^s x_{s'i}^s}{\sum_{s' \in \mathbf{S}} p_{s'i}^s x_{s'i}^s} = \frac{A_{s'i}^{s\sigma_{2i}} p_{s'i}^{s1-\sigma_{2i}}}{\sum_{s' \in \mathbf{S}} A_{s'i}^{s\sigma_{2i}} p_{s'i}^{s1-\sigma_{2i}}} \tag{9}$$

The production technology illustrated in Fig. 3 has a constant return to scale property. Hence, the price of commodities/services produced by industry i, P_i^s, equals the production cost for a unit output or the average cost:

$$P_i^s = a_{vi}^s \left[\sum_{k=1,2} w_k^s f_{ki}^s + w_{3i}^s f_{3i}^s \right] + \sum_{i' \in I} P_{i'}^s a_{i'i}^s \tag{10}$$

By rewriting the right-hand side of Eq. (10) in matrix form, we have a system of equations for P_i^s, $i \in \mathbf{I}$. We can then solve them to obtain a price vector of products,

$$\begin{bmatrix} P_1^s \\ \vdots \\ P_i^s \\ \vdots \\ P_I^s \end{bmatrix}^{\mathrm{t}} = \begin{bmatrix} a_{v1}^s \left\{ \sum_{k=1,2} w_k^s f_{k1}^s + w_{31}^s f_{31}^s \right\} \\ \vdots \\ a_{vi}^s \left\{ \sum_{k=1,2} w_k^s f_{ki}^s + w_{3i}^s f_{3i}^s \right\} \\ \vdots \\ a_{vI}^s \left\{ \sum_{k=1,2} w_k^s f_{kI}^s + w_{3I}^s f_{3I}^s \right\} \end{bmatrix}^{\mathrm{t}} \left[\bar{\mathbf{I}} - \bar{\mathbf{A}}^s \right]^{-1} \tag{11}$$

where $\bar{\mathbf{I}}$ is an identity matrix, $\bar{\mathbf{A}}^s$ is an input coefficient matrix associated with $a_{i'i}^s$, i, $i' \in \mathbf{I}$ and t for the column vector denote the transpose of the vector. If the commodity/service price vector satisfies Eq. (11), then a zero profit condition for each industry should hold.

2.3. Household Behavior

We assume that a representative household consumes to maximize its utility. The structure of household consumption is illustrated in Fig. 4. The household residing in region s, $s \in \mathbf{S}$ consumes commodities/services d_i^s, $i \in \mathbf{I}$ and self-produced composite goods Z^s. The self-produced composite goods include recreational activities such as skiing, golf, camping, sightseeing, and so on. Such activities should not be simply regarded as consumption of commodity/services. The household has to spend time by itself in combining several commodities/services such as rental of gear, hotel service, meals at restaurants, and making trips to sites. We then, most importantly, focus on such activities in relation to household trip behavior and interpret them as consumption of self-produced goods. Household behavior can be formulated in the same way as Becker's (1978) household production function. A household production function argues that a household produces some sort of product by inputting its own resources.

Self-produced composite goods Z^s is a single variable as the aggregation of a variety of self-produced goods $z_{s'}^s$ labeled by trip destination s', $s' \in \mathbf{S}$.

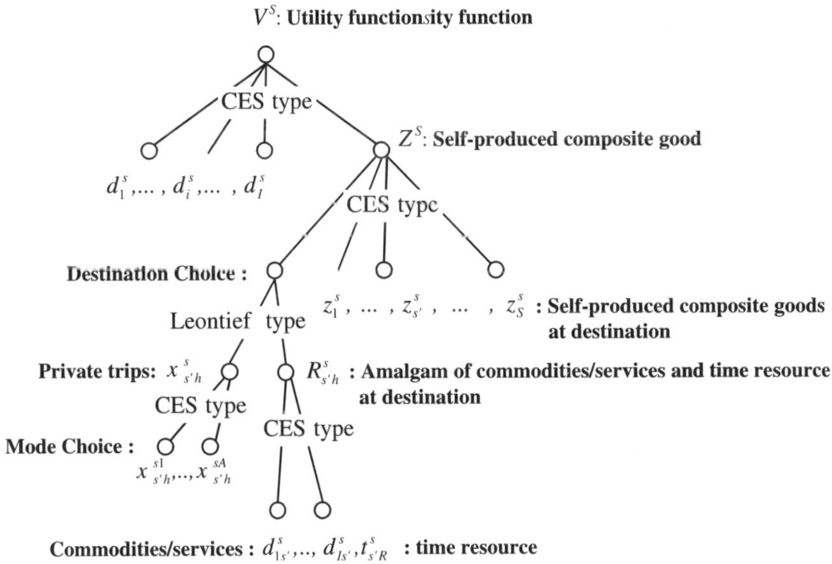

Fig. 4. Structure of Household Consumption and Self-production.

We introduce the label by destination because the self-production of $z^s_{s'}$ requires a trip to s'.

Utility function is specified as a nested CES type with the above elements. The self-production of $z^s_{s'}$ requires private passenger trips $x^{s^a}_{s'h}$ labeled by destination s' and mode a, the commodities/services at destination s', $d^s_{is'}$, and time resource $t^s_{s'R}$.

The household faces budget and time constraints. Expenditures for commodities/services of industry i consumed in residing region $P^s_i d^s_i$, $i \in \mathbf{I}$, expenditures for the commodities/services inputted to self-production at trip destination $P^{s'}_i d^s_{is'}$, $i \in \mathbf{I}$, $s' \in \mathbf{S}$, and expenditures for passenger transportation service $p^{s^a}_{s'} x^{s^a}_{s'h}$, $a \in \mathbf{A}$, $s' \in \mathbf{S}$ should be considered in the budget constraints. On the income side, the factor payment for endowments $w^s_k F^s_k$, $k \in \{1, 2\}$ and the dividend income from the transportation sector are included. Time constraints include the working time provided as an input factor F^s_1, the time for travel $t^{s^a}_{s'} x^s_{s'h}$, and the time spent for self-production $t^s_{s'R}$. Since the household also has to spend time for business trips, the total available time of a household in region s, T^s is defined as $T^s = \tau^s - \sum_{i\in\mathbf{I}} \sum_{s'\in\mathbf{S}} \sum_{a\in\mathbf{A}} t^{sa}_{s'} x^{sa}_{s'i} / N^s$, where τ^s is the endowment of time resource per household in region s and N^s is the number of households in region s.

We first formulate a utility maximizing problem with a utility function whose arguments are d^s_i, $i \in \mathbf{I}$, and Z^s. In the budget constraints at this level, the price of Z^s is denoted by P^s_Z for the representative price, and the time resources available to household F^s_1 are replaced with T^s. Then income is defined as gross income. The maximization problem is formalized as

$$V^s(P^s_{i(i\in\mathbf{I})}, P^s_Z, w^s_1 T^s + w^s_2 F^s_2 + \pi^s_0)$$

$$= \max_{d^s_i(i\in\mathbf{I}), Z^s} \left[\sum_{i\in\mathbf{I}} (\gamma^s_{1i})^{1/\rho_1} (d^s_i)^{(\rho_1-1)/\rho_1} + (\gamma^s_{1Z})^{1/\rho_1} (Z^s)^{(\rho_1-1)/\rho_1} \right]^{\rho_1/(\rho_1-1)} \tag{12a}$$

$$\text{s.t.} \quad \sum_{i\in\mathbf{I}} P^s_i d^s_i + P^s_Z Z^s = w^s_1 T^s + w^s_2 F^s_2 + \pi^s_0 \tag{12b}$$

where V^s is the indirect utility function, γ^s_{1i}, γ^s_{1Z}, $\rho_1 (>0)$ are parameters, and π^s_0 is the dividend income from the transportation sector per household in region s. The solutions of the programming in (12) give the demand functions

$$d^s_i = \frac{\gamma^s_{1i}(w^s_1 T^s + w^s_2 F^s_2 + \pi^s_0)}{P^{s\rho_1}_i \left[\sum_{i'\in\mathbf{I}} \gamma^s_{1i'} P^{s^{1-\rho_1}}_{i'} + \gamma^s_{1Z} P^{s^{1-\rho_1}}_Z \right]} \tag{13a}$$

$$Z^s = \frac{\gamma_{1Z}^s (w_1^s T^s + w_2^s F_2^s + \pi_0^s)}{P_Z^{s\rho_1} \left[\sum_{i' \in \mathbf{I}} \gamma_{1i'}^s P_{i'}^{s1-\rho_1} + \gamma_{1Z}^s P_Z^{s1-\rho_1} \right]} \tag{13b}$$

Since self-produced composite goods Z^s are composed of $z_{s'}^s$ and $s' \in \mathbf{S}$, the household determines their composition to minimize representative price P_Z^s. Next, we reformulate the minimization problem by specifying the aggregation structure of Z^s as a CES function

$$P_Z^s = \min_{z_{s'}^s (s' \in \mathbf{S})} \sum_{s' \in \mathbf{S}} P_{s'z}^s z_{s'}^s \tag{14a}$$

$$\text{s.t.} \quad (Z^s =) \left\{ \sum_{s' \in \mathbf{S}} (\gamma_{2s'}^s)^{1/\rho_2} (z_{s'}^s)^{(\rho_2 - 1)/\rho_2} \right\}^{\rho_2/(\rho_2 - 1)} = 1 \tag{14b}$$

where $P_{s'Z}^s$ is the price of the self-produced goods at destination $s' \in \mathbf{S}$ and $\gamma_{2s'}^s$ is a parameter specific to destination $s' \in \mathbf{S}$. The programming in (14) yields the demand share for self-produced goods at destination s' and the minimized value of P_Z^s, respectively:

$$z_{s'}^s = \left[\frac{\gamma_{2s'}^s P_{s'z}^{s1-\rho_2}}{\sum_{s' \in \mathbf{S}} \gamma_{2s'}^s P_{s'z}^{s1-\rho_2}} \right] \left[\frac{P_z^s}{P_{s'z}^s} \right] \tag{15a}$$

$$P_Z^s = \left[\sum_{s' \in \mathbf{S}} \gamma_{2s'}^s P_{s'z}^{s(1-\rho_2)} \right]^{1/(1-\rho_2)} \tag{15b}$$

To create self-produced goods at destination s', $z_{s'}^s$ for a share in Z^s, the household makes trips to s', $x_{s'h}^s$, purchases commodities/services, and spends time resources at destination s'. Label h stands for household hereafter. We make an amalgam of commodities/services and time resources denoted by $R_{s'}^s$. These two variables $x_{s'h}^s$ and $R_{s'}^s$ are assumed to be proportional to $z_{s'}^s$. Then the production function is the Leontief type. $R_{s'}^s$ is later disaggregated to a variety of commodities/services and time resources. The household chooses them to minimize the unit production cost of self-produced goods $P_{s'z}^s$. The cost minimization problem is formalized as

$$P_{s'z}^s = \min_{x_{s'h}^s, R_{s'}^s} \left[p_{s'h}^s x_{s'h}^s + P_{s'R}^s R_{s'}^s \right] \tag{16a}$$

$$\text{s.t.} \quad (z_{s'}^s =) \min \left[\frac{x_{s'h}^s}{b_1}, \frac{R_{s'}^s}{b_2} \right] = 1 \tag{16b}$$

where b_1 and b_2 are input coefficients for the passenger trip and the amalgam, respectively, and $P^s_{s'R}$ is the price of $R^s_{s'}$ or the representative price of ingredients in the amalgam. The minimization stated in (16) yields solutions:

$$x^s_{s'h} = b_1 \tag{17a}$$

$$R^s_{s'} = b_2 \tag{17b}$$

$$P^s_{s'Z} = p^s_{s'h} b_1 + P^s_{s'R} b_2 \tag{17c}$$

Amalgam $R^s_{s'}$ is composed of commodities/services $d^s_{is'}$, $i \in \mathbf{I}$ and time resource $t^s_{s'R}$. The household chooses them to minimize representative price $P^s_{s'R}$:

$$P^s_{s'R} = \min_{d^s_{is'}(i \in \mathbf{I}), t^s_{s'R}} \left(\sum_{i \in \mathbf{I}} P^{s'}_i d^s_{is'} + w^s_1 t^s_{s'R} \right) \tag{18a}$$

$$\text{s.t.} \quad (R^s_{s'} =) \phi_h \left[\sum_{i \in \mathbf{I}} \delta_{ih} d^{s(\sigma_h - 1)/\sigma_h}_{is'} + \delta_{0h} t^{s(\sigma_h - 1)/\sigma_h}_{s'R} \right]^{\sigma_h/(\sigma_h - 1)} = 1 \tag{18b}$$

where ϕ_h, δ_{ih}, δ_{0h}, and σ_h (>0) are parameters. The programming in (18) yields the composition of $d^s_{is'}$, $i \in \mathbf{I}$ and time resource $t^s_{s'}$, $s' \in \mathbf{S}$:

$$d^s_{s'i} = \phi^{s-1}_{s'h} \left[\frac{\delta^{\sigma_h}_{ih} P^{s'^{1-\sigma_h}}_i}{\sum_{i \in \mathbf{I}} \delta^{\sigma_h}_{ih} P^{s'^{1-\sigma_h}}_i + \delta^{\sigma_h}_{0h} w^{s^{1-\sigma_h}}_1} \right] \left[\frac{P^s_{s'R}}{P^{s'}_i} \right] \tag{19a}$$

$$t^s_{s'R} = \phi^{s-1}_{s'h} \left[\frac{\delta^{\sigma_h}_{0h} w^{s^{1-\sigma_h}}_1}{\sum_{i \in \mathbf{I}} \delta^{\sigma_h}_{ih} P^{s'^{1-\sigma_h}}_i + \delta^{\sigma_h}_{0h} w^{s^{1-\sigma_h}}_1} \right] \left[\frac{P^s_{s'R}}{w^s_1} \right] \tag{19b}$$

$$P^s_{s'R} = \left[\sum_{i \in \mathbf{I}} \delta^{\sigma_h}_{ih} P^{s'^{1-\sigma_h}}_i + \delta^{\sigma_h}_{0h} w^{s^{1-\sigma_h}}_1 \right]^{1/(1-\sigma_h)} \tag{19c}$$

On the other hand, passenger trips to s', $x^s_{s'h}$ should be disaggregated to trips labeled by mode, $x^{s^a}_{s'h}$, $a \in \mathbf{A}$. The household determines modal share to minimize the generalized and the representative prices of passenger transport service $p^s_{s'h}$. The minimization problem is defined as

$$p^s_{s'h} = \min_{x^{sa}_{s'h}(a \in \mathbf{A})} \sum_{a \in \mathbf{A}} (p^{sa}_{s'} + w^s_1 t^{sa}_{s'}) x^{sa}_{s'h} \tag{20a}$$

$$\text{s.t.} \quad (x^s_{s'h} =) \left[\sum_{a \in \mathbf{A}} (\gamma^{saP}_{s'})^{1/\rho_3} (x^{sa}_{s'h})^{(\rho_3 - 1)/\rho_3} \right]^{\rho_3/(\rho_3 - 1)} = 1 \tag{20b}$$

where $x^{sa}_{s'h}$ is household private trips s to region s' by mode a, $s, s' \in \mathbf{S}$, $a \in \mathbf{A}$; $\gamma^{saP}_{s'}$ is a parameter denoting the attractiveness level of mode a; and ρ_3 is an elasticity parameter of substitution between modes. After solving the programming in (20), we obtain the modal share for private household trips and the representative price:

$$x^{sa}_{s'h} = \left[\frac{\gamma^{saP}_{s'} (p^{sa}_{s'} + w^s_1 t^{sa}_{s'})^{1-\rho_3}}{\sum_{a \in \mathbf{A}} \gamma^{saP}_{s'} (p^{sa}_{s'} + w^s_1 t^{sa}_{s'})^{1-\rho_3}} \right] \left[\frac{p^s_{s'h}}{p^{sa}_{s'} + w^s_1 t^{sa}_{s'}} \right] \tag{21a}$$

$$p^s_{s'h} = \left[\sum_{a \in \mathbf{A}} \gamma^{saP}_{s'} (p^{sa}_{s'} + w^s_1 t^{sa}_{s'})^{1-\rho_3} \right]^{1/(1-\rho_3)} \tag{21b}$$

2.4. Transportation Sector's Behavior

The transportation sector in region s, $s \in \mathbf{S}$ produces transportation service Q^s_0 aggregated from several modes. We specify the production function of the transportation sector as a Leontief-type technology to input factors f^s_{0k}, $k \in \{1, 2\}$:

$$Q^s_0 = \min \left[\frac{f^s_{01}}{c^s_1}, \frac{f^s_{02}}{c^s_2} \right] \tag{22}$$

where c^s_k is the input coefficient for $k \in \{1, 2\}$. Based on (22), we obtain the input factor demand function.

$$f^s_{0k} = c^s_k Q^s_0 \quad \text{for } k \in \{1, 2\} \tag{23}$$

The price of transportation services is set as a part of policy scenario as well as a physical improvement of transportation infrastructure. The transportation sector has to supply an amount of service equal to the aggregated demand as stated in the following equation:

$$Q_0^s = \sum_{a \in \mathbf{A}} \sum_{s' \in \mathbf{S}} \left(\sum_{i \in \mathbf{I}} x_{s'i}^{s^a} x_{s'i}^{s} f_{3i}^{s} a_{vi}^{s} Q_i^s + x_{s'h}^{s^a} b_1 z_{s'}^{s} Z^s N^s \right) \tag{24}$$

Excess profit from the transportation sector is then expressed as

$$\Pi_0^s = \sum_{a \in \mathbf{A}} \sum_{s' \in \mathbf{S}} p_{s'}^{s^a} \left(\sum_{i \in \mathbf{I}} x_{s'i}^{s^a} x_{s'i}^{s} f_{3i}^{s} a_{vi}^{s} Q_i^s + x_{s'h}^{s^a} b_1 z_{s'}^{s} Z^s N^s \right) - \sum_{k=1,2} w_k^s f_{0k}^s \tag{25}$$

Excess profit per household π_0^s is

$$\pi_0^s = \frac{\Pi_0^s}{N^s} \tag{26}$$

2.5. *Equilibrium Conditions and Benefit Definition*

The system of equations for equilibrium first consists of supply and demand conditions for input factors. Assume that input factors can only be traded within a region. Then supply and demand balance conditions for time resources as labor and the capital are, respectively, as

$$\begin{aligned} &\sum_{i \in \mathbf{I}} (a_{vi}^s Q_i^s f_{1i}^s) + (c_1^s Q_0^s) \\ &\quad = \tau^s N^s - \sum_{s' \in \mathbf{S}} \sum_{i \in I} \sum_{a \in \mathbf{A}} t_{s'}^{s^a} x_{s'i}^{s^a}(w_1^s) x_{s'i}^{s}(w_1^s) f_{3i}^{s}(\mathbf{w}^s) a_{vi}^s Q_i^s \\ &\qquad - \sum_{s' \in \mathbf{S}} \sum_{a \in \mathbf{A}} t_{s'}^{s^a} x_{s'h}^{s^a}(w_1^s) b_1 z_{s'}^{s}(\mathbf{P}^{s'}, w_1^s) Z^s(\mathbf{P}, \mathbf{w}^s, \pi_0^s) N^s \\ &\qquad - \sum_{s' \in \mathbf{S}} t_{s'R}^{s}(\mathbf{P}^{s'}, w_1^s) b_2 z_{s'}^{s}(\mathbf{P}^{s'}, w_1^s) Z^s(\mathbf{P}, \mathbf{w}^s, \pi_0^s) N^s \quad \text{for all } s \in \mathbf{S} \end{aligned} \tag{27a}$$

$$\sum_{i \in \mathbf{I}} a_{vi}^s Q_i^s f_{2i}^s + c_2^s Q_0^s = F_2^s N^s \quad \text{for all } s \in \mathbf{S} \tag{27b}$$

where notations $\mathbf{w}^s = [w_1^s, w_2^s]$, $\mathbf{P}^{s'} = [P_1^{s'}, \ldots, P_i^{s'}, \ldots, P_I^{s'}]$, and $\mathbf{P} = [\mathbf{P}^1, \ldots, \mathbf{P}^s, \ldots, \mathbf{P}^S]$ denote endogenous price variables appearing in the equations.

The next important factors are supply and demand balance conditions for commodities/services. Production by industry or the aggregate supply should equal the total of the intermediate demand by other industrial sectors and the final demand by households. From the standard formulas in the input–output analysis, we can write production by each industrial sector in region s as

$$\mathbf{Q}^s = \begin{bmatrix} Q_1^s \\ \vdots \\ Q_i^s \\ \vdots \\ Q_I^s \end{bmatrix}^{\mathrm{t}} = [\bar{\mathbf{I}} - \bar{\mathbf{A}}^s]^{-1} \begin{bmatrix} d_1^s(\mathbf{P}, \mathbf{w}^s, \pi_0^s)N^s + \sum_{s' \in \mathbf{S}} d_{s1}^{s'}(\mathbf{P}, \mathbf{w}^{s'}, \pi_0^{s'})N^{s'} \\ \vdots \\ d_i^s(\mathbf{P}, \mathbf{w}^s, \pi_0^s)N^s + \sum_{s' \in \mathbf{S}} d_{si}^{s'}(\mathbf{P}, \mathbf{w}^{s'}, \pi_0^{s'})N^{s'} \\ \vdots \\ d_I^s(\mathbf{P}, \mathbf{w}^s, \pi_0^s)N^s + \sum_{s' \in \mathbf{S}} d_{sI}^{s'}(\mathbf{P}, \mathbf{w}^{s'}, \pi_0^{s'})N^{s'} \end{bmatrix}^{\mathrm{t}}$$

$$\text{for all } s \in \mathbf{S} \qquad (28)$$

On the RHS of (28), the final demand for each industry $i \in \mathbf{I}$ is listed in a column vector. The final demand is aggregated not only from individual demand d_i^s by household within region s but also from other regions $d_{si}^{s'}$, $s' \in \mathbf{S}$. This is because the household demands commodities/services at a trip destination.

Dividend income from the transportation sector per household π_0^s appears in (27a) and (28), and the production of transportation sector Q_0^s appears in (27a) and (27b). Recalling (24)–(26), we can rewrite them as

$$\pi_0^s = \frac{\Pi_0^s}{N^s} = \left(\frac{1}{N^s}\right)\left[\sum_{a \in \mathbf{A}} \sum_{s' \in \mathbf{S}} p_{s'}^{sa} \left(\sum_{i \in \mathbf{I}} x_{s'i}^{sa}(w_1^s) x_{s'i}^{s}(w_1^s) f_{3i}^s(\mathbf{w}^s) a_{vi}^s Q_i^s \right.\right.$$
$$\left.\left. + x_{s'h}^{sa}(w_1^s) b_1 z_{s'}^s(\mathbf{P}^{s'}, w_1^s) Z^s(\mathbf{P}, \mathbf{w}^s, \pi_0^s) N^s \right) - \sum_{k=1,2} w_k^s f_{0k}^s \right] \quad \text{for all } s \in \mathbf{S} \qquad (29a)$$

$$Q_0^s = \sum_{a \in \mathbf{A}} \sum_{s' \in \mathbf{S}} \left(\sum_{i \in \mathbf{I}} x_{s'i}^{sa}(w_1^s) x_{s'i}^{s}(w_1^s) f_{3i}^s(\mathbf{w}^s) a_{vi}^s Q_i^s \right.$$
$$\left. + x_{s'h}^{sa}(w_1^s) b_1 z_{s'}^s(\mathbf{P}^{s'}, w_1^s) Z^s(\mathbf{P}, \mathbf{w}^s, \pi_0^s) N^s \right) \quad \text{for all } s \in \mathbf{S} \qquad (29b)$$

The price of commodities/services in region s, P_i^s has been already decomposed with the prices of input factor in (11). Then we rewrite (11) with the notations used here

$$\mathbf{P}^s = \begin{bmatrix} P_1^s \\ \vdots \\ P_i^s \\ \vdots \\ P_I^s \end{bmatrix}^t = \begin{bmatrix} a_{v1}^s \left\{ \sum_{k=1,2} w_k^s f_{k1}^s(\mathbf{w}^s) + w_{31}^s(\mathbf{w}^s) f_{31}^s(\mathbf{w}^s) \right\} \\ \vdots \\ a_{vi}^s \left\{ \sum_{k=1,2} w_k^s f_{ki}^s(\mathbf{w}^s) + w_{3i}^s(\mathbf{w}^s) f_{3i}^s(\mathbf{w}^s) \right\} \\ \vdots \\ a_{vI}^s \left\{ \sum_{k=1,2} w_k^s f_{kI}^s(\mathbf{w}^s) + w_{3I}^s(\mathbf{w}^s) f_{3I}^s(\mathbf{w}^s) \right\} \end{bmatrix}^t [\bar{\mathbf{I}} - \bar{\mathbf{A}}^s]^{-1}$$

$$\text{for all } s \in \mathbf{S} \tag{30}$$

The system of equations for equilibrium should be fully described by (27a)–(30). Endogenous variables in the system are P_i^s, π_0^s, Q_i^s, w_1^s, w_2^s, $s \in \mathbf{S}$, $i \in \mathbf{I}$. However, we can simplify the system by eliminating variables P_i^s, π_0^s, and Q_i^s or by rewriting them as $\mathbf{w}^s = [w_1^s, w_2^s]$, $s \in \mathbf{S}$. Inserting (28)–(30) into (27a) and (27b), we obtain a reduced system of equations for $\mathbf{w}^s = [w_1^s, w_2^s]$, $s \in \mathbf{S}$. The unknown variables are $2S$, and the number of equations is also $2S$ from (27a) and (27b). Now the system is solvable. Because of the Walras law, one of the equations is redundant. We can therefore denote one unknown variable to be the constant of the price of numeraire goods.

We define the benefit of a project as equivalent variation (EV). Since the household utility level has been already formalized as the indirect utility function in (12a), EV satisfies the equation

$$\begin{aligned} &V^s(P_{i(i \in \mathrm{I})}^{sWO}, P_Z^{sWO}, w_1^{sWO} T^{sWO} + w_2^{sWO} F_2^{sWO} + \pi_0^{sWO} + EV^s) \\ &= V^s(P_{i(i \in \mathrm{I})}^{sW}, P_Z^{sW}, w_1^{sW} T^{sW} + w_2^{sW} F_2^{sW} + \pi_0^{sW}) \end{aligned} \tag{31}$$

where WO and W denote "without" and "with investment", respectively. EV, which should be measured per household, can be decomposed into several items of benefit/cost and aggregated to regional or national benefits.

3. APPLICATIONS TO HANEDA PROJECT

3.1. Sketch of the project

We analyzed the impact of expanding the capacity of Haneda airport in Tokyo by 10% in 1995 to increase the number of flight slots from 285 to 314 per day. The project's cost has been roughly estimated to be about 180 billion yen at 1995 yen values.

Slot increases lead to rescheduling of flight plans and finally to changes in levels of service. Airline competition in aviation markets may reduce the price of air tickets, and increasing the frequency of flights may also reduce waiting times and therefore the average travel time. These changes benefit or disbenefit airlines and users.

Haneda airport functions as a hub airport for domestic airlines in Japan. In a domestic aviation network, it is the central node so that the project can affect the spatial structure of Japan. In other words, the project not only benefits the users residing in Tokyo but also users in other regions.

3.2. Parameter setting

Parameters included in the production function of industry and the utility function of households should be identified by a calibration method (see Shoven & Whally, 1984; Ichioka, 1991). Space limitations do not allow us to show all values for parameters and exogenous variables. Table 1 shows only the source of values for parameters and exogenous variables. Each regional economy consists of seven industrial sectors, including agriculture, manufacturing, construction, energy, commercial, finance, and services other than the transportation sector.

3.3. Setting of level of service

We estimated the level of transportation service by price of service and travel time by an aviation network model based on Ohashi and Ando (1999), who collaborated with the authors to develop the model.

The aviation network model describes the airline's behavior as profit maximization, the user's behavior as multi-stage discrete choice, and equilibrium as a Nash-type non-cooperative game. This model also considers connecting passengers at such hub airports as Haneda to express the role of hub airport.

We input the capacity of Haneda for both cases with and without expansion into the aviation network model. The service levels outputted by the model were then transferred to the SCGE model.

Table 1. Parameters and Exogenous Variables.

Parameter	Source
$\bar{\mathbf{A}}^s$ (input coefficient)	Inter-regional input–output table (1995) (published by the Ministry of International Trade and Industry, Japan)
a^s_{vi} (production capacity rate)	Inter-regional input–output table (1995)
ϕ_{1i}, ϕ_{2i} (parameters for industry)	Estimated by using inter-regional input–output table (1995) and inter-regional net flow data (1995) (published by the Ministry of Transport, the Ministry of Construction, the National Land Agency, Japan)
δ_{ki} (parameters for industry)	Estimated by using inter-regional input–output table (1995)
σ_{1i} (elasticity of substitution)	Shoven and Whally (1992)
σ_{2i} (elasticity of substitution)	1.0
σ_{3i} (elasticity of substitution)	1.0
$A^s_{s'i}$ (destination share parameter)	Estimated by using inter-regional net flow data (1995)
γ^s_{1i} (household parameter)	Inter-regional input–output table (1995)
γ^s_{2i} (destination share parameter)	Estimated by using inter-regional net flow data (1995)
γ^s_{1Z} (household parameter)	Estimated by using inter-regional input–output table (1995), inter-regional net flow data (1995)
$\gamma^{saB}_{s'}, \gamma^{saP}_{s'}$ (mode share parameter)	Estimated by using inter-regional net flow data (1995)
ρ_1 (elasticity of substitution)	Ichioka (1991)
ρ_2 (elasticity of substitution)	1.0
ρ_3 (elasticity of substitution)	1.0
δ_{ih}, δ_{0h} (leisure share parameter)	Inter-regional net flow data (1995)
σ_h (elasticity of substitution)	1.0
b_1, b_2 (leisure share parameter)	Leisure and Resort Statistic (1996) (published by Japan Tourist Association)
c^s_k (input coefficient)	Inter-regional input–output table (1995)
Ω (initial endowment of time resources)	Inter-regional input–output table (1995)
F^s_2 (initial endowment of capital)	Inter-regional input–output table (1995)
N^s (number in household)	Population Census (1995) (published by Management and Coordination Agency, Japan)

3.4. Results of Application

3.4.1. Changes in Number of Trips

Tables 2–5 show changes in trips by air and rail. Tables 2 and 3 show that business and private air trips increased for many OD pairs, particularly trips from Kanto (Tokyo region), which increased the most. Tables 4 and 5 show

Table 2. Changes in Business Trips by Air.

O	D									
	Hokkaido	Tohoku	Kanto	Chubu	Kinki	Chugoku	Shikoku	Kyushu	Okinawa	Total
Rate of change (%)										
Hokkaido	−1.41	2.33	2.30	2.50	2.54	2.29	2.38	2.14	2.53	1.80
Tohoku	3.63	0.00	3.49	4.04	3.60	4.41	3.98	3.76	−0.33	3.57
Kanto	3.42	3.33	0.00	1.63	2.17	3.32	3.20	3.77	3.60	3.25
Chubu	3.89	4.15	1.74	−0.22	0.00	−0.22	−0.22	3.80	4.08	2.74
Kinki	3.87	3.64	2.36	0.00	0.00	−0.25	2.27	3.43	1.56	2.86
Chugoku	3.60	4.47	3.52	−0.28	−0.27	0.00	−0.27	−0.28	−0.28	2.86
Shikoku	3.18	3.49	2.88	−0.76	1.75	−0.76	0.00	3.85	−0.76	2.46
Kyushu	2.65	2.97	3.15	2.90	2.60	−1.07	3.48	−1.07	2.71	2.31
Okinawa	2.41	−1.75	2.27	2.48	0.04	−1.75	−1.75	2.03	0.00	1.41
Total	2.80	3.15	2.73	2.07	2.23	2.55	2.77	2.84	2.64	2.70
Trips per year										
Hokkaido	−12,265	13,508	74,313	9,330	13,231	2,428	1,654	5,619	794	108,612
Tohoku	16,696	0	19,133	5,153	10,318	5,024	3,046	8,070	−180	67,262
Kanto	122,464	32,403	0	18,109	31,998	42,610	44,933	179,402	20,098	492,017
Chubu	13,264	5,460	13,319	−88	0	−66	−220	20,655	3,214	55,538
Kinki	22,751	17,480	35,996	0	0	−260	17,931	50,470	2,837	147,204
Chugoku	3,391	4,872	31,221	−45	−167	0	−16	−335	−163	38,757
Shikoku	1,350	3,189	29,891	−586	9,601	−35	0	6,085	−300	49,194
Kyushu	4,286	9,499	131,324	13,841	33,385	−875	5,687	−14,191	9,337	192,293
Okinawa	496	−662	13,317	1,600	94	−1,956	−355	8,865	0	21,398
Total	172,433	85,749	348,513	47,315	98,460	46,870	72,660	264,639	35,637	1,172,276

Table 3. Changes in Private Household Trips by Air.

O	D									
	Hokkaido	Tohoku	Kanto	Chubu	Kinki	Chugoku	Shikoku	Kyushu	Okinawa	Total
Trips per year										
Hokkaido	−54	4,497	32,110	5,739	17,427	3,261	1,829	3,482	1,071	69,362
Tohoku	3,781	0	9,024	6,163	11,632	7,242	3,762	9,118	−66	50,656
Kanto	31,827	9,041	0	2,793	8,702	16,227	7,049	27,911	9,940	113,491
Chubu	10,961	8,390	5,184	−1	0	−6	−31	14,920	4,786	44,203
Kinki	14,454	11,260	10,758	0	0	−8	3,537	25,138	1,840	66,978
Chugoku	3,285	6,237	18,556	−11	−7	0	0	−16	−195	27,850
Shikoku	1,480	2,227	8,609	−31	3,125	0	0	2,693	−48	18,054
Kyushu	4,029	5,227	33,227	10,114	19,499	−42	1,906	−161	2,667	76,464
Okinawa	925	−267	10,764	8,040	1,382	−498	−89	2,000	0	22,256
Total	70,686	46,612	128,233	32,805	61,759	26,176	17,962	85,087	19,995	489,314
Rate of change (%)										
Hokkaido	−0.12	1.58	0.96	0.78	1.18	0.85	1.18	0.72	1.01	0.99
Tohoku	1.55	0.00	3.57	2.82	2.82	2.66	2.74	2.50	−0.04	2.44
Kanto	1.42	3.47	0.00	1.94	2.40	2.99	1.99	1.35	0.89	1.60
Chubu	1.48	3.45	1.81	−0.04	0.00	−0.04	−0.03	1.86	1.06	1.68
Kinki	1.21	2.45	2.42	0.00	0.00	−0.03	2.00	2.11	0.49	1.73
Chugoku	0.95	2.33	2.88	−0.03	−0.03	0.00	−0.03	−0.03	−0.03	1.43
Shikoku	1.46	1.49	1.88	−0.05	2.00	−0.04	0.00	2.87	−0.05	1.61
Kyushu	0.90	1.67	1.37	1.87	2.00	−0.08	3.65	−0.08	1.00	1.45
Okinawa	1.03	−0.19	0.83	1.75	0.20	−0.19	−0.19	0.68	0.00	0.68
Total	1.30	2.20	1.40	1.49	1.51	1.69	1.77	1.53	0.64	1.43

Table 4. Changes in Business Trips by Rail.

O	D									
	Hokkaido	Tohoku	Kanto	Chubu	Kinki	Chugoku	Shikoku	Kyushu	Okinawa	Total
Trips per year										
Hokkaido	−47,626	−1,952	−1,801	−199	−1,322	−119	0	−43	0	−53,063
Tohoku	−588	−16,114	−24,369	−978	−1,633	−281	−175	−815	0	−44,954
Kanto	−641	−39,438	−174,419	−54,675	−83,388	−11,241	−1,116	−2,313	0	−367,230
Chubu	−64	−783	−21,539	−3,006	−14,532	−2,124	−369	−719	0	−43,136
Kinki	−85	−814	−43,922	−16,550	−6,145	−12,045	−2,756	−4,559	−29	−86,904
Chugoku	−46	−181	−5,166	−1,827	−9,781	−7,897	−2,299	−4,728	0	−31,925
Shikoku	0	−62	−2,090	−1,517	−8,821	−6,499	−6,251	−1,133	0	−26,374
Kyushu	−62	−365	−4,651	−3,662	−17,784	−17,619	−2,324	−63,385	−93	−109,945
Okinawa	0	0	−36	−241	0	0	0	−327	0	−604
Total	−49,112	−59,709	−277,990	−82,657	−143,405	−57,827	−15,290	−78,023	−122	−764,135
Rate of change (%)										
Hokkaido	−1.42	−1.42	−1.42	−1.42	−1.42	−1.42	0.00	−1.42	0.00	−1.42
Tohoku	−0.33	−0.33	−0.33	−0.33	−0.33	−0.33	−0.33	−0.33	0.00	−0.33
Kanto	−0.44	−0.44	−0.44	−0.44	−0.44	−0.44	−0.44	−0.44	0.00	−0.44
Chubu	−0.22	−0.22	−0.22	−0.22	−0.22	−0.22	−0.22	−0.22	0.00	−0.22
Kinki	−0.25	−0.25	−0.25	−0.25	−0.25	−0.25	−0.25	−0.25	−0.25	−0.25
Chugoku	−0.27	−0.27	−0.27	−0.27	−0.27	−0.27	−0.27	−0.27	0.00	−0.27
Shikoku	0.00	−0.76	−0.76	−0.76	−0.76	−0.76	−0.76	−0.76	0.00	−0.76
Kyushu	−1.07	−1.07	−1.07	−1.07	−1.07	−1.07	−1.07	−1.07	−1.07	−1.07
Okinawa	0.00	0.00	−1.80	−1.80	0.00	0.00	0.00	−1.81	0.00	−1.80
Total	−1.30	−0.40	−0.36	−0.37	−0.41	−0.42	−0.44	−0.73	−0.60	−0.42

Table 5. Changes in Private Household Trips by Rail.

O	D									
	Hokkaido	Tohoku	Kanto	Chubu	Kinki	Chugoku	Shikoku	Kyushu	Okinawa	Total
Trips per year										
Hokkaido	−1,554	−3,813	−7,521	−684	−4,424	−413	−408	−58	0	−18,875
Tohoku	−2,081	−522	−10,000	−3,644	−6,470	−3,549	−1,921	−3,920	0	−32,109
Kanto	−10,006	−9,883	−3,838	−3,249	−7,446	−11,573	−3,803	−9,746	−44	−59,589
Chubu	−3,558	−4,848	−5,974	−178	−837	−185	−44	−7,550	−206	−23,380
Kinki	−3,480	−6,064	−8,926	−527	−540	−536	−2,637	−15,610	−195	−38,515
Chugoku	−382	−2,937	−12,797	−191	−559	−240	−116	−229	0	−17,451
Shikoku	−398	−636	−4,476	−45	−2,431	−175	−194	−1,472	−8	−9,835
Kyushu	−320	−1,901	−12,377	−5,534	−12,705	−509	−1,202	−2,139	−335	−37,022
Okinawa	0	−2	−221	−1,556	0	−1	0	−187	0	−1,966
Total	−21,780	−30,605	−66,130	−15,608	−35,413	−17,181	−10,325	−40,911	−788	−238,742
Rate of change (%)										
Hokkaido	−0.13	−2.15	−2.71	−3.07	−2.72	−2.80	−2.57	−2.78	0.00	−1.00
Tohoku	−2.33	−0.04	−0.25	−1.50	−1.08	−1.99	−1.52	−1.55	0.00	−0.46
Kanto	−2.36	−0.30	−0.04	−0.13	−0.20	−0.75	−1.61	−2.75	−3.04	−0.27
Chubu	−2.54	−0.89	−0.16	−0.04	−0.04	−0.04	−0.03	−2.09	−3.12	−0.29
Kinki	−2.81	−1.39	−0.19	−0.03	−0.03	−0.03	−0.52	−1.52	−1.30	−0.32
Chugoku	−2.82	−2.32	−0.89	−0.03	−0.03	−0.03	−0.03	−0.03	−0.03	−0.28
Shikoku	−2.41	−2.67	−1.73	−0.04	−0.51	−0.04	−0.04	−1.69	0.00	−0.53
Kyushu	−2.75	−2.32	−2.78	−2.05	−1.65	−0.08	−0.90	−0.08	−2.72	−0.73
Okinawa	0.00	0.00	−3.18	−2.50	0.00	−0.25	0.00	0.00	0.00	−2.58
Total	−1.07	−0.50	−0.26	−0.26	−0.31	−0.29	−0.50	−0.72	−1.49	−0.37

that business and private rail trips decreased because of the shift in demand to air transportation. Since changes in auto trips are small, we ignore them here.

Table 6 shows changes in the business trips by air and rail Table 7 and shows those in household's private trip. For almost all OD pairs, the trips increased. We can then assume that Haneda expansion may induce new trips, although the rate of total passenger trips is not high. Such conventional transport demand analysis as 4-step has not been able to estimate induced trips. In an SCGE model, since the microeconomic behavior of industrial sectors and households are explicitly formulated, trip demands can be consistently estimated between the mode, OD, and trip generation levels.

Table 7 also indicates that Haneda expansion may impact the distribution of household private trips at OD level because, particularly, trips to Hokkaido (the northern island) and Kyushu (the southern island) increase. Since more private household trips to such regions means an increase of tourists, we can assume that household consumption of commodities/ services during the trip may also increase.

3.4.2. Changes in Consumption at Private Trip Destinations

Such household consumption on trips and at the destination as hotels, souvenirs, and meals may increase about 11 billion yen per year. Such changes are especially significant in Hokkaido, Kyushu, and Okinawa as indicated in Fig. 5. We conclude that Haneda expansion stimulates the tourist industry in such peripheral regions.

3.4.3. Changes in Production

Production at the national economic level increases in all industries except the service sector, which includes communication, education, food service, etc. as shown in Fig. 6. The total value of the production increases or equals about 26 billion yen per year. Haneda expansion can significantly impact the Japanese national economy to such an extent.

3.5. Benefit Assessment

3.5.1. Social Net Benefit and Investment Criteria

Total annual benefit measured by equivalent variation amounts to about 33 billion yen per year. Assume that the service life of the expanded facilities is 30 years and that the social discount rate is 4%. Then the present net value of the project is roughly estimated at 400 billion yen, and the cost benefit

Table 6. Changes in Business Trips by Air and Rail.

O	D									
	Hokkaido	Tohoku	Kanto	Chubu	Kinki	Chugoku	Shikoku	Kyushu	Okinawa	Total
Trips per year										
Hokkaido	−59,892	11,556	72,513	9,131	11,910	2,308	1,654	5,575	794	55,549
Tohoku	16,108	−16,114	−5,236	4,175	8,685	4,743	2,872	7,254	−180	22,308
Kanto	121,823	−7,035	−174,419	−36,566	−51,390	31,369	43,817	177,088	20,098	124,787
Chubu	13,200	4,677	−8,220	−3,094	−14,532	−2,190	−589	19,937	3,214	12,401
Kinki	22,667	16,665	−7,926	−16,550	−6,145	−12,306	15,175	45,911	2,808	60,300
Chugoku	3,344	4,691	26,055	−1,871	−9,947	−7,897	−2,316	−5,063	−163	6,832
Shikoku	1,350	3,127	27,801	−2,103	780	−6,534	−6,251	4,952	−300	22,821
Kyushu	4,225	9,134	126,673	10,178	15,601	−18,494	3,363	−77,575	9,244	82,348
Okinawa	496	−662	13,282	1,358	94	−1,956	−355	8,538	0	20,794
Total	123,321	26,040	70,523	−35,342	−44,945	−10,956	57,369	186,617	35,515	408,141
Rate of change (%)										
Hokkaido	−1.42	1.61	2.16	2.35	1.94	2.02	2.38	2.10	2.53	0.57
Tohoku	2.52	−0.33	−0.07	0.98	1.11	2.37	2.21	1.57	−0.33	0.14
Kanto	3.26	−0.07	−0.44	−0.27	−0.25	0.81	2.64	3.35	3.60	0.13
Chubu	3.57	0.97	−0.08	−0.22	−0.22	−0.22	−0.22	2.31	4.08	0.06
Kinki	3.65	2.07	−0.04	−0.25	−0.25	−0.25	0.80	1.39	1.45	0.15
Chugoku	3.01	2.68	0.94	−0.27	−0.27	−0.27	−0.27	−0.27	−0.28	0.05
Shikoku	3.18	3.14	2.12	−0.76	0.05	−0.76	−0.76	1.61	−0.76	0.42
Kyushu	2.52	2.58	2.75	1.24	0.53	−1.07	0.88	−1.07	2.62	0.44
Okinawa	2.41	−1.75	2.26	1.74	0.04	−1.75	−1.75	1.88	0.00	1.34
Total	1.24	0.15	0.08	−0.15	−0.11	−0.07	0.94	0.93	2.60	0.18

Table 7. Changes in Private Household Trips by Air and Rail.

O	D									
	Hokkaido	Tohoku	Kanto	Chubu	Kinki	Chugoku	Shikoku	Kyushu	Okinawa	Total
Trips per year										
Hokkaido	−1,608	684	24,590	5,055	13,002	2,848	1,421	3,425	1,071	50,487
Tohoku	1,700	−522	−976	2,518	5,162	3,693	1,841	5,198	−66	18,548
Kanto	21,821	−841	−3,838	−456	1,255	4,654	3,247	18,165	9,896	53,902
Chubu	7,403	3,542	−790	−179	−837	−191	−75	7,370	4,581	20,822
Kinki	10,973	5,196	1,832	−527	−540	−544	899	9,529	1,645	28,463
Chugoku	2,903	3,301	5,759	−202	−566	−240	−116	−245	−195	10,399
Shikoku	1,082	1,591	4,132	−77	694	−176	−194	1,221	−56	8,218
Kyushu	3,708	3,326	20,850	4,580	6,794	−551	703	−2,300	2,332	39,443
Okinawa	925	−269	10,543	6,484	1,382	−499	−89	1,813	0	20,290
Total	48,906	16,007	62,103	17,196	26,345	8,995	7,637	44,176	19,207	250,573
Rate of change (%)										
Hokkaido	−0.13	0.15	0.68	0.66	0.79	0.72	0.83	0.71	1.01	0.57
Tohoku	0.51	−0.04	−0.02	0.54	0.51	0.82	0.70	0.84	−0.04	0.20
Kanto	0.82	−0.02	−0.04	−0.02	0.03	0.22	0.55	0.75	0.89	0.18
Chubu	0.84	0.45	−0.02	−0.04	−0.04	−0.04	−0.03	0.63	1.00	0.19
Kinki	0.83	0.58	0.04	−0.03	−0.03	−0.03	0.13	0.43	0.42	0.18
Chugoku	0.81	0.84	0.28	−0.03	−0.03	−0.03	−0.03	−0.03	−0.03	0.13
Shikoku	0.92	0.92	0.58	−0.05	0.11	−0.04	−0.04	0.68	−0.05	0.28
Kyushu	0.81	0.84	0.73	0.57	0.39	−0.08	0.38	−0.08	0.84	0.38
Okinawa	1.03	−0.19	0.81	1.24	0.20	−0.19	−0.19	0.60	0.00	0.61
Total	0.65	0.19	0.18	0.21	0.17	0.12	0.25	0.39	0.60	0.25

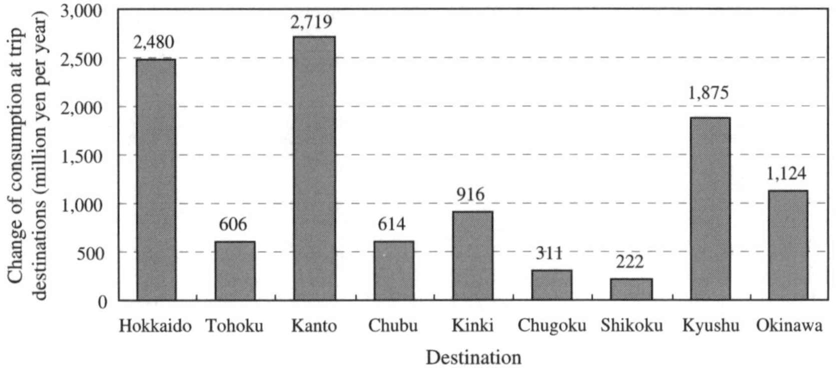

Fig. 5. Changes in Consumption at Private Trip Destinations (at 1995 yen values).

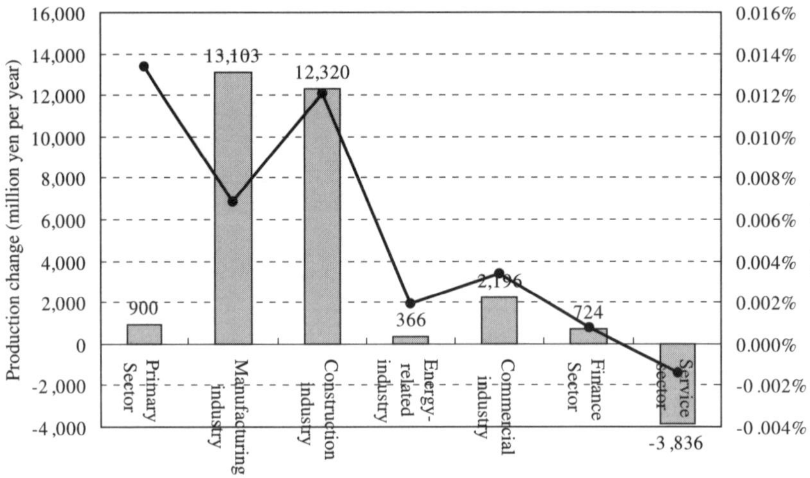

Fig. 6. Change in Production in Each Industry (at 1995 yen values).

ratio is about 3.3. From these figures, we conclude that the expansion of the Haneda airport passes the investment criteria in cost benefit analysis.

3.5.2. Benefit Incidence

For spatial incidence or distribution of benefit, Fig. 7 shows that the largest benefit accrues to Kanto (Tokyo metropolitan area). However, peripheral regions such as Hokkaido, Kyushu, and Okinawa may reap higher benefits per household than Kanto as shown in Fig. 8. Haneda expansion can benefit all of Japan.

A benefit incidence table (BIT) shows benefit distribution in spatial and industrial dimensions. The benefits measured for equivalent variation can be decomposed into several items of benefits/costs listed in the far left column in Table 8. The regions in the SCGE model are grouped into Kanto and other regions, and industrial sectors by the industry and the transportation firms. Needless to say, we can adopt a more detailed grouping of regions and industrial sectors.

BIT first shows that the total amount of user's benefits are about 82 billion yen per year. The competition between airlines in the aviation market

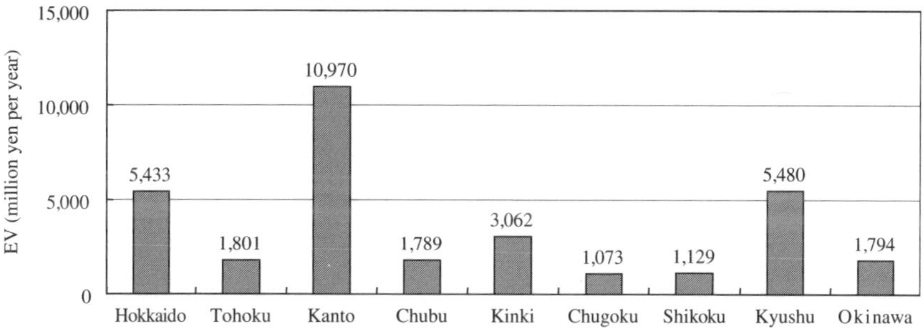

Fig. 7. Regional Benefits (at 1995 yen values).

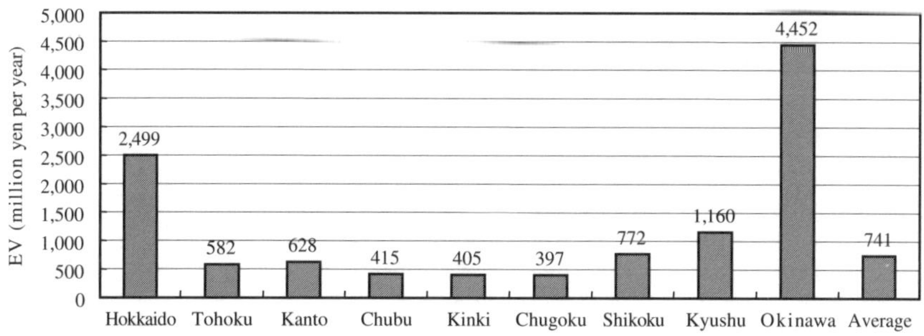

Fig. 8. Regional Benefits Per Household (at 1995 yen values).

Table 8. Benefit Incidence Table (at 1995 yen values).

	Kanto				Other than Kanto				National			
	Households	Industries	Transport firms	Total	Households	Industries	Transport firms	Total	Households	Industries	Transport firms	Total
Fare revenue			−13.6	−13.6			−35.1	−35.1			−48.7	−48.7
Users' benefit	8.5	16.0		24.5	30.9	26.3		57.1	39.4	42.2		81.6
Change in prices of commodities and services industries	12.3	−12.2		0.1	28.1	−28.2		−0.1	40.4	−40.4		0.0
Wage change	3.0	−3.0	−0.0	−0.0	1.9	−1.9	−0.0	−0.0	4.9	−5.0	−0.0	−0.0
Capital dividend change	0.9	−0.9	−0.0	−0.0	−3.6	3.6	0.0	−0.0	−2.7	2.7	0.0	−0.0
Dividend from transport firms change	−13.6		13.6	0.0	−35.1		35.1	0.0	−48.7		48.7	0.0
Total	11.2	−0.2	0.0	11.0	22.1	−0.2	0.0	21.9	33.3	−0.4	0.0	32.8

Note: Values in the hatched cell should theoretically be 0, but actually they are non-zero because of calculation residuals.

results in reduction of their profit, and therefore the dividend income of households decreases about 49 billion yen per year.

Indirect or spillovered benefits by changes in price of commodities/services, in wage rate and in capital rent are canceled out between demand side and supply side as long as markets are in perfect competition. The cancellation of indirect benefits was proved by Mohring (1976) and Kanemoto and Mera (1985). Since the SCGE model in this paper is built on the same theoretical basis, we have successfully showed that they can be canceled out numerically in an actual scale application of general equilibrium theory.

Indirect benefits in competitive markets don't contribute to the social net benefit (SNB) indicated in the right-bottom cell. However, the amount of the indirect benefits is not negligible. For example, households take about 40 billion yen per year at the national level as increments of consumer's surplus for commodities/services. Such a large amount of indirect benefit is so important that we should consider the benefit incidence in spatial and industrial dimensions.

The SNB amounts to about 33 billion yen per year, which is the sum of the user's benefit and the profit loss of transportation firms. The SNB can be decomposed to 11 billion yen for the Kanto region (Tokyo metropolitan area) and 22 billion for other regions. We should again remark that Haneda expansion will benefit the non-metropolitan regions much more than the metropolitan area. This suggests that the cost of Haneda expansion should be shouldered not only by Kanto but also by other regions; expansion could also be implemented as a national project supported by the central government.

4. CONCLUDING REMARKS

In this paper, we developed a SCGE model for evaluating a large-scale transportation project and applied it to the Haneda expansion question. In the application, we successfully evaluated the project from conventional cost benefit criterion, by simple SNB measurements, and from benefit incidence in spatial and industrial dimensions.

Our developed SCGE model can estimate induced trips and describe the consumption of commodities and services at a trip's destination. The analysis of induced trips is crucial for energy consumption and environmental emissions, although we did not focus on them in this paper. Consumption at trip destination is a key issue for impact on local industries, particularly on tourism sectors. These two points in the SCGE model would be useful when we assess impact from wider views than those in a conventional project evaluation.

We have seen the spatial distribution of the project benefits in BIT for the expansion of Haneda. Since it is a central node of domestic aviation network in Japan, its expansion will benefit non-metropolitan regions more than metropolitan areas. Such findings obtained by spatial benefit incidence analysis should form the basis for discussions of project cost sharing between local and central governments.

In further methodological development of the SCGE model, we will address the following problems:

(i) Comparison of projects of different transportation modes by SCGE that explicitly incorporate freight transportation into the model. Other SCGE models developed in Japan, as shown in Fig. 2, have been applied to freight transportation projects. We should employ the techniques in such models so that this paper's model can describe freight transportation.
(ii) To analyze long-term changes in spatial structure, we should describe interregional migration and capital flows. Many empirical models already exist for interregional migration. However, they are not always consistent with microeconomic foundations. In our line of thought, migration should be modeled that is also consistent with general equilibrium theory so that it can be linked to the SCGE model.

ACKNOWLEDGMENTS

This study is a part of a research project conducted by the Policy Research Institute, Ministry of Land, Infrastructure, and Transportation, Government of Japan. The research group was directed by Prof. Hisayoshi Morisugi, Tohoku University. The authors are grateful to all members in the group for their help, particularly to Dr. Tadahiro Ohashi for his collaboration in modeling and to Dr. Fumio Takuma for his insightful comments. However, only the authors are responsible for this paper.

REFERENCES

Becker, G. S. (1978). *The economic approach to human behavior*. Chicago: The University of Chicago Press.

Ichioka, O. (1991). *Applied general equilibrium analysis*. Tokyo: Yuhikaku (in Japanese).

Kanemoto, Y., & Mera, K. (1985). General equilibrium analysis of the benefits of large scale transportation investments. *Regional Science and Urban Economics*, *15*, 343–363.

Koike, A., & Ueda, T. (2001). SCGE model for passenger transport investment, *Proceedings of 9th World Congress of Transport Research*, CD-ROM

Mohring, H. (1976). *Transportation economics*. Boston, MA: Ballinger Publisher Co.

Morisugi, H., & Ohno, E. (1995). Proposal of a benefit incidence matrix for urban development projects. *Regional Science and Urban Economics*, *25*, 461–481.

Ohashi, T., & Ando, A. (1999). A study on the hub-spoke network and landing fee policy in the aviation market. *Journal of Infrastructure Planning and Management*, *611*(IV-42), 33–44.

Shoven, J. B., & Whally, J. (1984). Applied general equilibrium models of taxation and international trade. An introduction and survey. *Journal of Economic Literature*, *22*, 1007–1051.

Shoven, J. B., & Whally, J. (1992). *Applying General Equilibrium*. Cambridge University Press.

FURTHER READING

Akamatsu, T., Handa, M., & Nagae, T. (1998). Variational inequality approach to multi-regional computable general equilibrium modeling. *Infrastructure Planning Review*, *15*, 175–185.

Ando, A., & Mizokami, A. (1993). Equilibria in the context of infrastructure planning and applied general equilibrium analysis. *Infrastructure Planning Review*, *11*, 29–40.

Doi, M., Piyush, T., & Itoh, H. (2001). A computable general equilibrium analysis of efficiency improvements at Japanese ports. *Review of Urban & Regional Development Studies*, *13*(3), 123–142.

Fujiwara, T., Hasuike, K., & Kanemoto, Y. (2002). Welfare evaluation of tax policies to reduce CO_2 emissions by road transportation. *Journal of Applied Regional Science*, *7*, 39–50.

Hideshima, E., & Kobayashi, K. (1997). An applied general equilibrium model for urban systems within non-metropolitan areas. *Proceedings of Infrastructure Planning*, *20*(1), 259–262.

Ishiguro, K., & Inamura, H. (1999). Study on SCGE models for prediction of container cargo flow on international trunk line. *Proceedings of Infrastructure Planning*, *22*(2), 745–748.

Ishiguro, K., Ishikura, T., Kagawa, S., & Inamura, H. (2001). Development of international trade model taking account of ocean carriers' behavior. *Journal of the Eastern Asia Society for Transportation Studies*, *4*(1), 329–338.

Kobayashi, K., & Okumura, M. (1996). National land-axis development by providing rapid transportation systems. *Infrastructure Planning Review*, *13*, 57–66.

Koike, A., Ishikawa, Y., Ueda, T., & Kohno, M. (2002). Computable general equilibrium model for urban area. *Proceedings of Infrastructure Planning*, *26* CD-ROM.

Miyagi, T., & Honbu, K. (1993). A study on forecasting method for interregional trades based on the SCGE model. *Proceedings of Infrastructure Planning*, *16*(1), 879–886.

Miyagi, T., & Honbu, K. (1996). A study on an interregional trade model based on the applied general equilibrium analysis. *Journal of Infrastructure Planning and Management, 530*(IV-30), 31–40.

Miyagi, T., Honbu, K., Mizutani, A., & Oohashi, K. (1996). An economic evaluation of the effects of the TOKAI-HOKURIKU expressway and the TOKAI belt highway of SCGE model. *Proceedings of Infrastructure Planning, 19*(1), 189–192.

Miyata, Y. (1995). A general equilibrium analysis of the waste-economic system – a CGE modeling approach. *Infrastructure Planning Review, 12*, 259–270.

Miyata, Y. (1998). General equilibrium effects of waste abatement policies – a CGE-modeling approach. *Infrastructure Planning Review, 15*, 305–316.

Miyata, Y., Sato, Y., Takahashi, S., & Yamazaki, N. (1990). A regional computable general equilibrium model. *Proceedings of Infrastructure Planning, 13*, 45–52.

Miyata, Y., Shibusawa, H., & Chen, Z. (1998). An interregional general equilibrium analysis of energy demand/supply in China. *Infrastructure Planning Review, 15*, 359–368.

Mizokami, S. (1992). Interregional trades based on input–output and spatial price equilibrium framework. *Proceedings of Infrastructure Planning, 15*(1), 629–635.

Mizokami, S. (1994). An interregional trades model in consideration of input–output and price equilibrium. *Proceedings of JSCE, 494*(IV-24), 53–61.

Mizokami, S., Hayashi, Y., & Nakajima, Y. (1995). A multi-regional model on the general equilibrium theory for transport project evaluation. *Proceedings of Infrastructure Planning, 18*(1), 217–220.

Morisugi, H., Hayashiyama, Y., & Kojima, S. (1986). Time-saving benefit evaluation for transport projects – a simple practical method and its accuracy. *Infrastructure Planning Review, 4*, 149–156.

Morisugi, H., & Ohshima, N. (1985). A simple post-evaluation model for trunk transport network formation. *Proceedings of Infrastructure Planning, 7*, 125–132.

Morisugi, H., Ueda, T., Muto, S., & Kondo, Y. (1996). Constructing the computable general equilibrium model focusing the transport sector. *Infrastructure Planning Review, 13*, 349–360.

Mun, S. (1997). *Evaluation of transportation network from the viewpoint of population distribution – multiregional general equilibrium model focusing agglomeration economies (in Japanese)*. Tohoku Construction Association Report on Research.

Okuda, T. (1994). Multiregional general equilibrium model introduced from probabilistic approach. *Studies in Regional Science, 24*(1), 117–131.

Okuda, T. (1999). A computable general equilibrium model for an analysis of transport and communication activity. *Proceedings of Infrastructure Planning, 22*(2), 487–490.

Okuda, T., & Hayashi, Y. (1994). Applied general equilibrium analysis on effects of motorway improvement in Japan. *Studies in Regional Science, 25*(1), 45–56.

Shibata, T., & Ando, A. (1992). The effects of improving trunk-line transport facilities in China based on interregional equilibrium over commodities and their prices. *Infrastructure Planning Review, 10*, 167–174.

Shibata, T., Ando, A., & Okumura, T. (1990). Examining the feasibility of a multi-regional model for China from a statistical viewpoint. *Proceedings of Infrastructure Planning, 13*, 39–44.

Yamauchi, H., Ueda, T., & Kawai, T. (1999). A cost–benefit analysis of expressway with an applied general equilibrium model. *Expressways and Automobiles, 42*(5), 22–30.

AIRPORT CHOICE ANALYSIS OF INTERNATIONAL PASSENGERS USING TIME-SERIES DISAGGREGATE DATA

Tatsuo Takase and Takayuki Morikawa

ABSTRACT

Recently, large-scale hub airports have been developed at principal cities in Asia. Consequently, many direct air routes for Asian cities have been established from local airports in Japan. By December 2001, for instance, 22 airports in Japan offered direct flights to South Korea. These changes in air route settings seem to influence not only airport choice but also destination choice of Japanese passengers. This research investigates the airport and destination choices of Japanese passengers using repeated cross-section disaggregate air passenger data. Disaggregate nested logit models are employed to represent passenger's airport and destination choices. At the same time, a forecasting model of overseas travel generation demand is also estimated that uses a generalized least squares method. These models are used to evaluate the policies of increasing international flights at local airports in Japan.

Global Competition in Transportation Markets: Analysis and Policy Making
Research in Transportation Economics, Volume 13, 197–210

ISSN: 0739-8859/doi:10.1016/S0739-8859(05)13009-1

1. INTRODUCTION

International air transportation demand in Japan has steadily been increasing largely due to the rapid growth of Asian and European countries. This phenomenon can partly be explained by changes in the destination preferences of tourists, but it can also be attributed to the recent opening of direct flights to Asian countries from Japan's local airports. Until recently, international air travel demand was concentrated at Narita and Kansai International airports; it was impossible to increase the number of flights to/from these airports because of capacity. On the other hand, large-scale hub airports have been developing at principal cities in Asia. Consequently, many direct air routes for Asian cities have been established from local airports in Japan, enabling residents of rural areas to travel abroad from local airports, thereby increasing accessibility to foreign countries.

Airport administrators and airline marketers probably understand such evolving behavior of overseas passengers, especially the passenger's criteria for route choices. However, even researchers rarely have ready answers to such questions as: Why do passengers favor one route over another? Which key characteristic is more crucial for airport choice: ease of access or flight frequency?

To date, passengers' route/airport choice behavior has certainly been affected by such characteristics as accessibility to major cities, frequency of flights, ticket costs, and flight times. Such decisions may also be influenced by past experiences, word of mouth, or advertisements.

In this chapter, passengers' route/airport choice behavior is modeled by using not only engineering variables but also information accumulated by the experience of travelers themselves. Overseas passengers' survey data in Japan were used to estimate a nested logit model of travel route choice as a function of flight cost, flight frequency, access time, past choice probability, and other variables. A linear demand function for overseas travel is also estimated using a generalized least-squares method that forecasts changes in demand by travel destinations.

2. LITERATURE REVIEW

There is quite a large body of literature on international air demand analysis that describes total demand and route/airport choice. Koizumi and Yai (1993) analyzed the New Tokyo International Airport and estimated the direct demand for each of its routes, developing demand models for trip distribution and route choice by using index regression or an aggregate logit. Naohara, Yai, Hyodo and Morichi (1993) estimated an airport choice model by using an aggregate logit based on overseas travel survey data in Japan. The accessibility obtained from the airport choice model was used in a trip generation model, revealing that accessibility was closely related to airport choice. They also created an airport access mode choice model using multinomial logit with such variables. Oi, Yai, Morichi, and Yamamoto (1988) estimated a trip generation model with accessibility and exchange rates. They also estimated an airport choice model by trip purposes with access service and flight frequencies. Furuichi and Koppelman (1993) developed a multistage nested logit model to represent a joint choice of departure airports, destination regions, and overseas trip generations.

Harvey (1987) and Windle and Dresner (1995) worked on passengers' airport choice in multiple airport regions. By analyzing the airport choices of residents of the San Francisco Bay Area, Harvey found that a simple logit model with two variables, airport access time and flight frequency to destinations, provided a good approximation of airport choice behavior, especially for business passengers. Windle and Dresner found that passengers' experience with a particular airport is generally an important determinant of airport choice and suggested using it for future airport choice models. At the same time, in keenly competitive markets they revealed that the importance of airport access time decreases dramatically while the importance of flight frequency increases.

In an aggregate demand analysis of overseas passengers, Murayama and Morikawa (1992) forecasted the number of overseas travelers by applying the Diffusion Model. Kawai, Murakami, and Hasegawa (1995) used a gravity model with economic levels and traffic resistance to forecast long-term overseas travel demand.

3. CONCEPTUAL FRAMEWORK

To analyze the demand of overseas trips based on passenger's behavior in Japan, we developed an overseas travel generation demand model. The choice model of the departure airport and the air route was made in this research.

The first feature of the model made in this research hierarchically showed the passenger's behavior for each destination. In concrete terms, a nested logit model for the three hierarchies (route choice, airport choice, and access mode choice) was made. Airport choice and access mode choice have been discussed in a number of previous works. However, the increasing rate of

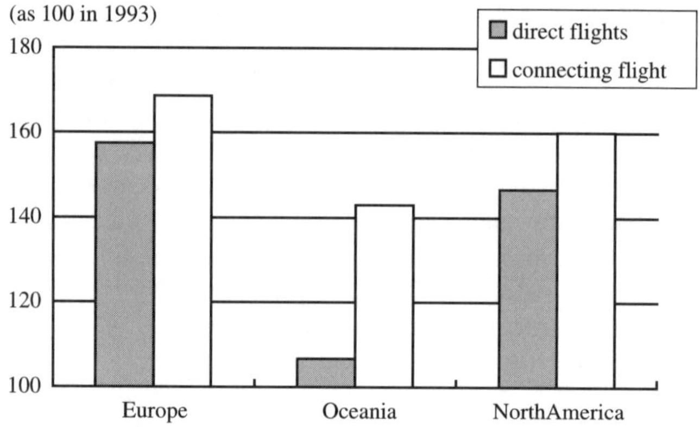

Fig. 1. Number of Passengers to Each Destination in 1999.

passengers using connecting flights via Asia is larger than the number of direct flight passengers in recent years, as is shown in Fig. 1. Moreover, the number of air routes from local airports to the Asian region has increased due to the development of major Asian airports and the capacity limit of domestic central airports. Therefore, route choice level was inserted in the structure of the nested model.

The next feature used airport choice probability estimated by a one term previous model as an explanatory variable. Except for frequent business travelers and those who have repeatedly visited the same region, many passengers may not have direct information based on experience. If consumers could obtain such information as destinations, which airport to choose, what access mode to use, and so on from people with experience, they may investigate past tendencies. Hence, airport choice probability obtained from the previous survey is used as an independent variable of the nested logit model for airport choice level.

4. DATA AND MODEL

4.1. Data

The data used in the airport choice analysis were obtained from overseas air passenger surveys in Japan conducted every 2 years at the peak travel season (August) and the off-peak season (November) by the Civil Aviation Bureau of the Ministry of Land, Infrastructure, and Transport. This paper used the 1977 and 1999 survey data. Questionnaires were conducted in the departure lounges of all international airports in Japan of overseas passengers. The raw data include about 21,500 samples. Note that the data were collected in a choice-based manner from the viewpoint of the airport choice mode. The passengers were asked about the following aspects of their trip: destination, flight-number, purpose, access to the airport, origin of trip and access mode, reasons for airport choice, distance from home, residents or non-residents, and group or individual tour.

Additional data required to estimate the choice models were obtained from the following sources; access characteristics (e.g. access time and fares) to each airport from all origin zones were obtained from timetable, route, fare search software, and digital maps. Airport service characteristics (e.g. flight frequency) at each airport were obtained from timetables. Airline service characteristics (e.g. flight times and fares) of both direct and connecting flight fares for the best flight routes available on each destination were obtained from the Official Airline Guide. Y-class (economy class) was used for each fare. The data sets used in the forecasting model of overseas travel generation demand came from the gross provincial products of the System of National Account by The Cabinet Office.

4.2. Model

It is assumed that a passenger's decision making is hierarchically structured, as illustrated in Fig. 2. The nested model (NLM) of passenger's choice behavior, shown in Fig. 2, is composed of route and airport choice levels and access modes to the airport choice level. Table 1 shows a detailed composition of alternatives. At the airport choice level, the number of airports available from each residence ranges from two (e.g. Kanto and Kinki regions) to four (e.g. Hiroshima, Okayama, and Kagoshima). This choice behavior model is estimated by using a nested logit model, and a primary factor that influenced the passenger's choice behavior at each choice level is used as an independent variable. The major independent variables of each level are flight time and flight fare (route choice level), flight frequency and choice proportion of each airport in the last survey (airport choice level), access time and access cost (access mode choice level).

A linear regression model was used to generate overseas travel demand. However, a serial correlation often exists between regression models from time-series data. Therefore, if there is a significant correlation between error

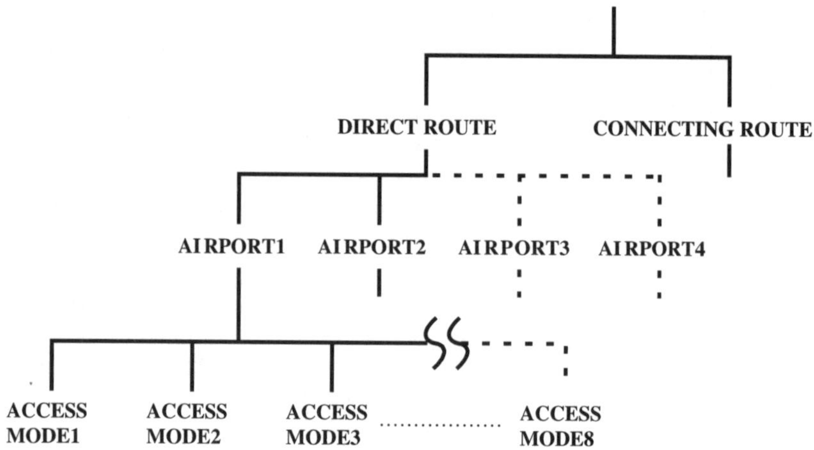

Fig. 2. Nested Logit Model Structure.

Table 1. List of Alternatives.

Airport 1	Narita
Airport 2	Kansai
Airport 3	Sendai (only Tohoku region) Nagoya (only Chubu region) Fukuoka (only Kyushu region)
Airport 4	Other international airports (each prefecture)
Access mode i ($i = 1-8$)	Auto, rail, bus, air, auto, and hotel, rail and hotel, bus and hotel, air and hotel

terms, a generalized least-squares method is used for the model estimation as

$$y_t - \rho y_{t-1} = \beta_1(1-\rho) + \beta_2(x_t - \rho x_{t-1}) + \varepsilon_t \qquad (1)$$

where y_t is the number of overseas travelers in t term, x_t the gross provincial product (GPP) in t term, ρ the autocorrelation coefficient, and β_1, β_2 are the regression coefficients.

4.3. Estimation Results

The estimation results from the nested logit model of the passenger's route and mode choices from origin to each destination are shown in Tables 2–4.

Table 2. Estimation Results of NLM to European Region.

	Units	Coefficient Estimate	*t* Statistic
First stage (mode choice)			
Air constant	0/1	−2.00	−12.2
Rail constant	0/1	−2.61	−17.2
Bus constant	0/1	−3.51	−23.3
Auto constant	0/1	−3.59	−24.6
Air and hotel constant	0/1	3.43	58.0
Rail and hotel constant	0/1	1.48	28.9
Bus and hotel constant	0/1	0.176	2.87
Access cost	100 yen	−0.0175	−34.2
Access time	10 min	−0.0720	−39.7
$L(0)$		−79,616.8	
$L(\hat{\beta})$		−57,648.4	
ρ^2		0.276	
Second stage (airport choice)			
Narita constant		−1.59	−18.0
Another main airport constant (Sendai, Nagoya, or Fukuoka)		0.445	6.68
Local airport constant		0.868	5.69
Flight frequency	Flights/week	0.0010	28.1
Airport choice probability based on last survey		1.90	31.6
Mode scale parameter		0.418	22.0
$L(0)$		−32,715.4	
$L(\hat{\beta})$		−14,281.5	
ρ^2		0.563	
Third stage (route choice)			
Direct flight constant	0/1	−4.08	−4.25
Flight time	10 min	−0.0997	−6.16
Business trip dummy	0/1	0.176	3.25
Airport scale parameter		0.370	7.52
$L(0)$		−32,715.4	
$L(\hat{\beta})$		−10,701.3	
ρ^2		0.673	
Number of observations		47,198	
Number of cases		2121	

Table 3. Estimation Results of NLM to North America.

	Units	Coefficient Estimate	*t* Statistic
First stage (mode choice)			
Air constant	0/1	0.0353	0.25
Rail constant	0/1	−0.314	−2.44
Bus constant	0/1	−1.24	−9.78
Auto constant	0/1	−1.02	−8.24
Air and hotel constant	0/1	2.73	40.9
Rail and hotel constant	0/1	1.47	25.6
Bus and hotel constant	0/1	0.402	5.94
Access cost	100 yen	−0.0094	−22.4
Access time	10 min	−0.0548	−36.7
$L(0)$		−92,363.1	
$L(\hat{\beta})$		−61,333.7	
ρ^2		0.336	
Second stage (airport and route choice)			
Direct flight constant	0/1	0.8692	38.8
Flight frequency	Flights/week	0.0037	45.0
Flight time	10 min	−0.0127	−9.95
Business trip dummy	0/1	1.05	17.5
Individual tour dummy	0/1	0.36	9.95
Airport choice probability of the last survey		2.35	89.0
Mode scale parameter		0.464	38.8
$L(0)$		−76,484.1	
$L(\hat{\beta})$		−30,695.8	
ρ^2		0.599	
Number of observations		51,512	
Number of cases		2,833	

All estimated parameters have correct signs, and most are statistically significant. In the first stage, the values of time calculated from estimated coefficients in the airport access are: 3,500 JPY/h to North America, 2,470 JPY/h to Europe, and 1,790 JPY/h to Oceania. Perhaps the distinction of values of time between the three destinations is closely related to the difference of the proportion of business trips between them. Business passengers may be relatively more sensitive to travel time because they often travel under time restrictions. In fact, the proportion of business trips to the number of passengers to Europe and North America is about twice that of passengers to

Table 4. Estimation Results of NLM to Oceania Region.

	Units	Coefficient Estimate	*t* Statistic
First stage (mode choice)			
Air constant	0/1	−11.8	−32.7
Rail constant	0/1	−12.7	−38.9
Bus constant	0/1	−13.5	−41.9
Auto constant	0/1	−12.0	−37.7
Air and hotel constant	0/1	0.801	6.78
Rail and hotel constant	0/1	−1.02	−13.1
Bus and hotel constant	0/1	−2.45	−20.0
Access cost	100 yen	−0.0514	−49.2
Access time	10 min	−0.153	−36.7
$L(0)$		−27,308.1	
$L(\hat{\beta})$		−16,802.7	
ρ^2		0.385	
Second stage (airport choice)			
Narita		−2.46	−16.5
Another main airport constant (Sendai, Nagoya or Fukuoka)		0.305	3.46
Local airport		3.90	7.35
Flight frequency	Flights/week	0.121	19.9
Mode scale parameter		0.338	73.2
$L(0)$		−12,376.6	
$L(\hat{\beta})$		−5,118.6	
ρ^2		0.586	
Third stage (route choice)			
Direct flight constant	0/1	−7.13	−12.4
Flight time	10 min	−0.198	−11.8
Flight cost	100 yen	−0.00425	−4.81
Airport scale parameter		0.0915	1.28
$L(0)$		−10,923.0	
$L(\hat{\beta})$		−3,688.3	
ρ^2		0.662	
Number of observations		15,759	
Number of cases		692	

Oceania. Consequently, the value of time by airport access of passengers to Europe and North America is higher than that of passengers to Oceania.

In the upper level model (the second and third stages), the most effective model for each destination was decided by trial and error. As a result, the selected model of passengers' route and airport choice behavior has a different structure between destinations. In particular, the selected route and airport choice model of passengers to North America is not a nested logit but a joint logit model. Moreover, because the proportion of business trips to the number of passengers to Europe and North America is high, as previously described, a business trip dummy was added to the utility function of direct flights to Europe and North America. As the proportion of individual tour trips to the number of passengers to North America is higher than that of passengers to other places, an individual tour trip dummy was also included into the utility function of direct flights to North America.

Flight fare was omitted from the utility function of the route and airport choice of passengers to Europe and North America because the estimated coefficient was not statistically significant (e.g. t statistic equals about -0.3) in cases that included flight fare as an dependent variable in the utility function. This is probably strongly related to the business trip ratio. Business passengers may be comparatively insensitive to cost because they often do not pay for their own travel expenses.

Another feature of the estimated results of the nested logit model is that the coefficient of flight frequency is smaller than the other coefficients. The estimated coefficient of the alternative specific constant of the local airport is larger than the others. This is especially noticeable in the model for Europe and North America directions, which shows that many passengers choose to fly from their local airport regardless of flight frequency. Moreover, the coefficient of airport choice probability from the last survey has the same significance as other major independent variables, and it is positive. Therefore, if the proportion of passengers who use local airports increases, or if positive information for using local airports is given, other passengers may also choose to fly from closer to home.

Finally, Table 5 presents the estimation results for the forecasting model of overseas travel generation demand by examining whether the model had a significant correlation between error terms.

The Durbin–Watson d-statistic of the model of each direction was as follows: 0.19 to North America, 0.21 to Europe, and 0.20 to Oceania. This outcome indicates a high correlation between error terms, and the generalized least-squares method was used for the forecasting model. But the ρ of Eq. (1) used the $\hat{\rho}$ calculated by employing the Cochrane–Orcutt method.

Table 5. GLS Estimation Results.

	Units	Coefficient Estimate	t Statistic
To European region			
$GPP_t - \rho GPP_{t-1}$	Million yen	0.00437	17.1
$1-\rho$		16,009.4	1.5
ρ		0.913	
R^2		0.612	
To North American region			
$GPP_t - \rho GPP_{t-1}$	Million yen	0.00482	15.6
$1-\rho$		22,492.4	1.5
ρ		0.930	
R^2		0.568	
To Oceania region			
$GPP_t - \rho GPP_{t-1}$	Million yen	0.00180	16.5
$1-\rho$		−3,514.4	−0.9
ρ		0.929	
R^2		0.695	
Term of observations		4	
Number of cases		47	

The results from Table 5 indicate that the degree of change in the number of travelers to Europe is similar to the number of travelers to North America because the estimated coefficients of both models are similar. Another feature is that the number of travelers to Oceania decreases if the increasing rate of the gross provincial product is small. As for travel to Oceania, there are a significant number of non-business trips and package tours. Therefore, it is conceivable that some people choose destination by the falling of a trade barometer.

5. SOME POLICY EXAMPLES

In this section we analyze the following policy examples by using a nested logit model of the passenger's route and mode choice from origin to destination. Two examples are shown.

Case 1. The number of flights from all local airports (except Okinawa) to Asian destinations doubles (the total number of flights increases by 60) and connecting times decreases by 2/3.

Case 2. The number of direct flights from Narita to North American and European destinations is increased by 30 flights.

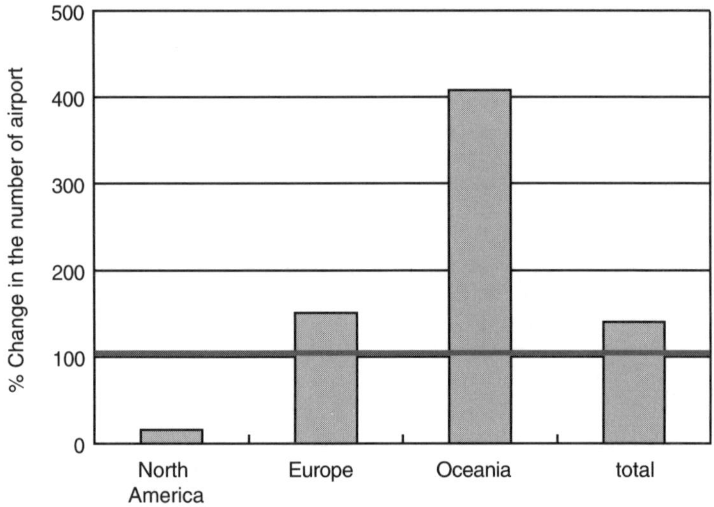

Fig. 3. Policy Analysis (Case 1).

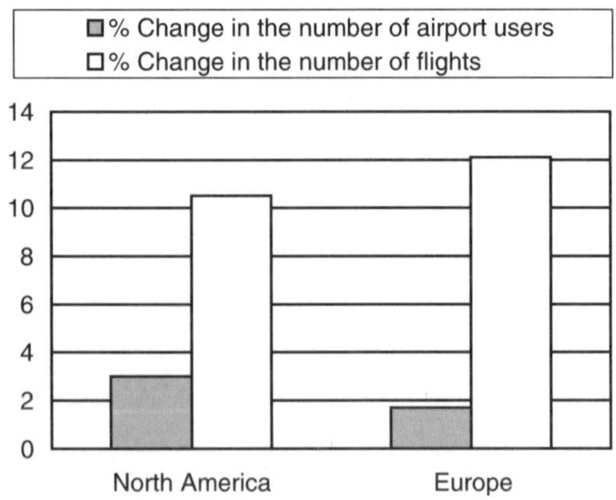

Fig. 4. Policy Analysis (Case 2).

The forecasting results of each case are shown in Figs. 3 and 4. In Case 1, the number of passengers with connecting flights to Europe and Oceania increases significantly, whereas the increase in the number of passengers with connecting flights to North America is constant because connecting flights to North America via Asia seem roundabout when compared to connecting flights to Europe. In Case 2, the number of passengers of direct flights from Narita in both directions barely increases, and the increase in flights has little effect on the number of direct flight passengers.

6. CONCLUSION

This paper analyzed the choice behavior of overseas passengers in Japan. Nested logit models of route and airport choices were developed, and a forecasting model of overseas travel generation demand was also estimated by using a linear regression model with a generalized least-squares estimation. This work is distinct from previous works because a series of passenger's choice behavior (access mode, airport, and air route choices) was modeled. This model enabled the evaluation of such policies as increasing international flights at local airports in Japan. The analysis points to a number of specific conclusions:

- The choice behavior of passengers to Europe and North America, in the route and access mode choice levels, is mainly determined by travel time rather than cost. In particular, business trips to Europe place the highest value on time. However, flight frequency in the route choice level is very small when compared with other coefficients.
- The choice behavior of passengers to Oceania is different from that of Europe and North America. The majority of travelers to Oceania are tourists, who have joined package tours, heavily influenced by economic situations.

The conclusions presented above show that passengers will choose to fly from their local airport if more direct air routes to Asian large-scale hub airports are established from local airports in Japan and if transfer time at those hub airports is reduced. Consequently, increasing international flights from local airports, especially to Asian major cities, is more effective rather than expanding major airports in Japan.

REFERENCES

Furuichi, M., & Koppelman, F. S. (1993). Integrated forecasting model for international air travel demand. *Infrastructure Planning Review of JSCE*, *11*, 239–246.

Harvey, G. (1987). Airport choice in a multiple airport region. *Transportation Research A, 21A*(6), 439–449.

Kawai, T., Murakami, K., & Hasegawa, T. (1995). Development of a super-long range demand forecast model for international air passengers. *Proceedings of Infrastructure Planning of JSCE, 17*, 335–338.

Koizumi, Y., & Yai, T. (1993). A study on the demand of the airline in Pacific area. *Proceedings of Annual Conference of JSCE, 4*(48), 550–551.

Murayama, K., & Morikawa, T. (1992). Demand analysis on overseas tourism using diffusion models. *Proceedings of Infrastructure Planning of JSCE, 15*(1)-1, 205–210.

Naohara, S., Yai, T., Hyodo, T., & Morichi, S. (1993). Analysis of international passengers from non-major airports. *Proceedings of Annual Conference of JSCE, 4*(48), 548–549.

Oi, T., Yai, T., Morichi, S., & Yamamoto, S. (1988). Demand analysis model for local airport internationalization. *Proceedings of Annual Conference of JSCE, 4*(43), 492–493.

Windle, R., & Dresner, M. (1995). Airport choice in multiple-airport regions. *Journal of Transportation Engineering, 121*(4), 332–337.

EFFECTS OF DEREGULATION ON LOCAL AIR PASSENGER DEMAND

Naoto Ida and Tohru Tamura

ABSTRACT

In Japan, supply–demand adjustments for aviation regulation were abolished under the revised Aviation Law of February 2000. This act of deregulation provided airline companies the freedom to set airfares and enter into or withdraw from any route. Competition after deregulation brought many benefits to the consumers of the major routes as well. On the other hand, it inconvenienced the consumers of minor routes by widening the gaps in airfares and services. The purpose of this study is to determine how domestic aviation networks have been affected by airline deregulation. Moreover, the possibility for revitalizing regional airports is suggested.

1. INTRODUCTION

Deregulation of the Japanese airline industry started in 1986. In principle, the enactment of the revised Aviation Law in February 2000 led to full liberalization, providing airlines the freedom to set airfares and enter into or withdraw from any route.

In 1998 two airline companies, Skymark Airlines and Hokkaido International Airlines, were born. They were the first new entrants in the

Global Competition in Transportation Markets: Analysis and Policy Making
Research in Transportation Economics, Volume 13, 211–231

ISSN: 0739-8859/doi:10.1016/S0739-8859(05)13010-8

Japanese airline industry in 35 years. This triggered an increase in airfare discount rates offered by the other existing three major airlines: Japan Airlines, All Nippon Airways, and Japan Air System. This provided more benefits to passengers of major routes. On the other hand, however, this also widened gaps of service levels, such as fares, for passengers using minor routes connecting local areas. Moreover, this also provoked anxieties about possible withdrawals from unprofitable routes.

In June 1994 at the Council for Civil Aviation, the Ministry of Transport argued that "even after discontinuation of unprofitable routes, small-scale commuters will emerge in due time, among which optimum ones that meet respective regional conditions will fulfill transportation functions in accordance with market-economy principles." Furthermore, a report was submitted, which claimed that "a withdrawal of the existing airlines from unprofitable routes is unavoidable." It predicted the beginning of "inter-regional competition" for municipalities located near regional airports (qtd. in Tozaki, 1995, p. 112).

This research aims to improve understanding of the impact of airline deregulation on local airports in Japan. In particular, this chapter focuses on the impact to passengers of airline deregulation, centering on changes in civil aviation networks. The following four issues are specifically addressed and clarified:

- Changes in the number of passengers and seats provided by airline companies by studying individual domestic airline routes;
- Comprehension of post-deregulation changes in civil aviation networks;
- Analysis of the gap between the preferences of local airport users and levels of service provided by airline companies; and
- Discussion on the feasibility of introducing regional aviation into Hokkaido.

2. AIRLINE DEREGULATION IN JAPAN

2.1. Policy Changes after Deregulation

2.1.1. Establishment of the "70/72 Scheme"

Since the Japanese aviation industry was constrained by general headquarters (GHQ) occupation policies after World War II, it had a much slower start than other developed countries. Currently, however, it has made great strides in air transportation, spurred by high economic growth and innovations in aviation technology. On the other hand, the effect of distinctive Japanese aviation policies cannot be neglected, especially the "70/72 Scheme" that greatly shaped the current air transport system.

The 70/72 Scheme's framework was based on the outcome of the cabinet meeting in November 1970 entitled "Approval on Management Regimes of Airline Companies" and the reply by the Minister of Transport, Notification in July 1972. This scheme determined the business spheres of respective domestic airlines by specifically determining the following: Japan Airlines would operate international and major domestic routes that connected Sapporo, Tokyo, Osaka, Fukuoka, and Okinawa-Naha with each other; All Nippon Airways would operate major domestic and local routes; and Toa Domestic Airlines, which is presently known as Japan Air System, would operate domestic local routes. This plan stabilized the management of the three respective companies by cross-subsidization, leading to the expansion of route networks and increases in the number of passengers.

2.1.2. Shift in Aviation Policies

As the airline industry matured, strict regulations became unnecessary. On the other hand, strong, prosperous airlines felt fettered by the regulations. Therefore, they sought to expand into new fields and revitalize business as well as their right to realize them.

In September 1985, the Minister of Transport referred to the Council for Transportation Policy a report entitled "About Basic Policies on Ideal Administrative Systems of Airlines in Japan." The Council submitted its final report in June 1986. The "70/72 Scheme" was abolished in 1986 at a cabinet meeting that agreed with interim reports submitted back in December 1985. This final report in 1986 recommended the adoption of new aviation policies centered on the following three points: (1) a shift to a multi-company operation system of international routes; (2) the promotion of competition on domestic routes; and (3) the complete privatization of Japan Airlines.

The above interim reports were intended to promote competition under flexible administrative management that considered the chronic shortage of slots at Haneda and Itami Airports and the differences in strength among companies. Deregulation in Japan proceeded gradually and realistically.

One example is the measures that promoted competition on domestic routes: first, in 1986, double-track standards were established for routes with annual demands of more than 700,000 passengers, while triple-track standards were established for routes with annual demand of more than 1,000,000 passengers. These standards were lowered in stages in 1992 and 1995 and completely abolished in 1997.

2.1.3. Advent of Policies to Promote Competition

Deregulation of airfare policies proceeded in stages, announcing policies that introduced principles of market mechanism function among companies. First in 1995, a change was made in the "notification procedure" of discount airfares and charges of less than 50%. In 1996, all fares were able to set within $\pm$25% range from the standard operation cost, which introduced a so-called "specified range airfare system." With the abolition of supply and demand adjustment regulations in 2000, all matters related to fares shifted from an approval system to a prior notification procedure. This deregulation in fare system led to diversified airfare settings and promoted healthy competition.

2.2. Civil Aviation Networks and Feasibility of a Hub-and-spoke System in Japan

After deregulation of the American airline industry in 1978, major airline companies rapidly established hub-and-spoke systems that enabled airline companies to centralize crew assignments and equipment maintenance. It also enhanced occupancy ratios and improved operational efficiency.

Such a system, however, requires exclusive securing of slots to reduce transit times, especially during peak times. For this reason, after deregulation airline companies tried to monopolize hub airports in the U.S. (see Table 1, Steven & Clifford, 1995).

In Japanese domestic aviation networks, such a hub-and-spoke system proved impossible for the following reasons:

- National land cannot be expanded. → Spoke routes are not feasible.
- Most air routes are short. The longest route in Japan is only 2,250 km between Sapporo and Naha.
- The number of airlines is small. → There is no competition between airlines.
- There is strong demand by local users for direct flights.
- Many local demands are centered on Tokyo or Osaka.
- Equipment operations are centered on large aircraft. → The operation of smaller airplanes is not feasible for the establishment of a spoke.

Table 1. Passenger Shares by Dominant Airline Companies at American Hub Airports.

Airport	1978		1993	
	Share (%)	Airline	Share (%)	Airline
Atlanta	49.7	Delta	83.5	Delta
Denver	32.0	United	51.8	United
Detroit	21.7	American	74.8	Northwest
Memphis	42.2	Delta	76.3	Northwest
Minneapolis = St. Paul	31.7	Northwest	80.6	Northwest
Pittsburgh	46.7	Allegheny	88.9	US Airways

There are also problems resulting from competition with alternative means of transportation.

- *Shinkansen* and expressway networks have been improved.

Japan does not have the appropriate environment for providing civil aviation services through a U.S.-style hub-and-spoke system.

3. POST-DEREGULATION DOMESTIC AVIATION NETWORK

The following two issues were analyzed to improve understanding of the changes in aviation networks for domestic flights since deregulation. First, the number of passengers and seats were tabulated according to major and minor routes to better understand the overall trends of domestic flights. Next, the behavior of airline companies was studied based on the number of passengers and seats through load factor classification to comprehend fluctuations at local airports in Japan.

3.1. Data Used

The number of passengers and seats was tabulated using the "Monthly Operation and Transportation Results by Scheduled Domestic Airline Routes and Sections" included in the "Annual Report on Air

Transportation Statistics" published by the Ministry of Land, Infrastructure, and Transport. The data included the number of flights in operation, mileage, number of hours in operation, number of passengers and seats, passenger-kilometer and seat-kilometer, seat occupancy ratios, etc. Data were also tabulated by route, airline, and month for all scheduled domestic airline routes. In these statistics, however, transit behavior was divided into two groups, making it impossible to perform analysis by extraction of flight transit.

October flights were tabulated because it is the month that experiences the smallest degree of seasonal fluctuations. In addition, the tabulation covered only flights in daily operation from October 1 to 31.

3.2. Supply and Demand Fluctuations of Domestic Flights

Figs. 1 and 2 indicate the transition of the number of passengers and seats, respectively, between 1992 and 2001 for all domestic routes including island routes. Fig. 1 describes the number of passengers (bar chart; vertical left axis; unit: millions) and the ratio of number of passengers with 1992 assumed to be 100 (line graph; vertical right axis). Fig. 2 indicates the number of seats (bar chart; vertical left axis; unit: millions) and the ratio of number of seats with 1992 assumed to be 100 (line graph; vertical right axis). Major routes and minor routes are described separately.

Fig. 1 indicates that the number of passengers for domestic flights has been increasing. Since 1996 in particular, when the impact of deregulation began to take shape, significant increases have been witnessed on major routes, indicating that demand increase trends shifted from minor to major routes. The increase in demand for minor routes before 1996 was caused by growing size of aircraft with extending runways at local airports. As shown in Fig. 2, supplies from airline companies indicate that the number of seats on major routes drastically increased after deregulation because almost all major routes increased in frequency. On the other hand, for the most part minor routes remained constant.

Based on the above, shifts occurred from minor to major routes in both user supply and demand by airline companies.

3.3. Supply and Demand Fluctuations at Local Airports

This section covers 37 routes arriving in and departing from Tokyo, excluding island routes, among airline routes in continuous operation from

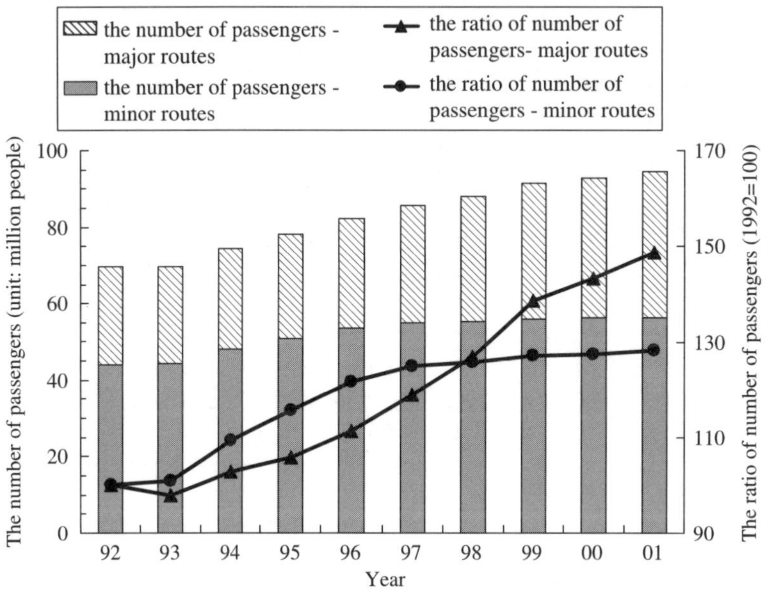

Fig. 1. Numbers and Ratio of Passengers for Domestic Flights in Japan.

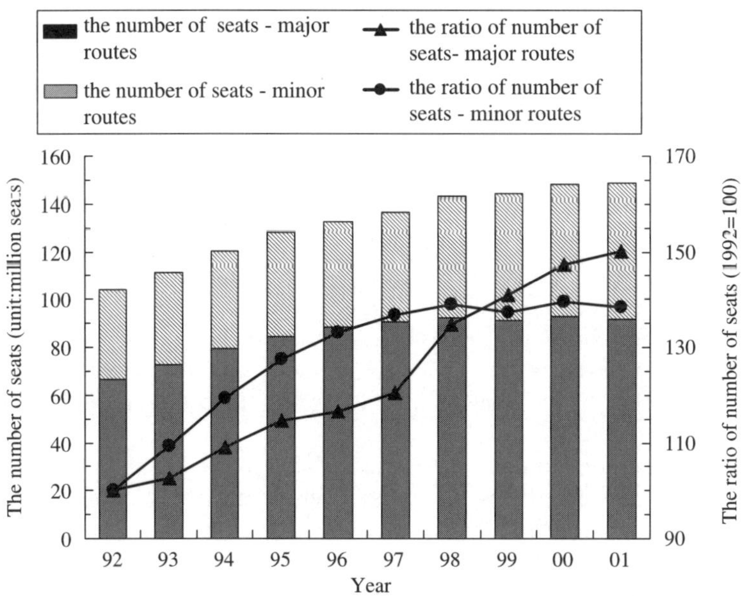

Fig. 2. Numbers and Ratio of Seats for Domestic Flights in Japan.

1997 to 2001. Haneda Airport has the largest number of routes in operation, making it the logical choice to study national trends.

For these 37 routes, load factors by route, the ratio of number of passengers, and the ratio of the number of seats in 1997 were assumed to be 100. The number of flights in operation were compared and verified.

In Japan over the past several years, the average break-even load factor has been about 65% (Institution of Fiscal and Monetary Policy of the Ministry of Finance, 2000). In target routes where load factor exceeded 65%, few declines were observed in the number of seats. For this reason, 65% was considered the break-even point. That is, airline companies are more likely to take actions to improve load factors by reducing the number of seats when the load factor is below the break-even point due to excessive supply. Table 2 classifies routes arriving at and departing from Haneda based on whether the load factor is 65% or higher.

Table 2 indicates those few routes in Hokkaido and Kyushu whose load factor exceeds the break-even point.

3.4. Supply and Demand Fluctuation for Routes Below Break-even Point

Routes below the break-even point can be divided into two groups. The differences are explained below with the typical two routes as examples. One is the route between Haneda and Wakkanai. Wakkanai Airport is located in the northernmost part of Hokkaido. Another is between Haneda and Nemuro-Nakashibetsu. Nemuro-Nakashibetsu Airport is located in eastern Hokkaido (see Fig. 5). The information on these airports is shown in Tables 3 and 4.

Figs. 3 and 4 indicate the ratio of the number of passengers and the ratio of seats using the 1997 load factor data, which is assumed to be 100, for the Haneda–Wakkanai and Haneda–Nemuro-Nakashibetsu routes, respectively.

First, type (A), for example, the Haneda–Wakkanai route, the number of passengers has tended to increase slightly, even though the load factor is below the break-even point. Airline companies, however, have decreased the number of seats. They seem intent to improve load factors and maintain their current routes.

For type (B), for example, the Haneda–Nemuro-Nakashibetsu route, not only is the load factor below the break-even point, but the number of passengers has also decreased drastically. Under such circumstances, it is possible for airline companies to reduce the number of routes or withdraw completely. It has actually been decided that one of the two round trip direct

Table 2. Classification by Load Factors of Routes Arriving/Departing from Haneda Airport.

			Hokkaido	Tohoku, Hokuriku	Kinki, Chugoku, Shikoku	Kyushu
Load factor > 65%	Major routes		New Chitose		Itami Kansai	Okinawa Naha
	Minor routes	Profitable		Toyama Komatsu	Okayama Yamaguchi Ube Takamatsu Kochi	Kitakyushu
		Near the break-even point	Obihiro	Akita Syonai	Tottori Izumo	
Load factor < 65%	Major route					Fukuoka
	Minor routes	(A)	Wakkanai Memambetsu Hakodate		Nanki Shirahama Hiroshima Yonago Tokushima Matsuyama	Oita Kumamoto Kagoshima
		(B)	Nakashibetsu Kushiro Asahikawa	Aomori	Iwami	Nagasaki
		Exception		Misawa Yamagata		Miyazaki

Table 3. The Information of Wakkanai and Nemuro-Nakashibetsu Airports (in 2002).

Airport	Destination	The number of flights (round-trip flight/day) October 2001	The number of passengers (persons/year) 2001
Wakkanai	Tokyo	4	164,026
	Sapporo (New Chitose and Okadama)	6	88,509
Nemuro-Nakashibetsu	Tokyo	4	118,888
	Sapporo (New Chitose and Okadama)	4	108,784

Table 4. Flight Schedule Presented in Questionnaire.

Haneda	→	Asahikawa		→	Haneda
Aircraft	Departure	Arrival	Aircraft	Departure	Arrival
A3	8:30	10:05	*Increase*	*7:20*	*9:05*
A321	9:05	10:40	A3	10:50	12:35
A3R	11:30	13:05	A321	11:35	13:20
A321	13:40	15:15	A3R	14:00	15:45
M90	15:30	17:05	A320	15:50	17:35
A3	17:05	18:40	M90	17:40	19:25
Increase	*20:55*	*22:30*	A3	19:30	21:15
Increase	*21:55*	*23:30*			

flights on the Haneda–Nemuro-Nakashibetsu route will be changed into a nondirect route via the New Chitose Airport.

The aforementioned analyses in this section indicate the following:

- In Kyushu and Hokkaido, where no competition with *Shinkansen* exists, many Haneda routes have load factors below the break-even point.
- For routes in parts of Hokkaido whose load factor is stagnant, there are plans to reorganize the network by consolidating routes into the Haneda–

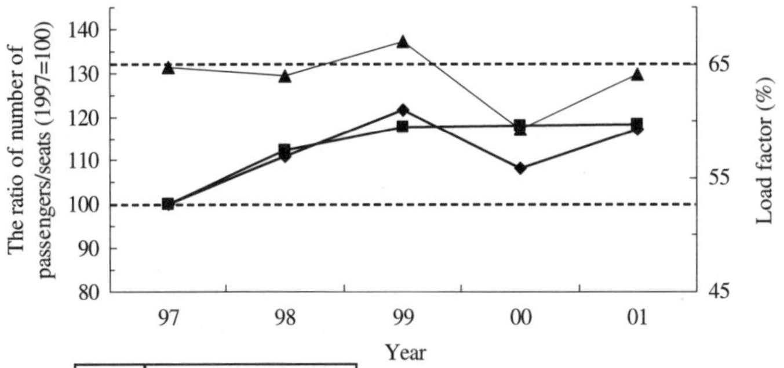

year	the number of flights
1997	4
1998	4
1999	4
2000	4
2001	4
2002	4

Fig. 3. Ratio of Passengers, Seats, Load Factors, and Flights for Haneda–Wakkanai Route.

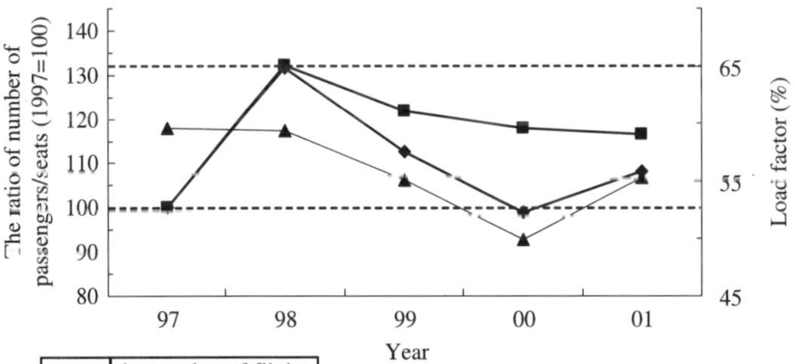

year	the number of flights
1997	4
1998	4
1999	4
2000	4
2001	4
2002	4

Fig. 4. Ratio of Passengers, Seats, Load Factors, and Flights for Haneda–Nemuro-Nakashibetsu Route.

New Chitose route, the world's top trunk route.

4. COMPREHENSION OF DEMANDS OF LOCAL AIRPORT USERS

This section studies the gap between the preferences of local airport users and the levels of service provided by airline companies. A survey was conducted near a local airport to better understand the views of local citizens regarding changes in airline networks. The content of the survey was designed to determine and evaluate the willingness of local citizens to bear the burden of measures to be implemented by airport administrators to revitalize Asahikawa Airport using a contingent valuation method (CVM).

4.1. Reasons for Selection of Asahikawa Airport

The population of Japan is approximately 127 million in an area of 377,873 km^2. Hokkaido, located in the northernmost part of the Japanese Archipelago, has less than 5% of the country's total population; yet its area accounts for over 20% of the country's entire land. Its center is Sapporo, where 32.1% of Hokkaido's population resides. Although the New Chitose Airport was originally developed to serve the urban area of Sapporo, it is currently poised as a gateway to Hokkaido. A particularly key industry in the prefecture is tourism, and over 30% of tourists to Hokkaido use the New Chitose Airport.

Meanwhile, Asahikawa Airport is located approximately 200 km from the New Chitose Airport. Asahikawa Airport was selected for our case study because it was affected by airline deregulation and because air travelers defected to New Chitose Airport. Fig. 5 shows the location of the two airports.

In 2000 at Asahikawa Airport, the annual number of domestic passengers stood at 949,089. Meanwhile, at New Chitose Airport, the annual number of domestic passengers numbered 17.2 million in 2000, which in Japan was second only to Haneda Airport.

First, correlations between changes in the usage conditions of Asahikawa Airport and deregulation policies are verified below. Next, the possibility of airport revitalization by region is examined based on questionnaire results conducted in Asahikawa by the author and others.

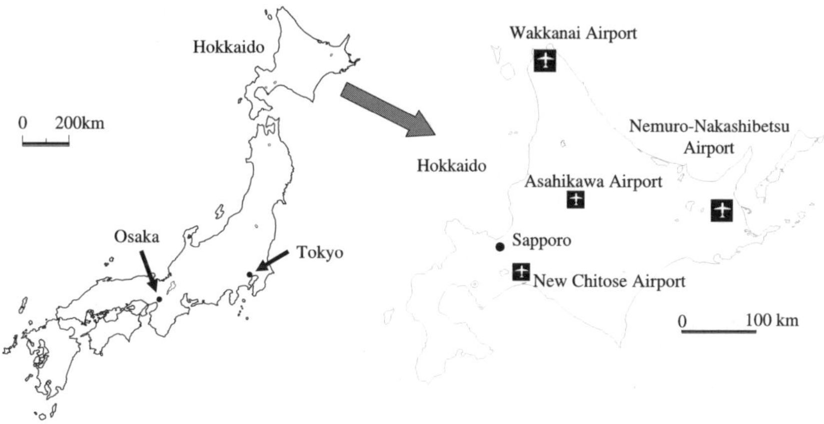

Fig. 5. Locations of Asahikawa and New Chitose Airports.

Asahikawa Airport has experienced stagnant demand due to economic declines in the airport area, causing airlines to reduce flights. In addition, as described in Fig. 6, the load factor is below the break-even point, and demand has also been decreasing. The airport is of the Nakashibetsu-type mentioned in Section 3.

4.2. Environment Surrounding Asahikawa Airport

4.2.1. Competition with New Chitose Airport

Hokkaido International Airlines started a route between New Chitose and Haneda Airports in December 1998. It improved service levels due to reduced fares. In addition, enhanced railroad transportation between Asahikawa and New Chitose Airports is another possible cause of the decreasing number of passengers who use Asahikawa Airport.

4.2.2. Changes in the Number of Passengers Using Asahikawa Airport

Fig. 7 indicates the number of residents of Asahikawa and the surrounding area (from a regional population of 10,000) who selected Asahikawa and New Chitose Airports based on survey results of passengers in 1993 and 1999. Fig. 7 reveals a decrease in the number of passengers who used Asahikawa Airport, from both residents of Asahikawa and the area around Asahikawa Airport. On the other hand, the number of passengers who chose New Chitose Airport clearly rose, reflecting the flow of passengers into that airport.

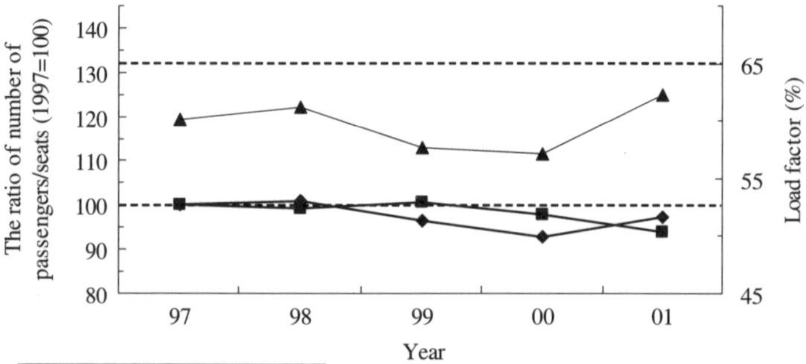

year	the number of flights
1997	20
1998	14
1999	14
2000	14
2001	14
2002	12

Fig. 6. Ratio of Passengers, Seats, and Load Factors for Haneda–Asahikawa Route.

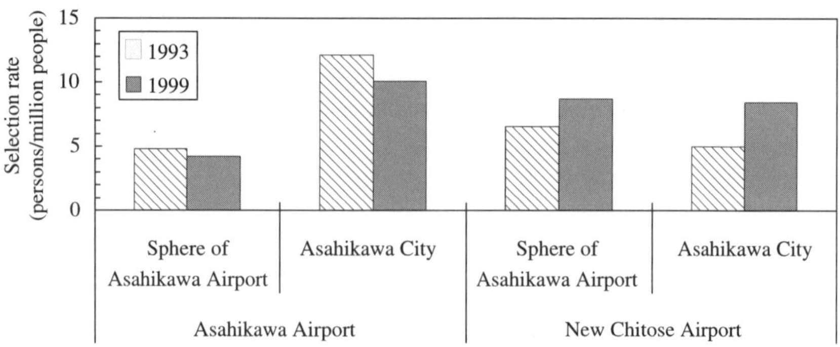

Fig. 7. Selection of Airports by Residents Living near Asahikawa Airport.

4.3. Questionnaire

In December 2001, a distribution and collection method questionnaire was mailed that targeted 60 residents of Asahikawa that appraised local residents' intentions to bear the costs of upgrading Asahikawa Airport. Airport administrators had implemented measures based on a CVM. Fig. 8 shows this questionnaire sheet.

Fig. 9 shows the approval rate of willingness to pay (hereafter WTP) by increase in the number of flights between Asahikawa and Tokyo. Fig. 9 shows that WTP is determined at ¥24,568 ($196.80 ($1 = ¥124.84: the average dollar–yen exchange rate during the survey period)).

4.4. Discussion

The questionnaire revealed that local citizens were willing to pay ¥24,568 ($196.80) for the project. Compared with the airfare between Asahikawa and Tokyo at the time the questionnaire was conducted, i.e. ¥33,500

The following programs are fictitious scenarios. These programs are not real at all.

Here is the program that is managed by a municipal body established in the vicinity of the Asahikawa Airport including the Asahikawa municipal government. The municipal body will promote the use of Asahikawa Airport.

There are six round trip flights from Asahikawa to Tokyo per day in November 2001. There is a program to increase three flights (see Table 4) so that it can be one day trip from Asahikawa to Tokyo. This program can be come to realization soon only if you pay contribution. If not, this program is postponed indefinitely. This contribution should be collected from every household in Asahikawa only once a year.

If this program comes true, the number of flights between Asahikawa and Tokyo increases that you can cut travel time for Tokyo. However, notice that if you pay the contributions then what's left from your income will you limited goods and services.

1. Do you agree to this program?

1) Yes 2) No 3) Cannot decide

2. If your answer is YES in question number 1, please circle your answer in the following table.

willingness to pay (yen)		answer	
1)	0	1. agree	2. oppose
2)	5,000	1. agree	2. oppose
3)	10,000	1. agree	2. oppose
4)	15,000	1. agree	2. oppose
5)	20,000	1. agree	2. oppose
6)	25,000	1. agree	2. oppose
7)	30,000	1. agree	2. oppose
8)	35,000	1. agree	2. oppose
9)	40,000	1. agree	2. oppose
10)	45,000	1. agree	2. oppose
11)	50,000	1. agree	2. oppose
Fill in maximum payable amount (,000)			

Fig. 8. Questionnaire Sheet.

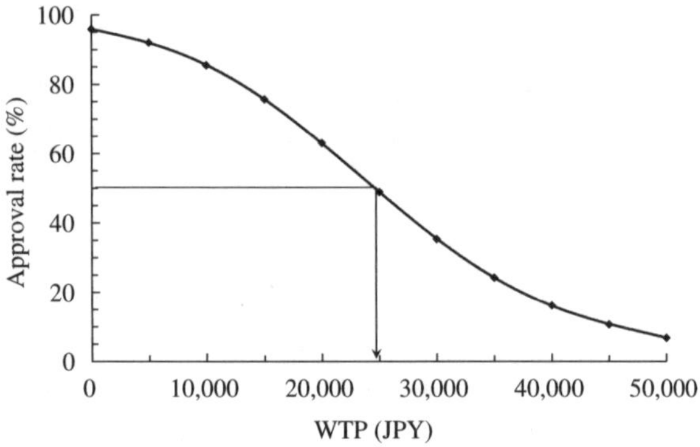

Fig. 9. Willingness to Pay Approval Rate.

($268.34), this amount was fairly high. The results clarified that local citizens were strongly dissatisfied with the present air routes provided by Asahikawa Airport as well as those between Haneda and Asahikawa. In other words, local citizens near Asahikawa Airport did not favor reorganizing the network and making New Chitose Airport into a hub airport.

5. SUPPLY AND FEASIBILITY OF REGIONAL AVIATION IN HOKKAIDO

5.1. Present Transportation Network Situation in Hokkaido

Approximately 1,000 km from the nation's capital, Tokyo, Hokkaido is located at the northern tip of the Japanese archipelago. Although Hokkaido and Honshu are separated by the sea, there are established railways (no *shinkansen* bullet trains, however), a ferry, and civil aviation networks. Civil aviation services account for 85% of the total passenger demand between Hokkaido and Honshu, the main island of Japan.

Administrative, economic, and other central functions are concentrated in Sapporo. On the other hand, in areas besides the metropolis of Sapporo, population is sparsely distributed, resulting in low passenger demand. Therefore, the marketability of public transportation projects has little potential. In fact, automobiles account for approximately 90% of transportation in Hokkaido. Nevertheless, construction of expressways is lagging behind other areas in the nation. In fact, many major cities are not yet completely linked to the expressway network. As the distances covered by intercity traffic tends to increase in Hokkaido, higher transportation costs are required, accordingly.

Hokkaido has 13 airports, including three on outlying islands as shown in Fig. 10. The Japanese government colonized Hokkaido 130 years ago as a defensive stronghold against Russia. At that time Hokkaido was neither a territory of Russia nor Japan, but an independent area inhabited by the Ainu, the island's indigenous people. Many airfields were constructed before World War II, most of which are still used as airports. The challenge we now face is to utilize such a transportation infrastructure to revitalize the region.

5.2. Regional Challenges and Aviation Roles in Hokkaido

Currently, Japan is on the verge of transforming its infrastructure development policies. Conventional policy used to ensure that, regardless of where people lived, they could enjoy a certain degree of urban services, such as administrative, medical, educational, etc. To that end, transportation and infrastructure facilities were designed and constructed to demonstrate urban functions. However, the Japanese people will no longer accept the cost of further construction, increasing the probability that in some areas facilities and transportation infrastructure will remain insufficient.

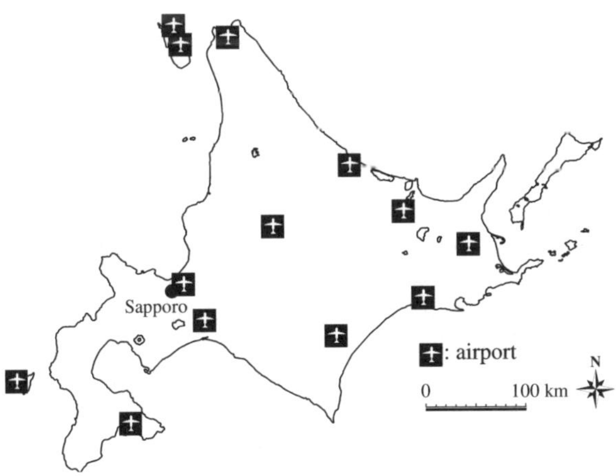

Fig. 10. Location of Airports in Hokkaido.

On the other hand, a globalization trend has been relentlessly sweeping over regional cities. Regions are encouraged to become more independent by promoting competitive local industries. Although Hokkaido is fully capable of independence based on tourism, agriculture, and fisheries, inferior market accessibility has created bottlenecks in some regions. In some of these regions, airports have already been constructed. As mentioned in Section 3, however, it is difficult to say that functions are fully utilized under the existing air transportation system. Therefore, the introduction of a regional aviation system is thought to be a solution.

Based on the above facts, the introduction of regional aviation into Hokkaido has the potential to address the region's transportation woes. The sections below review examples of advanced regional aviation and then study the feasibility of introducing it into Hokkaido.

5.3. Examples of Advanced Regional Aviation

In the United States, "commuter air transportation" was legally defined for the first time in 1969, signaling the start of regional aviation. After deregulation of civil aviation in 1978 in the United States, airlines introduced a hub-and-spoke system network (indicated in Section 2). Such networks significantly developed and expanded regional airlines as a part of spoke routes. Even now, primarily in Europe and the United States, regional airlines play an important role in air transportation. Over the past decade in particular, the introduction of regional jet services has progressed in various countries, enabling regional airlines to make great strides.

In the United States, the airlines own more than 10% of all small aircraft. Between 2001 and 2002, 123 additional regional jet air routes were added, approximately half of which were new routes. The development of regional airlines has improved civil aviation networks. Such airlines as Atlantic Coast Airlines and American Eagle, for example, operate routes that also serve as feeders for mega carriers. Moreover, these routes have maintained a steady demand since September 11, 2001 and are still growing. In Europe, routes operated by regional jets have continued to expand at various airports. For example, at Munich Airport in Germany, regional jets operate on 43 of the 209 routes; of 640 flights, 195 are by regional jets, with an average flight distance of approximately 700 km.

In particular, this section cites Cross Air (presently Swiss International Airlines) as an example of a European regional airline. Based in Zurich, Cross Air was launched in 1978 by an airline operating on three routes.

Since 1982, the company has expanded its network by adding Basel, Switzerland to its base. Around 1990, the company also assumed the role of a feeder to Swiss Airlines, achieving further development. However, the major target of market development was the niche market. Compared with the United States, where regional airlines developed as spoke routes in a hub-and-spoke system, in Europe such regional airline network formation is based on a round-robin system. That is, the company ventured into routes that no other airline operated and developed the market. Initially, Cross Air introduced a hub-and-spoke system by regional jets. The realities were, however, that demand for transit was not generated because an "economy of scale by mass transportation between hubs," which is a salient feature of a hub-and-spoke system, did not occur (Yai, 2003, pp. 23–25). This is also the reason that such a hub-and-spoke system failed in Japan as stated in Section 2.2. Furthermore, the aircraft operation method is characteristic of a network expansion strategy. Since flights usually begin at a hub airport, the first flights at the other airports depart later. Cross Air, however, operated the first flights from both airports simultaneously, enabling both regions to equally enjoy transportation links. Demand was also established around business passengers. Therefore, they made profits even when regional jets were used, whose operational costs per passenger are high.

5.4. Feasibility of Introducing Regional Aviation into Hokkaido

As stated in Section 5.2, the introduction of regional aviation into Hokkaido will support regional independence from a transportation aspect. Based on Section 5.3, however, the aforementioned introduction faces the following two major challenges:

(1) Since a hub-and-spoke system network is difficult to form in Japan, would it be possible to establish a European-style round-robin system network? That is, is it possible for regional airlines to create new demand by developing a niche market and playing the role of feeder to mega carriers?
(2) Will business users, who are willing to pay slightly high airfares, become the mainstay?

These two issues are matters of revenue and expenditure. One airline has overcome these issues and created a successful operation in Hokkaido: a commuter air service company named Hokkaido Air System (hereafter HAC).

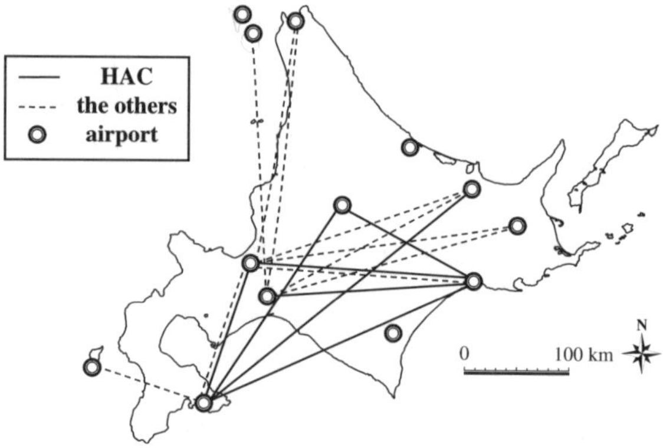

Fig. 11. Air Transportation Network in Hokkaido (as of February 2004).

Fig. 11 shows the current air transportation network operated by HAC and the other airlines in Hokkaido.

HAC began operations on three routes in Hokkaido in 1998, using one SAAB 340 aircraft. Currently, the company operates three SAAB 340 aircraft on six routes in Hokkaido. HAC became profitable in 1999, admittedly aided greatly by local government subsidies. The local government made a 49% investment when the airline was established. It also provided subsidies for the initial purchase of aircraft and other related expenses. However, revenue grants were not provided; after inauguration of services, management on market mechanisms has been conducted. As a result, HAC developed a niche market and improved transportation services as a whole.

Section 4 revealed that local residents were willing to pay for necessary upgrades to civil aviation services to improve flight frequencies and schedules, according to the aforementioned analysis. Since the survey subjects were different, a conclusion cannot be made without thorough consideration, but it may be possible to introduce a regional aviation in a form that satisfies the people's expense.

6. CONCLUSION

This research has made the following conclusions:

(1) The establishment of a U.S.-type airline network based on a hub-and-spoke system is not yet feasible in Japan.

(2) After deregulation, airline companies tended to change the number of available seats according to load factors. In fact, even for routes whose load factors were below the break-even point, decisions to maintain or reduce routes were based on demand fluctuations.
(3) CVM surveys showed a wide gap between the level of services desired by local airport users and those actually provided by airline companies.
(4) On the other hand, local residents were willing to pay to improve civil aviation services. Therefore, the introduction into Hokkaido of a regional aviation network is feasible that satisfies people's expenses.

Modeling is scheduled in the future to quantitatively comprehend airline behavior after deregulation. In addition, communities in which local airports are located are going to propose frameworks that will enable airport operations to reflect the wishes of passengers.

REFERENCES

Institution of Fiscal and Monetary Policy of the Ministry of Finance (2000). *Japanese economical efficiency and policy to recover*. Tokyo.

Steven, A. M., & Clifford, W. (1995). *The evolution of the airline industry*. Washington, DC: The Brookings Institution.

Tozaki, H. (1995). *Aviation deregulation*. Tokyo: Keiso Syobo.

Yai, T. (2003). New development of regional aviation – future of air commuter transportation and airports situation. *Symposium on regional aviation*. Nagoya, Japan: Regional Aviation Promotion Organization.

AIR SERVICE DEVELOPMENT OF LOCAL AIRPORTS AND ITS INFLUENCE ON THE FORMATION OF AVIATION NETWORKS

Hideyuki Kita, Atsushi Koike and Keishi Tanimoto

ABSTRACT

This chapter focuses on strategic actions of local governments to enrich aviation services to their local airports, and analyzes the influence of these actions on the formation process of aviation networks. Air service development action gives an influence not only on the airport but also on other airports because of the existence of network externality. This study models such an interdependent and path-dependent situation as a repeated n-person non-cooperative game. Numerical analysis shows that competition between local governments using subsidies for petition flights forms a regional hub endogenously under certain conditions.

1. INTRODUCTION

In the competitive situation of the air transportation market, mega-carriers are eager to attract users by supplying lower fares and convenient services

Global Competition in Transportation Markets: Analysis and Policy Making
Research in Transportation Economics, Volume 13, 233–249

ISSN: 0739-8859/doi:10.1016/S0739-8859(05)13011-X

based on bundling transport between major airports and rich feeder services. Enriching feeder service to/from local airports at a low cost is essential for this purpose. However, low-cost carriers are eager to attract users by offering convenient point-to-point service with lower fares. Developing aviation services that connect local airports at a lower cost is also essential to them. Since major airports tend to be congested, the role of local-to-local aviation services that bypass major airports becomes more important. That is to say, developing new connections to local airports at a low cost is a key issue for airlines.

It is not only a key issue from the airlines' viewpoint, since enriching local-to-local service is also a big interest of local governments that have local airports. From the viewpoint of such local governments, it is desirable that their airports have much richer aviation services to increase the level of welfare. Many local governments are eager to take a range of actions to develop air services.

This being the case, it is necessary to take into consideration the behavior of local airports or local governments in air service development, in addition to the airlines' behavior, for understanding future transitions of aviation networks. However, few studies have discussed the influence of air service development actions of local governments on the formation of aviation networks. This study focuses on the air service development actions of local governments, and models the network formation process by taking network externality and the interdependency of development actions into consideration.

2. MODELING APPROACH

There exists a rich accumulation of studies that deal with the nature of an air transport network and its formation process (see e.g. Hendricks, Piccione, & Tan, 1999; Button & Stough, 2000). O'Kelly (1987) pointed out that the location of a hub is determined endogenously, and Campbell (1996) proved that several hubs are generated through a self-organizing process on a network having a scale of economy. Brueckner and Spiller (1991) analyzed the characteristics of a hub-and-spoke network from the point of view of airline competition, and obtained some findings by comparing the total surplus of aviation services on a hub-and-spoke-type network in cases of both monopoly and competitive conditions, and direct and connecting services. Hong and Harker (1992) studied the influence of frequency, price, and slots on the network equilibrium. Panzar (1979) discussed equilibrium and welfare in the air transportation market. Oum, Zhang, and Zhang (1996) investigated the influence of airport pricing at a major airport and some feeder airports under several pricing rules.

In many airports, especially in Japan, even though the development of facilities has been carried out at an adequate level, they do not necessarily completely fulfill their transportation functions, due to poor air services connecting them with other airports. Local governments that operate these airports are eager to increase the levels of activity and social welfare of the jurisdictions that may be carried by the enriched air services. However, it is not easy to develop a new aviation service. For example, in Japan, there exists a competitive situation among local governments, in air service development, to obtain a higher number of flights and new air routes, because of capacity restraints at major airports, such as Haneda, Tokyo International Airport.

Because of network externality, i.e. changing fare or frequency of a route affects demand, not only on that route but also on other routes, air service development action by a local government has an influence on other local governments, and vice versa. This means that there exists an interdependency between air service development actions by local governments. Hence, the resultant level of services due to an action varies depending on the actions of other local governments. Each local government acts to maximize their level of welfare by taking this fact into account.

Air service development is not a one-shot action, but one that is repeated for a long time for enriching air services. There exists also a path-dependency in developing air service, in that earlier actions restrict the capacity to modify the following actions. For investigating the transition of aviation network, it is necessary to evaluate the formation process over the long run.

With this as its background, this study considers the above situation to be a game among local governments that seek more flights and new air routes, and models the transition of network formation resultant from air service development actions by local governments as a repeated n-person non-cooperative game. We model the behavior of local governments that is often seen in reality, of giving incentives to airlines for starting new flights and air routes by offering subsidies to a certain extent. In the modeling, the difference between major airports facing capacity restrictions and local airports with no capacity restrictions is taken into consideration in order to analyze the influence of capacity restrictions at major airports.

This study analyzes the above situation by using the partial equilibrium analysis to focus our attention on their strategies in the air transport networks based on the earlier works by Kita, Sakata, and Yoshimura (1999) and Kita, Sakata, and Tanimota (2000). The payoff of each local government is defined as the difference between the increment of equivalent variation (EV) of consumers in the jurisdiction and the amount of subsidy for petition flights. EV will be estimated by using the network equilibrium model based on the description of user and airline behavior. Though this model takes into account the competition between air and railway service, the fare and frequency of railway service are fixed for focusing on the formation process of a transportation network. Each of air and railway transportation markets is assumed to be monopolistic, for simplicity.

3. THE MODEL

3.1. Player and Strategy

Denote players, local governments that hold airports, as $\boldsymbol{I} = \{1,\ldots,i,\ldots,j,\ldots,I\}$, the set of strategies of each player $i \in \boldsymbol{I}$ as $\boldsymbol{g}_i = \{g_{i1},\ldots,g_{ii-1},g_{ii+1},\ldots,g_{ij},\ldots,g_{iI},g_{ih}\}$, and the set of strategy sets as $\boldsymbol{G} = \{\boldsymbol{g}_1,\ldots,\boldsymbol{g}_i,\ldots,\boldsymbol{g}_j,\ldots,\boldsymbol{g}_I\}$, where subscript h means the major airports that do not take air service development action. A strategy set of player i to player j, g_{ij}, can be written as $g_{ij} = \{2x_{ij}\}$, where x_{ij} is the number of round trip petition flights between i and j. In the case of increase in the number of flights, there exists the following restriction on the strategy in each period of the game because of the limited availability of aircraft operation:

$$\sum_{j \in K} g_{ij} \leqslant R, \quad g_{ij} \in g_i \text{ and } g_{ij} \geqq 0 \tag{1}$$

where R is the maximum number of incremental flights in each period. There is no restriction in the case of reducing the number of flights.

3.2. The Behavior of Users

A user chooses a route, ϕ_{ij}, and a number of trips, $X^m_{\phi_{ij}}$, which maximize the user's utility determined by income, y_i , and the generalized cost, $p'^m_{\phi_{ij}}$. Generalized cost consists of fare, $p^m_{\phi_{ij}}$, frequency, $F^m_{\phi_{ij}}$, and travel time, $T^m_{\phi_{ij}}$. The desired time of trip generation is assumed to be uniformly distributed over the range of [0, E], and the route for a return trip is assumed to be the same as the outward trip.

Define the indirect utility function of a resident at the city i as follows:

$$V_{ij} = \sum_{m \in M} \ln\left(\mu \frac{D_j}{p^m_{\phi_{ij}}} + \nu y_i\right) \tag{2}$$

where D_j denotes the socio-economic factors of the destination city j, M the set of the transportation services m, and μ and ν are parameters. The generalized cost of the trip from i to $j, p^m_{\phi_{ij}}$, along the optimal route, ϕ_{ij}, is formulated as

$$p^m_{\phi_{ij}} = \sum_{\phi_{ij}} P^m_{\phi_{ij}} + \alpha \sum_{\phi_{ij}} \left(T^m_{\phi_{ij}} + \frac{E}{\min\ F^m_{\phi_{ij}}}\right) \tag{3}$$

where $\sum_{\phi_{ij}} P^m_{\phi_{ij}}$ denotes the sum of the fares of all the links forming the route ϕ_{ij}, and $\sum_{\phi_{ij}} T^m_{\phi_{ij}}$ the sum of the travel time on these links. The third term denotes the average waiting time of a passenger at the airports, where $\min\ F^m_{\phi_{ij}}$ is the minimum frequency of service on the route. $E/\min\ F^m_{\phi_{ij}}$ is the average waiting time, and E the operating hours in a day. α is a parameter to the passenger's value of travel time.

The demand function for the service between the cities i and j is given as

$$X^m_{\phi_{ij}} = \frac{\mu D_j N_i}{\nu (p^m_{\phi_{ij}})^2} \frac{(\mu D_j / p^m_{\phi_{ij}} + \nu y_i)^{-1}}{\sum_{m \in M} (\mu D_j / p^m_{\phi_{ij}} + \nu y_i)^{-1}} \tag{4}$$

where N_i denotes the population of the city i.

The link demand function is formulated as

$$X^m_{ij} = \sum_{k \in K} \sum_{l \in K} \delta^m_{ijkl} X^m_{\phi_{kl}} \tag{5}$$

where K denotes the set of cities. δ^m_{ijkl} is a dummy variable describing whether the optimal route between k and l passes the link between i and j or not, and is defined as follows:

$$\delta^m_{ijkl} = \begin{cases} 1 & \text{for passing} \\ 0 & \text{otherwise} \end{cases} \tag{6}$$

3.3. The Behavior of the Airline

Airlines are assumed not to change their fares or aircraft types, and to select the frequency F^n_{ij}, which maximizes their profit for a given aviation network.

The profit maximization problem of airline n can be formulated as

$$F^n_{ij} = \arg\max_{F^n} \sum_{i\in K}\sum_{j\in K}(P^n_{ij}X^n_{ij} - C^n_{ij} - C^n_F)$$
$$\text{s.t.}\quad X^n_{ij} \leqslant F^n_{ij}S_{ij} \tag{7}$$

where P^n_{ij} denotes the airfare, S_{ij} the passenger capacity of aircraft, C^n_{ij} the variable cost, and C^n_F the fixed cost of the flight from i to j, respectively.

The variable cost, C^n_{ij}, has the following structure, similar to a Cobb–Douglas-type production technology, with the distance L_{ij} and the frequency F^n_{ij},

$$C^n_{ij} = \sigma(L_{ij})^{\tau}(F^n_{ij})^{v^n_{ij}} \tag{8}$$

where σ and τ are parameters, and v^n_{ij} is given as follows:

$$v(F^n_{ij}) = \rho_1(F^n_{ij})^2 + \rho_2 F^n_{ij} + \rho_3 \tag{9}$$

where ρ_1, ρ_2, and ρ_3 are parameters.

Because the demand, X^n_{ij}, in the profit maximization problem of airlines shown in Eq. (7) includes the generalized cost of other airlines and the railway company, each airline takes the optimal behavior for the conditions of other transportation services.

3.4. Subsidy for Petition Flights

In a case where the profit of the airline company on the route ij is reduced resulting from an action for air service development by a local government, the amount of subsidy, PS_{ij}, for route ij to the airline paid by the local government concerned, is

$$PS_{ij} = -(P_{ij}X_{ij} - C_{ij}) \tag{10}$$

and $PS_{ij} = 0$ for the increase in the case, where P_{ij} is the fare between ij, X_{ij} the number of passengers between ij, and C_{ij} the variable cost for supplying air service for route ij.

In a case where both the local governments simultaneously wish to have new flights, both the governments share the subsidy to the airline. We assume that the shared amount of subsidy is based on the requested number of flights f_i and f_j by each government. In case of $f_i > f_j$, the costs for petition flights c_i and c_j are given as follows:

$$c_i = PS^n_{ij} \frac{\max[f_i, f_j] - (1/2)\min[f_i, f_j]}{\max[f_i, f_j]} \quad (11)$$

$$c_j = PS^n_{ij} \frac{(1/2)\min[f_i, f_j]}{\max[f_i, f_j]} \quad (12)$$

3.5. Payoff

For focusing on the feasibility of petition flights, the level of welfare in each jurisdiction is evaluated in the EV of passengers in changing aviation service.

Indirect utility function of a passenger is

$$V_{ij} = \sum_{m \in M} \ln\left(\mu \frac{\mathrm{D}_j}{p^m_{\phi ij}} + v y_i\right) \quad (13)$$

where, $m \in M$ is mode, V_{ij} is utility level, D_{ij} is destination weight, $p^m_{\phi ij}$ is generalized cost, and μ, v are parameters.

From the definition of EV,

$$V_{ij}(p^{m\mathrm{wo}}_{\phi ij}, y^{\mathrm{wo}}_j + \mathrm{EV}) = V_{ij}(p^{m\mathrm{w}}_{\phi ij}, y^{\mathrm{w}}_j) \quad (14)$$

where, superscripts w and wo mean with and without, respectively.

Substituting Eq. (14) into Eq. (13),

$$\sum_{n \in M} \ln\left(\mu \frac{\mathrm{D}_j}{p^{m\mathrm{wo}}_{\phi ij}} + v(y^{\mathrm{wo}}_j + \mathrm{EV})\right) = \sum_{m \in M} \ln\left(\mu \frac{\mathrm{D}_j}{p^{m\mathrm{w}}_{\phi ij}} + v y^{\mathrm{w}}_j\right) \quad (15)$$

Assume $y^{\mathrm{wo}}_j = y^{\mathrm{w}}_j$, and extend it only on the routes with changing fares,

$$\mathrm{EV}^m_{ij} = \frac{\mu}{v} \mathrm{D}_j \left(\frac{1}{p^{m\mathrm{w}}_{\phi ij}} - \frac{1}{p^{m\mathrm{wo}}_{\phi ij}}\right) \quad (16)$$

Here, considering $p^{m\mathrm{wo}}_{\phi ij} = \infty$,

$$\mathrm{EV}^m_{ij} = \frac{\mu}{v} \mathrm{D}_j \left(\frac{1}{p^{m\mathrm{w}}_{\phi ij}}\right) \quad (17)$$

Then, the following will be obtained by aggregating it over the jurisdiction.

$$\mathrm{EV}_j = \sum_{m \in M} \sum_{j \in J} \xi_{ij}^{m} \frac{\mu}{\nu} \mathrm{D}_j \left(\frac{1}{p_{\phi ij}^{m\mathrm{w}}} \right) \tag{18}$$

where ξ_{ij}^{m} is a dummy variable to describe whether the destination becomes reachable,

$$\xi_{ij}^{m} = \begin{cases} 1 & \text{reachable} \\ 0 & \text{unreachable or no change} \end{cases}$$

3.6. Air Service Development Actions of Local Governments

Local governments try to maximize the sum of payoff in each period in air service development. Uncertainty in forecasting future environmental change will be taken into account by introducing a discount rate, δ, in evaluating the total amount of payoff. We assume that there exist two types of local government in their strategies; governments that behave myopically, and governments that behave by optimizing their actions for n periods. In a case in which multiple equilibria exist, the latter-type government, seeking long-run optimization, can select an equilibrium to maximize total payoff during n periods by paying the difference between the payoff under the desired equilibrium and others to the concerned local government:

The total payoff, $\Pi_{i'}$, is given as

$$\Pi_{i'} = \sum_{t=1}^{n} \Pi_i^t \delta^{t-1} - Q \tag{19}$$

where Q denotes the sum of payment shown in Eq. (22) to the concerned local governments for selecting the best equilibrium for them,

$$Q = \sum_{t=1}^{n} \varsigma^t q_i^t \delta^{t-1} \tag{20}$$

ς^t is a dummy variable shown below,

$$\varsigma^t = \begin{cases} 1 & \text{multiple equilibria exist in } t\text{th period} \\ 0 & \text{no multiple equilibria exist in } t\text{th period} \end{cases} \tag{21}$$

q_i^t is the payment by local government i in tth period to make other local government take an action that realizes the desired equilibrium of i:

$$q_i^t = \sum_{j \in I,\ i \neq j} \{\max[\pi_j^t] - \pi_j^t\} \tag{22}$$

where $\max[\pi_j^t]$ is the maximum payoff of local government j among multiple equilibria in the tth period, and π_j^t is the payoff of government j under the desired equilibrium of government i.

Air service development actions of local governments have interdependency with each other; Assume that local governments choose their actions corresponding to a Nash equilibrium.

4. NUMERICAL ANALYSES

4.1. Setting

In this section, we investigate the process of network formation resulting from air service development actions by local governments with and without forecasting. The aviation network used for this analysis connects four cities, a metropolitan city like Tokyo and three local cities. Railway service is available for all city pairs. The fare and the frequency of railway service for all city pairs are fixed. The fare–distance relationship and the time–distance relationship of aviation services are estimated from statistical data (see, Kita, 1999). The length of time period for each game is set to 1 year, and the game is repeated till equilibrium is reached. Assume that only one local government, airport 4, has long-term prediction ability, and that the other three airports do not have it and behave in a myopic manner. Assume that the maximum number of incremental flights in each period is six, and airport 1 can be reconsidered once in four periods because of capacity restriction. The parameters of demand function and cost function are estimated from statistical data, and set as $\mu = 0.230$, $\nu = 2.371$, $\sigma = 3751.4$, $\tau = 0.808$, $\rho_1 = 0.040$, $\rho_2 = -0.498$, and $\rho_3 = 3.020$.

4.2. Numerical Example 1

Fig. 1 shows the dimension of the hypothetical network for analysis, and Fig. 2 shows the population of the service area of each airport and the initial number of flights.

Fig. 3 shows a part of the analysis. The symbol "•" shows the airport that developed a new air service, and the number beside the symbol means the number of incremental flights. In this case, multiple equilibria occurred in the 2nd period. We can find a result that shows that having formation process 1 in the 2nd period carries a higher payoff in the short run; however, selecting formation process 2 carries a higher total payoff in the long run.

4.3. Numerical Example 2

The network formation process is affected by the population of the service areas of local airports. For the hypothetical network shown in Fig. 4, we analyzed two cases in the combinations of population; case 1 = {10,000, 300, 600, 600} and case 2 = {10,000, 300, 400, 1,000}, where the numbers within parentheses show the population of cities 1–4 in thousands. In both the cases, development actions fall into a stable situation in the 6th periods. However, in case 2, the number of flights becomes 14 each among cities 1–2, 1–3, and 1–4, and 12 between cities 3 and 4, which is different from the stable situation of the formation process 3 in case 1 (Fig. 5). Many similar results tell us that the network formation process is rather sensitive to the population of service areas. From another analysis, we got a result that the network formation processes are identical in formation processes 1 and 3;

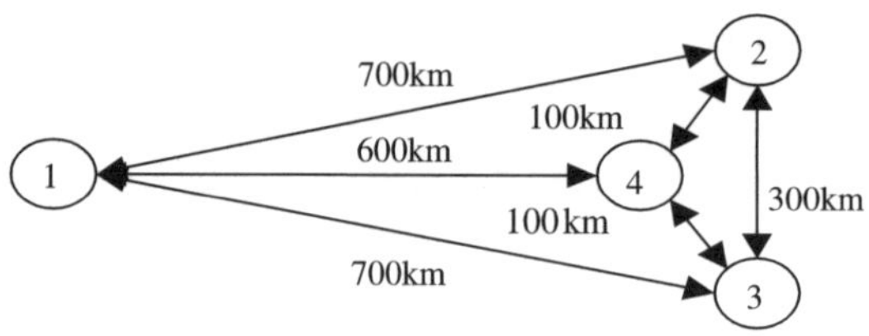

Fig. 1. Hypothetical Network for Numerical Example 1.

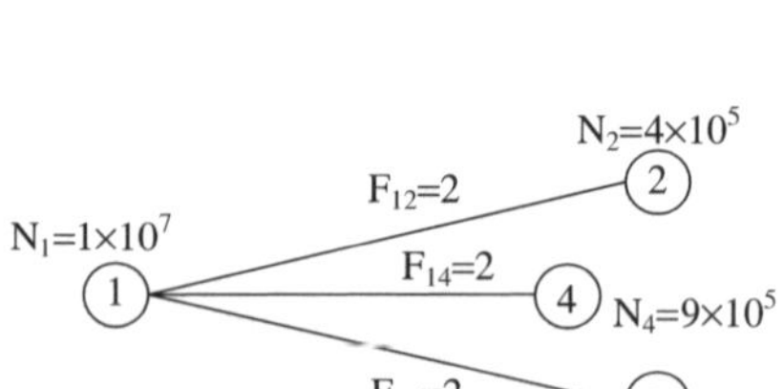

Fig. 2. Population and the Initial Number of Flights.

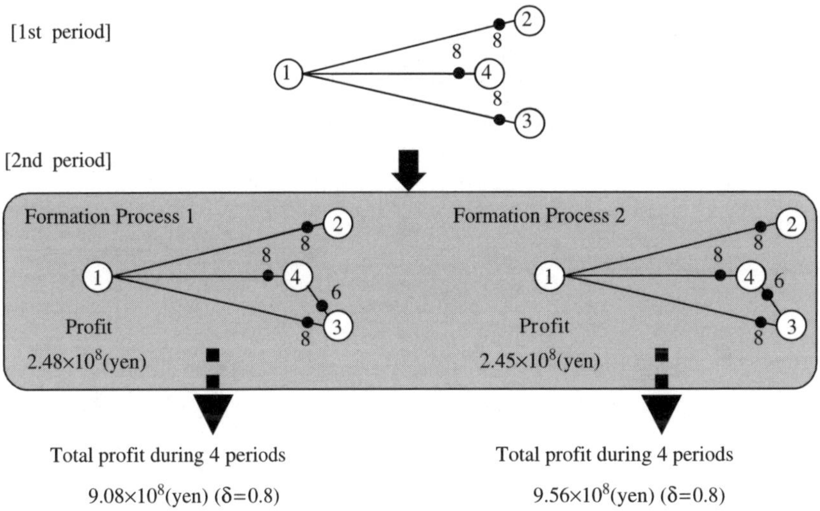

Fig. 3. Network Formation Process in Numerical Example 1.

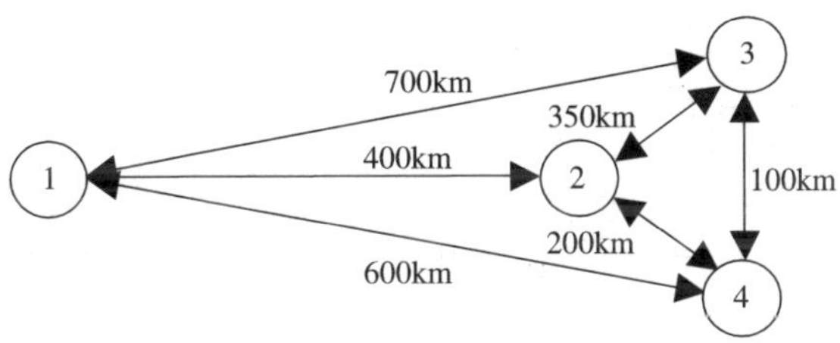

Fig. 4. Network for Numerical Example 2.

however, the total amount of payoff in formation process 3 during all the periods is bigger than formation process 1. This implies a possibility of increase in the level of service by selecting the route to be developed appropriately, and the importance of selecting network formation process by making another local government choose the best equilibrium for the local government having long-term forecasting ability.

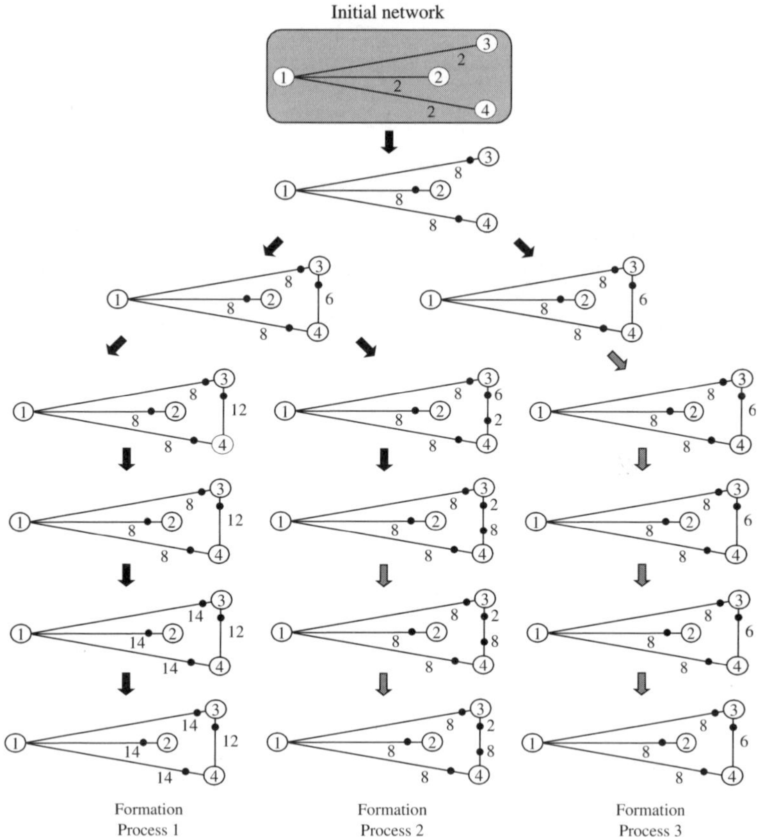

Fig. 5. Network Formation Process in Numerical Example 2 (case 1).

4.4. Numerical Example 3

The result of air service development actions is affected also by the location of other local airports and the distance to the major airport. The networks shown in Fig. 6 are identical in the distances between local airports and differ only in the distances between major and local airports. Fig. 7 shows

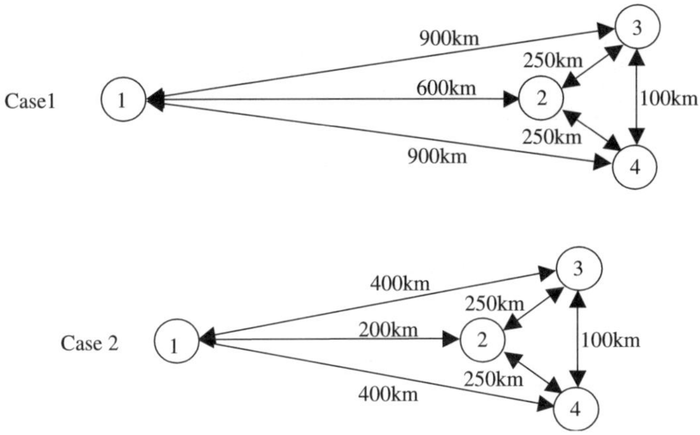

Fig. 6. Hypothetical Network for Numerical Example 3.

the network formation processes of each network. This result demonstrates that the shorter the distance to the major airport, the greater the convenience of having direct routes. The longer the distance to a major airport, the greater the advantage in connecting airports indirectly via another airport, that is to say a regional hub system. In such a case, the advantage as a regional hub airport depends greatly on the location of other airports and the population of the service area of the airports. Through this analysis, it can be recognized that the network formation process is strongly dominated by the strategy for air service development of each airport.

4.5. Numerical Example 4

In this section, we analyze the influence of insufficient capacity of the major airport on the behavior of air service development by local governments. Suppose an aviation network that consists of a major airport at the center and three local airports that are located surrounding the major airport (Fig. 8). The network configuration, initial number of flights based on the profit maximization behavior of the airline company, and the population of service areas of the airports are shown in Fig. 9.

The estimated network formation process is shown in Fig. 10. Under the setting conditions, no multiple equilibria occurred in any period of development. Where there was no capacity restriction at the major airport, the formation process became stable after three periods. If there was existing

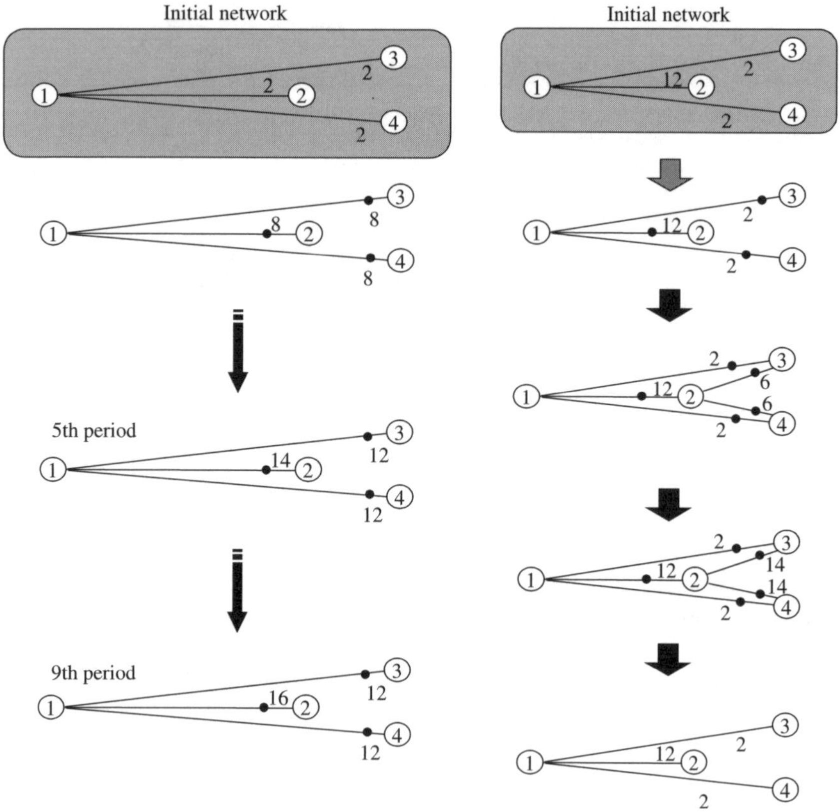

Fig. 7. Network Formation Process in Numerical Example 3.

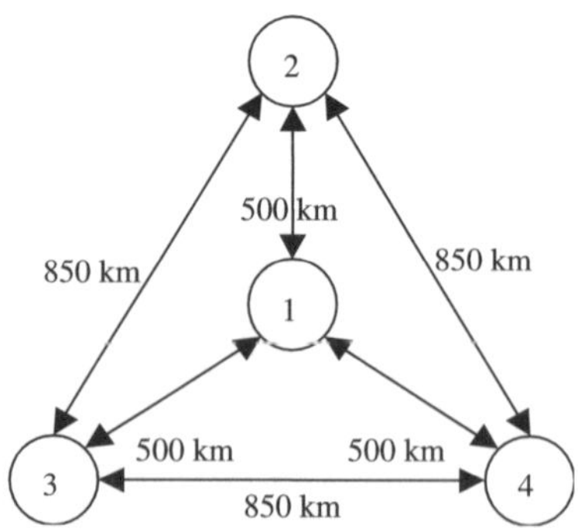

Fig. 8. Network for Numerical Example 4.

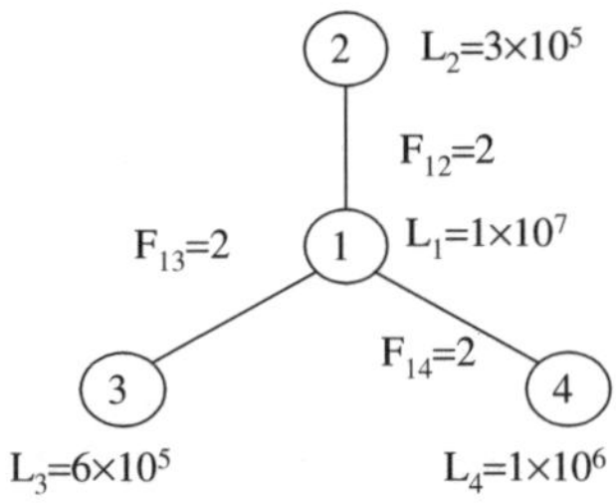

Fig. 9. Initial Condition of Flights in Numerical Example 4.

capacity restraint at the major airport, no action for air service development between local airports was made in the periods when development of new flights to the major airport could not be started. No change was observed in the shape of the network and the number of flights. Network changes occurred only in the periods when air service development to the major airport commenced. Though the final shapes of both the networks are identical,

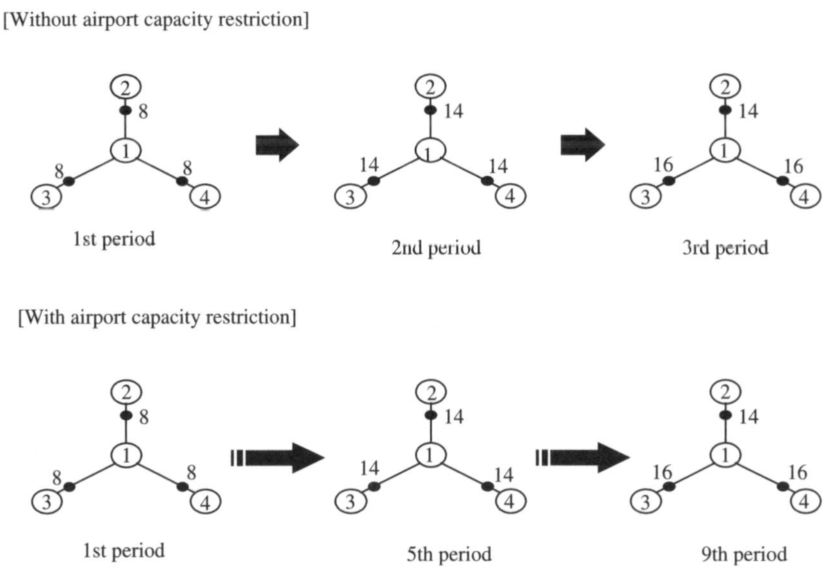

Fig. 10. Network Formation Process in Numerical Example 4.

development under capacity restraint took three times longer than in cases of no capacity restraint. From this result, it can be said that the existence of capacity restraint becomes a big obstacle for local governments in taking action for development of air service.

5. CONCLUSIONS

This study analyzed air service development actions by local governments that have local airports, by using *n*-person non-cooperative repeated game based on demand–supply equilibrium model. Numerical analysis shows that competition between local governments using subsidies for petition flights forms a regional hub endogenously under certain conditions. In addition, the following findings are obtained through numerical examples:

(1) It is effective to take strategic action based on forecasting and evaluation of the aviation network formation process for enriching air service for local airports.
(2) Myopic actions may lead us to a situation of not being able to come back onto the optimal path.
(3) Capacity restriction of major airport may slow down the network formation process of local aviation services and become a barrier under certain conditions, implying that capacity expansion of major airport may help to vitalize local aviation services.
(4) The role of equilibrium selection, given the existence of multiple equilibria, implies the importance of cooperation between airports.

However, many assumptions and simplifications remain in this study. We assumed that the strategy of local government consists of the number of petition flights only, and the type of aircraft and airfare are fixed. Changing these factors is also an important strategy, and should be taken into consideration in the analysis. Payment of a subsidy to an airline by the government may reduce the utility level of the region. The effect of this reduction is considered small enough compared with the scale of government finance. However, conducting applied general equilibrium analysis should be considered for evaluation by measuring the effect of the payment on the welfare. Assumptions on the monopolistic situation of the aviation market and the prediction ability of local governments should also be reconsidered in further studies.

REFERENCES

Brueckner, J. K., & Spiller, P. T. (1991). Competition and mergers in airline networks. *International Journal of Industrial Organization, 9*, 323–342.

Button, K., & Stough, R. (2000). *Air transport networks – Theory and policy implications*. Northampton, MA: Edward Elger.

Campbell, J. F. (1996). Hub location and the *p*-hub median problem. *Operations Research, 44*(6), 923–935.

Hendricks, K., Piccione, M., & Tan, G. (1999). Equilibria in networks. *Econometrica, 67*(6), 1407–1434.

Hong, S., & Harker, T. (1992). Air traffic network equilibrium; toward frequency, price, and slot. *Transportation Research, 26B*, 307–323.

Kita, H. (1999). A model analysis of flight subsidization for sustaining local aviation services. *Proceedings of the international symposium on local knowledge and innovation: Enhancing the substance of non-metropolitan regions*, MARG, pp. 279–288.

Kita, H., Sakata, H., & Tanimoto, K. (2000). A model analysis of aviation services with multi-sector flights. *Infrastructure Planning Review, 17*, 67–74 (in Japanese).

Kita, H., Sakata, H., & Yoshimura, S. (1999). A feasibility analysis of subsidized flights for local aviation services. *Infrastructure Planning Review, 16*, 195–200 (in Japanese).

O'Kelly, M. E. (1987). A quadratic integer program for the location of increasing hub facilities. *European Journal of Operations Research, 32*, 393–404.

Oum, T. H., Zhang, A., & Zhang, Y. (1996). A note on optimal airport pricing in a hub and spoke system. *Transportation Research*, 30(1)-B, 11–18.

Panzar, J. C. (1979). Equilibrium and welfare in unregulated airline market. *American Economic Review, 69*, 92–95.

PART II: MARITIME TRANSPORT

WHAT FUTURE THE MARITIME SECTOR?: SOME CONSIDERATIONS ON GLOBALISATION, CO-OPERATION AND MARKET POWER

Eddy E. M. Van de Voorde

ABSTRACT

Competition between maritime companies is fierce, as, for that matter, is competition between port authorities or indeed between maritime companies on the one hand and port authorities on the other. This continuing struggle may be considered a consequence of a changed world economy, characterised by an international redistribution of labour and capital, and an integration and globalisation of markets. At the same time, a variety of co-operation agreements have materialised. These agreements assume different forms. However, the real objectives, the tools used and the impact realised are not always clear. The purpose of this paper is to take away some of the above uncertainties by examining the economic consequences of possible evolutions in the market in which shipping companies and ports operate.

Global Competition in Transportation Markets: Analysis and Policy Making
Research in Transportation Economics, Volume 13, 253–277

ISSN: 0739-8859/doi:10.1016/S0739-8859(05)13012-1

1. INTRODUCTION

Competition between maritime companies is fierce, as, for that matter, is competition between port authorities or indeed between maritime companies on the one hand and port authorities on the other. This continuing struggle may be considered a consequence of a changed world economy, characterised by an international redistribution of labour and capital, and an integration and globalisation of markets. Nothing is the same. Recent examples of this evolution are the loss in market share suffered by three giants in the container business, namely the ports of Rotterdam, Singapore and Hong Kong. In the case of Rotterdam, competition comes from most other ports in the Hamburg-Le Havre range. Singapore, for its part, has recently seen two major clients, Maersk and Evergreen, move operations (or part of operations in the case of Evergreen) to the Malaysian port of Tanjun Pelepas. Hong Kong, finally, has lost market shares to certain ports in Southern China, including Shenzhen.

At the same time, a variety of co-operation agreements have materialised. These agreements assume different forms, including alliances and mergers between shipping lines, the involvement of shipping companies in terminal management, and extending interests in inland transport on the part of shipping companies (for an overview see Heaver et al., 2000). However, the real objectives, the tools used and the impact realised are not always clear to see.

There can be no doubt that this rapid evolution in the marketplace has important consequences for the maritime and port business. The key maritime players have all become critical nodes in complex logistics chains, each of which needs to be continuously optimised. The result of such optimisation can be sudden and significant loss of market power.

Therefore, insight into what the future holds for the maritime business is becoming increasingly important. Indeed, some crucially important questions have arisen: What consequences might a further increase in the scale of containerships have? Will the co-operation process between shipping companies continue? What kind of alliances will port authorities have to accept? Finding the correct answer to these questions can help us formulate economic strategies for boosting market share, for increasing profits, and indeed for surviving in a negative marketplace.

The purpose of this paper is to take away some of the above-mentioned uncertainties by examining the economic consequences of possible evolutions in the market in which shipping companies and ports operate. Fig. 1 illustrates our approach. We shall take as our starting-point the current market structure and formulate a number of (possible) future scenarios,

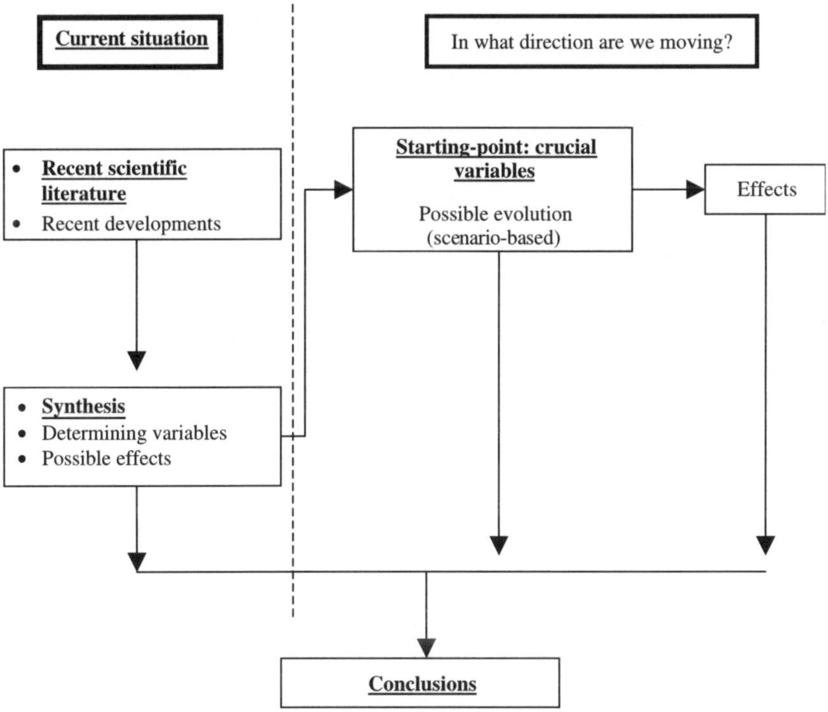

Source: own composition

Fig. 1. The Framework.

each of which may entail costs and benefits for the key players. Hence, an alternative title for this paper might have been "Dealing with Uncertainty".

2. RECENT SHIFTS WITHIN THE MARITIME SECTOR

Maritime transport, especially of containers, has expanded enormously in the recent past. The driving force behind this development has been growth in international trade, combined with the ongoing process of globalisation. The key question is therefore whether this growth will continue in the future.

According to certain consultants, it definitely will. By way of illustration, Table 1 provides an overview of forecasts by Ocean Shipping Consultants

Table 1. Forecast North European Container Port Throughput by Port Region up to 2015: Base Case (million TEU).

Year	North Continent West	North Continent East	British Isles	Scandinavia	Other Baltic	Total
2000	12.70	7.05	7.50	2.71	0.80	30.77
2001	12.53	7.66	7.40	2.74	1.07	31.41
2005	15.31	10.29	8.95	3.30	2.03	39.88
2010	19.77	14.42	11.21	4.20	3.41	53.01
2015	25.28	19.38	14.09	5.23	5.22	69.19

Source: Ocean Shipping Consultants (2002).

(2002) of developments in the Northern European market for containerised cargo, on the basis of a number of macroeconomic, container-specific and regulatory factors. In addition, due account was given to some important trends:

- the continuing development of the transshipment sector, including as a result of the policy of large shipping companies to transform the direct freight flows into transshipment flows;
- the continuing trend towards larger vessels;
- the increase in investment by shipping companies in container terminals;
- the ongoing internationalisation of the container stevedoring business;
- the increase in intermodality;
- the growing competitive pressure from Mediterranean ports;
- environmental pressures;
- the attempts by the European Commission to create a single regulatory framework for all European ports.

Two scenarios of macroeconomic growth were worked out as a basis for container traffic demand forecasts up to 2015.

Depending on the economic growth scenario applied, total North European container port throughput is expected to grow between 2001 and 2010 by 58–69% to a level of between 49.5 and 53.0 million Twenty foot equivalent unit (TEU). For the period 2010–2015, a further growth is predicted in the order of 25–31%, to a level of between 62.1 and 69.2 million TEU (Table 2).

Despite considerable volatility, the North European transshipment market has grown by an average 9.7% per annum since 1995. Under the base-case scenario, the average annual growth in container transshipment throughput is predicted to amount to 8.2% up to 2005, followed by 6%

Table 2. Forecast North European Container Port Throughput by Port Region up to 2015: Low Case (million TEU).

Year	North Continent West	North Continent East	British Isles	Scandinavia	Other Baltic	Total
2000	12.70	7.05	7.50	2.71	0.80	30.77
2001	12.53	7.66	7.40	2.74	1.07	31.41
2005	15.20	10.08	8.63	3.25	1.74	38.91
2010	19.01	13.52	10.45	4.02	2.51	49.51
2015	23.55	17.58	12.55	4.90	3.48	62.05

Source: Ocean Shipping Consultants (2002).

growth on average between 2010 and 2015. Under the low case, predicted average annual growth declines from 7.5% to 5%.

The above aggregate forecasts apply only to the region indicated and they should therefore be considered to be illustrative. In other regions, circumstances will be different, and so too will growth figures.

From a business perspective, things would not appear to be going well in the maritime sector. A shipping company such as P&O Nedlloyd was anticipating a loss of US$220 million for 2002, and the underlying indicators speak for themselves. There was growth in terms of TEU compared to the previous year, among other things, thanks to a second-quarter increase of no less than 13% over the comparable period in 2001. The average load factor in that same quarter was 83%, which was significantly higher than the year before. The problem was, however, that the average yield per TEU dropped in the second quarter of 2002 by a very substantial 15%. NOL/APL from Singapore found itself in a similar situation and was anticipating a loss of between US$185 and 208 million. The French group CMA CGM, by contrast, realised a modest profit in the first half of 2002. The volume transported rose spectacularly in that period (+33.5%), but at significantly lower freight rates, so that the increase in turnover amounted to only 6.8%.

This implosion of freight rates was clearly a direct consequence of excess container capacity. This, in turn, was due to the fact that, in times of growth and euphoria, shipping companies tend to commission additional capacity. They forgot that the maritime market is cyclical and that growth is therefore inevitably followed by a downturn.[1] Even with a constant level of demand, prices will drop if capacity increases. The opposite can also occur. Despite the recovery of transport volumes, the price increase that should ensue materialises agonisingly slowly.[2]

How are shipping companies responding? The main problem is clearly overcapacity within the sector, but paradoxically, large shipping companies are continuing to commission additional ships. Indeed, virtually all these companies are still making sizeable investments in fleet expansion. Remarkably, these additional orders are justified as a response to low freight rates. It is argued that the introduction of ever-larger containerships will result in a lower operational cost per slot.

Shipping companies are also increasingly confronted with exogenous factors that have an upward effect on operational costs. Examples are plenty: increased transit dues for the Panama Canal; more stringent safety requirements imposed by the U.S. Coast Guard for vessels calling at U.S. ports; increased protection of load boxes in container shipping; incidents such as the West Coast ports lockout (mid-2002); higher insurance premiums because of war risks and the threat of terrorism in some regions.

Financial deficits inevitably prompt cost savings measures. P&O Nedlloyd, for example, have taken measures that would save US$350 million by the end of 2003. Of course, the question remains how many degrees of freedom are retained. With losses looming and the prospect of unhappy shareholders a reality, commercial partners and service providers are put under heavy pressure. This has consequences for stevedores in particular, many of whom are already operating at relatively narrow margins, while they are, moreover, being forced by shipping companies to invest heavily in order to accommodate increasingly large ships (and thus greater freight volumes). For the time being, there are no indications that we are entering the era of 10–12,000 TEU containerships.

Consequently, a substantial expansion of port capacity also seems inevitable. By way of example, Table 3 provides an (incomplete) overview of major expansion plans in the North European Hamburg-Le Havre range.

In its totality, this represents quite a significant capacity increase, which in view of the growth forecast presented in Tables 1 and 2, would nevertheless appear to be absorbable. However, one should realise that there are also examples of overcapacity. In Zeebrugge, for example, the Flanders Container Terminal (FCT) is underused. And in the Mediterranean range, the 1 million TEU Cagliari International Container Terminal, which is partly owned by P&O Ports and Gruppo Investment Portuali, is used merely sporadically.

While it may appear as if growth in port handling capacity can continue unbridled, this is by no means the case. On the contrary, port expansion is no longer self-evident, if only because of the growing influence of the environmentalist movement. Furthermore, some ports are confronted with

Table 3. Planned Expansion of Container Capacity in North European Ports.

Port	Terminal	Unused Capacity/Planned Introduction
Amsterdam	CERES terminal	950,000 TEU/year of unused capacity
Antwerp	Deurganckdock terminals	5,785,000 TEU/year from 2005
Bremen	CTIV	950,000 TEU by 2007, 950,000 TEU more by 2009
Flushing	Westerschelde Container Terminal	2,250,000 TEU/year from 2006 at the earliest
Hamburg	Altenwerde Terminal	800,000 TEU by 2005
	Burchardkai	700,000 TEU by 2007, 700,000 TEU by 2010
Le Havre	Port 2000	4 quays by 2005, 2 more quays scheduled, 6 more quays to be built upon shipping companies' demand
Rotterdam	EUROMAX Terminal	From 2007 onwards
	Tweede Maasvlakte	No date specified, about 3,000,000 TEU
Rouen	Grand Couronne	200,000 TEU by 2005
Wilhemshaven	Eurogate Deep Sea Terminal	900,000 TEU by 2007, 900,000 TEU more by 2009

Source: Own processing of data from various port authorities. See also Meersman, Van de Voorde, and Vanelslander (2003).

physical barriers to expansion, so that they can only continue to increase capacity for specific traffic (e.g. containers) by redeveloping or renovating older parts of the port or through a reconfiguration of terminals (e.g. stacking six high instead of two high).

One can also discern certain response patterns among container handlers in ports. First and foremost, it is noticeable that, largely in response to the concentration trend among shipping companies, container handling is now also becoming more concentrated. These firms, ranging from traditional stevedores to terminal operating companies, would rather not run the risk of being pushed aside in negotiations over rates and volumes with (strong) shipping companies. Moreover, concentration provides a buffer against possible verticalisation.

Theoretically, there is of course always a danger that concentration will result in less competition, lower productivity growth and thus longer handling times and less favourable rates. For the time being, this has not happened, as competition continues to be fierce at the global player level (PSA Holdings, Hutchison Holdings, P&O Port Holding).

In this respect, much can be learnt from the change of strategy on the part of PSA, a global player that used to be impossible to ignore. However, its position, especially in the basic container handling market, suddenly came under pressure when its main customer, Maersk Sealand, relocated to Tanjun Pelepas, where handling rates are exceptionally low. Evergreen soon followed suit with part of its traffic. PSA responded instantly with aggressive rates and service policy. On 12 July 2002, it slashed handling rates for empty containers by 50% for a period of 12 months (note that empty containers account for 15% of total volume), while handling rates for loaded containers were cut by 10%. In the future, shipping companies might be able to participate in PSA. This more flexible position has meant that 75% of container throughput in Singapore is now fixed in contracts of at least 5 years.

From the perspective of the goods handler, the era of the supply-driven economy would appear to be a thing of the past. Demand, in this case from shipping companies, is now paramount. On the other hand, we appear to have entered a period marked by shorter production cycles and much steeper learning curves. After all, a commercial rotation of vessels requires speedy freight handling, as faster rotation times means fewer ships to deploy. This explains the additional response pattern whereby shipping companies have tried to gain greater control over the goods-handling sector. The manner in which they have tried to achieve this goal has varied: through long-term partnerships, by means of direct minority stakes in the capital, or by acquiring total control. Recent examples have mostly concerned so-called dedicated terminals, whereby a port authority grants an exclusive concession to a particular shipping company, often in combination with a goods-handler. This type of concession is primarily inspired by a growing desire for security (cf. the rather footloose nature of shipping companies).

The question arises how the various players in the shipping market, ranging from shipping companies to port authorities to goods-handlers, will respond in the future. Heaver et al. (2000) have studied the various forms of co-operation and concentration that occur in the maritime sector. The configuration put forward in their study remains intact. Only, some players are presently more active than others in seeking partnerships. Table 4 provides an overview of the variety of co-operation agreements that have occurred in the sector over time, specifically among shipping companies, stevedores and port authorities. The table does not offer an overview of evolutions in the types of partnerships, nor does it provide insight into the intensity with which important market players opt for particular strategies.

A number of co-operation agreements between shipping companies have clearly been occasioned by the fact that these companies now have new, larger, vessels at their disposal. We have already indicated that one of the underlying motives was to reduce operational costs. Large containerships of

Table 4. Strategic Agreements in the Maritime Sector.

Market players	Shipping companies	Stevedores	Port authorities
Shipping companies	Vessel sharing agreements (148 agreements being filed with the European Commission in 2001) Joint ventures (e.g. Lloyd-Triestino and Zim (2003) in AUX-service) Consortia (e.g. Cosco, K-Line, Yangming at PSW-service) Alliances (Grand Alliance, New World Alliance etc.) Mergers/acquisitions (e.g. OMI's bid on Stelmar in the tanker sector) Conferences (Italy alone e.g. is served by 19 conferences)		
Stevedores	Joint ventures (e.g. Maersk/APM 50/50 in Xiamen Port) Dedicated terminals (e.g. HPH in four Mexican ports) Share (COSCO 17.5% in Shekou terminal) Consortia (e.g. SAGT terminal Sri Lanka with a.o. P&O Ports, P&O Nedlloyd and Evergreen)	Mergers/acquisitions (e.g. PSA in HesseNoordNatie) Joint venture (e.g. P&O Ports and modern terminals in Shekou together with China Merchant an Swire Pacific)	
Port authorities	Concessions for dedicated terminals (e.g. MSC and HesseNoordNatie in Antwerp).	Concessions (vb. Deurganckdock Antwerp PSA and P&O) Joint ventures (e.g. Ningbo Port Authority and HPH in local port)	Alliances (e.g. Port of Rotterdam and Humber Trade Zone)

Source: Own processing of data from various shipping companies, stevedoring companies and port authorities.

over 5,000 TEU are particularly popular in this respect. These vessels are presently sold very cheaply.[3]

Nevertheless, they are capital goods that need to be deployed efficiently. Certainly such large containerships enable one to slash operational costs per slot by up to 30% as compared with 3,500–4,000 TEU vessels, which is

Table 5. Overview of new vessel orders placed during June–September 2004[a].

Shipping company	Registration	Number and type of ships	Date of Delivery
Maersk	2/8/04	12 of more than 8,000 TEU, 7 of 6,070 TEU	By 2009
MSC	1/9/04	5 of 8,100 TEU, five more?	
Evergreen	31/8/04	10 of 7,032 TEU, 4 of 4,200 TEU, ? of over 9,000 TEU	
P&O Nedlloyd	2002	4 of 7,500 TEU, 4 more on option?	
CMA-CGM	4/6/04	A number of 8,200 TEU	
NYK	31/8/04	A number of over 9,000 TEU	
COSCO	2004	8 of 8,500 TEU	2006
K-Line	2004	4 of over 8,000 TEU, 5 of over 5,000 TEU	2006–2007
CSCL	31/8/04	4 of 9,600 TEU?	
IRISL	31/8/04	5 of 5,500 TEU, 3 of 5,500 TEU in option	By 2007/2008
MOL	6/7/04	8 of 6,350 TEU and 4 of 8,100 TEU on option	By 2005 and option By 2007/2008

Source: Own processing of data from various shipping companies.
[a]In the case of P&O Nedlloyd, figures relate to 2002.

useful in a market environment with declining rates. By way of illustration, Table 5 provides an overview of a number of ship commissions from July 2002. This table is incomplete and it does not cover vessels that were delivered earlier in the year.

It is quite clear that these vessels are not merely intended as replacements for older ships, but that they will actually add further tonnage to the market. In 2002, China Shipping Container Line (CSCL), for example, suddenly substituted 5,450 TEU vessels for 2,500–2,700 TEU vessels on its European run. In doing so, the shipping company effectively skipped the intermediate phase of 4,000–5,000 TEU vessels.

It is generally accepted that there is an urgent need for remedying the market. The most drastic measure would of course consist in the scrapping of tonnage. Alternatively, one could withdraw from the market, as the Chinese company Sinotrans did.

Another less painful course of action is the establishment of joint services, coupled with a reorganisation of services. Consider the following example: three brand new 4,100 TEU containerships, delivered to Contship

Containerlines, are to be deployed as part of a reorganisation of various services that were previously operated jointly by Contship and P&ONL. Among other things, the two shipping companies decided to provide a weekly service with 10 containerships of this capacity within the scope of the eastbound round-the-world shipping schedule. Under this scheme, Australia and New Zealand are served from Europe via the Suez Canal. On the return journey via the Panama Canal, the ships first call at the East Coast of the U.S. This is an example of rationalisation and optimisation that ties in with market developments. In fact, the new vessels are to replace 12 2,500–2,700 TEU ships that were originally distributed over two distinct services.

3. IN WHICH DIRECTION ARE WE MOVING?

In the previous sections, we tried to outline how the maritime sector has evolved in recent years towards the present situation. After all, knowledge about the past and the present forms the basis for sound predictions about the future. However, in the volatile sector that maritime transportation clearly is, forecasting is invariably subject to many uncertainties.

What will the future maritime market look like? What kind of measures may we expect from the principal maritime players (shipping companies, stevedores) with a view to enhancing their competitive position? Which measures will be imposed upon those players by the regulatory bodies? It is quite clear that the future strategies of shipping companies and stevedores will be affected by their own decisions (endogenous variables) and by (responses to) decisions by others (exogenous variables).

In the following paragraphs, we shall keep this distinction in mind. We shall first concentrate on a number of steps that may be expected from the maritime companies themselves. We shall deal consecutively with further rationalisations (including through the deployment of even larger vessels and a new wave of mergers and acquisitions) and with the evolution towards a new port landscape. We shall also consider the derived effects of these types of developments for, among other things, the configuration and use of transshipment terminals.

Subsequently, we shall deal in greater detail with a number of measures, mostly politically inspired, that are exogenous to the sector, yet may exert a considerable influence. We refer specifically to environmental and safety aspects, and to the economic consequences of certain types of incidents or events.

3.1. New Rationalisation Measures by Shipping Companies

As we have already mentioned, 2002 was one of the most disastrous financial years ever for the maritime sector. Few if any shipping companies reported a profit, certainly in the liner shipping business. This was a major worry for any shipping company, especially as load factors have been relatively good but rates have failed to follow suit.

We may assume that the present trend towards a far-reaching rationalisation of the market will continue. It will involve the deployment of new, larger vessels which are expected/hoped to boost market share through a lower operational cost per slot. This rationalisation may also occur through mergers and forms of co-operation. Furthermore, one may expect landside technological improvements to have an impact on the cost structure and port landscape. Finally, some shipping companies may redirect or diversify operations. We shall deal with each of these possible developments separately.

3.1.1. Deployment of Larger Vessels

Round-the-world container services are undergoing drastic reorganisation that will continue in years to come. This rationalisation is translating into new partnerships, the re-routing of vessels, the deployment of larger vessels and shifts in direct port calls.

Each rescheduling is preceded by a careful analysis of market demand and cost structure. By way of illustration, we shall briefly consider two schedules that became operational from the end of November 2002. The westbound route was to be served by CMA CGM, Marfret, Contship and the newcomer P&ONL. It involved a 10-day frequency, with a fleet of eight 2,200 TEU containerships. The choice of vessel was determined by conditions of access and use at a number of Pacific ports. The eastbound route, on the other hand, would be operated by P&ONL and Contship, with CMA CGM as co-leader. In this instance, two different services were reorganised into one. The original services each involved a 14-day frequency with six 2,500–2,700 TEU vessels. The new service would involve 10 newly built ships of 4,100–4,200 TEU.

Where will the scale increase in containership sizes end? Wijnolst, Scholtens, and Waals (1999) have tried to demonstrate that the concept of a Malacca-Max containership (18,000 TEU) is already technologically and economically feasible.[4] The main advantage of such large vessels lies in potential benefits of scale. The authors calculate that they would yield a cost reduction of at least 16% compared with the most cost-effective ships available now. This cost reduction would be realised mainly in fuel consumption and in Suez Canal dues.

At first glance, it would seem then that a further scale increase is inevitable, perhaps towards vessels that are twice as large as those of the present generation.[5] However, in considering such technological developments, one tends only to take account of direct vessel costs. Terminal costs and costs of hinterland transportation tend to be ignored. Yet, the latter have often proven decisive in the choice of vessel type and port of call.

Moreover, there is an additional industrial and economic problem. These kinds of super vessels will inevitably be limited in the ports they can serve. This explains why the ports of Rotterdam and Singapore have been approached to contribute to the funding of a necessary widening of the Suez Canal. Shipping companies will want to prevent a situation where they are held hostage by the few ports they can enter. With the example of the transportation of crude by ULCCs still fresh in their memories (i.e. the dependency of this type of vessels upon the port of Rotterdam for serving the European continent), they will realise that price inelasticity due to a lack of alternatives can/will encourage port authorities to increase dues.

3.1.2. Mergers and Partnerships among Shipping Companies

In fact, the deployment of new and increasingly large vessels on the part of leading shipping companies such as Maersk Sealand and Evergreen could be interpreted as an effort to continue to put pressure on their competitors. Indeed, they might well regard it as a possible way of further reducing the cost per slot. The ensuing overcapacity is accepted as a matter of course.

Some shipping companies will clearly be tempted to pursue such a strategy (Slack, Comtois, & McCalla, 2002). They will also aim at vessels of larger dimension in order to reduce the cost per slot and spread risks. Alternatively, this could be achieved through acquisitions and mergers, as other shipping companies are confronted with an identical market environment. Mergers and/or takeovers also provide an opportunity to realise additional savings plans and, possibly, to reduce capacity.[6]

One may expect shipping companies to co-operate even more closely in the future, e.g. within the scope of alliances.[7] A recent example is the alliance between Cosco Container Lines Ltd, K Line, Yang Ming and Hanjin, which has allowed the participating shipping companies to provide a new range of joint services, for which they hire and exchange slots. In the short term, this type of alliance provides some breathing space for each of the partners, while in the long-term it may even lead to mergers (Song & Panayides, 2002).

This trend towards mergers may/will be further enhanced by strategic and financial considerations within the holdings that control certain shipping companies. A strategy that considers the liner shipping section as a profit centre will later provide selling opportunities.

However, one thing is very clear. In the short and middle term, the pressures of rationalisation will result in a thorough rearrangement of services provided and give rise to new alliances. In the first instance, these alliances will seem as brittle as alliances have been in the recent past, because they will be affected by strategic moves on the part of shareholders and management of the controlling holding. Stability will only be attained when (some of) the alliances evolve further into mergers or acquisitions.[8] At that moment, we will have a market with fewer players so that it will be easier to stabilise prices through sizeable capacity restrictions.

3.1.3. Landside Technological Improvements

It is also important that one should gain adequate insight into (necessary) landside technological and organisational improvements. The economic benefits that shipping companies seek through further scale increases and a corresponding cost reduction should not be wasted through landside bottlenecks. The ongoing trend towards further automation may be expected to persist. This is a consequence of, among other things, a shortage of skilled personnel, the high cost of labour and customer expectations with regard to availability of service (i.e. 24 hours a day, 7 days a week).

We may assume that, at the actual terminals, the trend towards the deployment of automatic stacking cranes (ASCs) will continue. ASCs are primarily used in the stacking area. They are fed on one side by automatic guided vehicles (AGVs) that operate to and from the transshipment quay. On the other side, there are straddle carriers that connect storage zones with the zone where trucks (and other vehicles or vessels) load and unload.

The most significant point of uncertainty, however, concerns the consequences that far-reaching policy of co-operation and acquisitions will have in terms of goods-handling operations. To what extent might a purchaser (e.g. PSA from Singapore) impose technologies that are used at the home terminal upon terminals controlled in other continents (e.g. HesseNoord Natie in Antwerp, Voltri in Genoa)?

3.1.4. Diversification and Reorientation

The rationalisation strategies pursued by some shipping companies may include other methods than the deployment of larger vessels or the realisation of mergers or partnerships, some of which may even be extreme (cf. Lorange, 2001). Consider the example of the Chinese liner shipping company Sinotrans, a relatively small player in the liner shipping market. Under pressure from low freight rates on the East-West Route, this company has decided to terminate its liner shipping operations altogether from the end of 2002 and to concentrate instead on basic forwarding activities.

Others will opt for a strategy of diversification. A recent example in this respect was NYK Line, which transformed itself from a shipping company that did not consider logistics as a core activity into an integrated logistics service provider. This kind of diversification policy makes companies less vulnerable to extreme fluctuations in freight rates.

Other shipping companies may opt to establish complementary activities, possibly within a partnership. Maersk Sealand and P&ON, for example, are now more closely involved in hinterland transportation, including by organising their own freight carriage by rail. To this end, ERS Railways, a spin off from European Rail Shuttle (ERS).

In short, as in other sectors of the economy, a trend may/will emerge in the shipping industry towards greater diversification and/or the development of complementary activities. In this manner, a company is made strategically and financially less vulnerable to the economic cycle in one particular sector. Moreover, such a strategy could, in the longer run, be incorporated into a striving for further 'verticalisation', whereby the aim is to achieve as much control as possible over the logistics chain. Ultimately, this is also the striving of a number of shipping companies who have transformed into integrated logistics service providers.

3.2. Towards a New Port Landscape

In recent years, a trend has been unfolding towards vertical and horizontal integration of international maritime transport chains. Often, shipping companies have taken the initiative in this integration drive. As a consequence of this evolution, these companies have gained in market power over other service providers, including port authorities and stevedores (Heaver, Meersman, & Van de Voorde, 2001).[9]

The other players were left with a choice of two possible strategies: to undergo the evolution and continue to be dominated, perhaps even assimilated, or, on the contrary, to respond quickly and effectively. Most opted for the latter (see e.g. Terada, 2002). Port authorities took new initiatives, while the traditional structure of the stevedoring sector was transformed by a series of mergers, acquisitions, and regional and global expansion projects. Notteboom (2002) asserts in this respect that "market concentration at the port demand side, port privatisation and the fact that investors finally consider the port industry as a mature business, are elements that explain the present consolidation in the cargo handling business".

These structural changes herald a new phase in port competition.[10] The large shipping companies no longer face relatively small partners in negotiations over volumes and rates, but are confronted instead with large terminal operating companies that are active in different ports and that can therefore negotiate over larger packages at once. This implies that port competition is gradually shifting from the level of port authorities to that of private terminal operators, who are presently providing regional networks that may nevertheless develop in the longer term into global networks (Notteboom, 2002).

Thus, in the first instance, the negotiating power of the shipping companies will be undermined. At the same time, however, these companies will become increasingly interested in acquiring dedicated terminals, perhaps in partnerships with terminal operating companies. Or they might simply want to acquire partial (or total) shareholding in such terminals. This evolution will make shipping companies less footloose (Heaver et al., 2001). After all, acquiring a financial stake in a terminal operating company (cf. negotiations between PSA and a number of shipping companies) will inspire shipping companies to develop long term relationships and to refrain from relocating to a competing port.[11]

So what precisely will be the role of the port authorities in this process? In the present negotiating "game" between shipping companies and terminal operators, their trump card is that they have the power to grant concessions. One may assume that, eventually, a number of port authorities will be tempted to use this power as a lever to acquire a more influential role in port competition and in the port throughput process. A good example in this respect is Rotterdam, where in 1999 the port authority acquired a 35% stake in the capital of the important terminal operating company ECT, which was otherwise controlled by Hutchison. This may ultimately lead to a conflict of interest, e.g. in the granting of new concessions. The question therefore arises to what extent the position of the port authority as the regulating power and interface with the government may be compromised in the eyes of third parties.

In the short to medium term, a number of significant changes may be expected to occur in relation to port competition and the management of large terminals.[12] However, these developments go hand in hand with a considerable degree of uncertainty that will only disappear when adequate answers have been formulated to a number of pertinent research questions (in this context, see among others Heaver et al., 2001). One typically important question is how a further concentration of terminal operating companies and an active participation of port authorities will affect the pricing of port services. A further point of interest is how this development will impact on the dynamics of competition in the sector.

3.3. Exogenous Influences and their Price Effect

Shareholders expect a private company to strive towards profit maximisation. This clearly also holds for companies that are active in the maritime and port sector. Of course, profits are influenced not only by a company's revenues, but also by the costs incurred. It is, for that matter quite noticeable that in times of crisis, cost control or even reduction inevitably occupies an important place in the operational approach of managers.

Management can influence many cost items, either directly (e.g. number of employees) or indirectly (e.g. wage levels). Others, however, are exogenous. That is to say, they are imposed from outside the company or sector. Managers must, nevertheless, respond alertly to such costs, as the speed and manner of this response can influence market share in this hugely competitive sector.

In what follows, we shall deal in some greater detail with three exogenous factors that may strongly effect the maritime sector in the future: safety aspects, environmental aspects and miscellaneous incidents.

3.3.1. Safety Aspects

Since the terrorist attacks of 11 September 2002, safety has certainly become a major issue in the maritime industry in the United States. Maritime freight flows, and especially containers, are after all a potential target or they could even be used in future attacks.

Some European ports immediately shared this American concern. They reached an agreement with American customs over the so-called "Container Security Initiative". The European Commission, for its part, was opposed to such bilateral agreements, arguing that it alone was empowered to sign up to the initiative. Furthermore, it expressed fears that the bilateral approach would distort competition between (European) ports.

The latter concern is not entirely unfounded. The largest terminal operators (Hutchison Port Holdings, PSA Port Holdings and P&O Port Holdings), who together account for about 70% of global maritime container flows, signed up in 2002 to the Smart and Secure Tradelanes (SST). The SST is a large-scale initiative by the U.S. authorities aimed at guaranteeing the safety of sea freight. By signing up to the SST, the ports of these operators were almost automatically granted green-lane status. The specific technological requirements (including special tags, electronic seals, radio frequency identification, etc.) necessitate additional investment, but, at the same time, they constitute a barrier to entry for other operators.

These safety costs come on top of other, often vessel-related, safety costs that were already in place. Under European regulations, each member state must, for example, check at least 25% of vessels calling at its ports (so-called Port State Control). But these measures, too, hold a risk of affecting the competitive balance between ports. In 2000, authorities in France checked only 12.2% of vessels calling at French ports, or less than half the agreed percentage. The proportion subsequently declined to just 9.6% in 2001.

While the maritime sector generally understands this closer focus on safety in American and European ports, some shipping companies have criticised the random security measures imposed at U.S. ports. Often, crews are required to stay aboard. Moreover, shipping companies are made to pay for safety measures that they have not requested. The bill for new, additional safety measures may be expected to rise further. From a purely economic perspective, these additional costs could probably be passed on to customers. However, it remains to be seen to what extent safety regulations (or failure to comply with such regulations) might distort competition between shipping companies and ports. Perhaps one can learn from the experience gained in the fight against the deployment of substandard vessels.

3.3.2. Environmental Aspects

The maritime sector is also increasingly subject to environmental regulations. On 28 December 2002, a European guideline became effective concerning port waste reception facilities for cargo residues and ship-generated waste such as domestic refuse and engine room slop.[13]

The purpose of this regulation is to deal effectively with illegal discharges of ship-generated waste into the open sea or surface waters. However, the regulation alone will not suffice: it will also need to be enforced. This implies that a system needs to be devised whereby ships are obliged to deliver their waste. Ports, on the other hand, must work out a scheme whereby all costs associated with the reception and processing of waste are passed on to the ships. The European Directive does not specify how these costs may be passed on. One possibility is incorporating these costs in the port dues, but this might raise concerns regarding possible distortion of competition between seaports.

Alternatively, one could introduce a deposit system under which ships would be required to pay a deposit according to vessel type, size and journey. This sum would be returned on submission of a certificate stating that waste has been delivered.[14]

This is an example of an additional cost item for most shipping companies. From the perspective of transport economics, the introduction of such a system would clearly represent a justifiable internalisation of externalities.[15]

In other words, it confronts the producer with the full socioeconomic cost. External costs fundamentally disturb the functioning of the market mechanism. If the producer sells a product or service at a price that does not contain all externalities, overconsumption will ensue.

One may therefore safely assume that, sometime in the future, other externalities will also be internalised. Examples that come to mind are other environmental cost factors such as discharges, external accident costs, external infrastructure costs and external safety costs. Inevitably, someone will need to pay for the internalisation of these external costs. It seems reasonable to assume that they will be passed on to the freight, i.e. to the user. At first glance, this would not appear to be much of a problem in the maritime sector. However, we have no clear picture at the present of the price-tag attached to such an internalisation, nor of the proportion it represents in the total cost of maritime transport and the logistics chain. Furthermore, we have insufficient empirical information about price sensitivity.

3.3.3. The Economic Cost of Incidents

The transport sector, including the maritime sector, is very sensitive to incidents that affect normal operations. This is not only the case for transport performances that need to be concluded according to a strict schedule (e.g. in liner shipping), but also for other types of carriage. Any delay implies a short-term loss of time and therefore money, and in the long-term it might even result in loss of load. This economic cost can be calculated quite accurately.

Recent examples of incidents are rife. The strike that paralysed American West Coast ports (second half of 2002) caused a direct loss of revenue, not only for the ports affected, but at a global scale. It also had a negative effect on productivity, it created an acute shortage of empty containers and ships, and it resulted in the imposition of congestion tolls in some places. Around the same time, the Ivory Coast was hit by a politico-military crisis, which had direct consequences in terms of port throughput in Abidjan, an important hub in that part of West Africa.

The real cost of such incidents is quite significant. First and foremost, it encompasses direct costs associated with the cargo and with operating the vessel. In the longer term, it may also involve damage claims that could translate into higher insurance premiums or in higher risk behaviour. Finally, it includes an indirect cost associated with damage to the maritime mode's reputation in terms of reliability and damage risks. Demand for (maritime) transport and throughput is also affected by this type of variables.

4. BY WAY OF CONCLUSION: A SCENARIO-BASED APPROACH

The entire maritime sector, i.e. shipping companies, port authorities and terminal operating companies, has undergone a real transformation in recent times. Each market player separately is confronted with the rapid emergence of new technologies and indeed new customs. In other words, they are confronted with a totally new market structure. Each party has tried to take strategic steps in this rapid evolution with the purpose of acquiring greater market share and, more importantly, realising greater profits. It seems self-evident that predicting the future in this sector should involve great uncertainties.

In this contribution, we have tried to channel this uncertainty. A detailed study of the past indicates a number of trends. The question arises whether these trends will persist in the future. Moreover, a number of market players (especially shipping companies and terminal operating companies) have, in the recent past, taken a number of long-term decisions that will have important repercussions in the future. Examples that come to mind are decisions with regard to the acquisition of ships (number, capacity, speed etc) and regarding mergers and partnerships. If one takes due account of these decisions, it is possible to consider a number of alternative scenarios.

As far as future container traffic is concerned, all forecasts agree. For the period up to 2015, they predict substantial growth. While the growth rate does depend on the scenario retained and on the origin and destination under consideration, it is noticeable that, even in the worst-case scenario, one expects average per annum growth to amount to around 5%. This clearly explains why most port authorities focus primarily, sometimes even exclusively, on the container market. Investments are, to a large extent, geared towards the container business, so that the rest of the maritime sector is unjustifiably neglected.

The growth figures cited are aggregate forecasts. From a business perspective, however, matters look a lot less rosy. Some shipping companies have reported figures that are simply catastrophic, so that shareholders feel they must put pressure on management. Management will then in turn pressurise other links in the logistics chain, e.g. terminal operating companies. Consequently, the latter are also working at margins that are (too) narrow, and they likewise feel compelled to make heavy, risky investments. The limits have, most probably, already been reached. Things will simply have to change in the future.

Shipping companies urgently need rate increases. While this may be achieved in different ways, rates are ultimately a matter of supply and demand. In order for rates to increase, demand for container traffic must, in any case, increase more rapidly than supply (i.e. than capacity). Sharp increases in demand should not prompt shipping companies to provide additional capacity. On the other hand, if demand grows more slowly, capacity should best be reduced.

Of course, certain strategies are already being pursued. At the moment, there is, for example, a tendency to invest in new, additional capacity. Some shipping companies seem convinced that low freight rates are here to stay and that the only solution therefore lies in reducing the operational cost per slot. If this strategy is not combined with the scrapping of older vessels, even greater overcapacity may ensue, which would translate into even lower freight rates. This may be good news for forwarders, but it leaves shipping companies with no alternative but to economise at company level, through mergers, partnerships and other cost-saving operations. Moreover, the forwarder will, quite probably, be confronted with the passing on of additional costs imposed upon the shipping companies. Examples that spring to mind are higher transit dues (e.g. Panama Canal), more stringent safety requirements, rising insurance premiums because of a greater risk of war, and the internalisation of external costs.

Port authorities and terminal operators, too, will feel the effects of this evolution and they will need to take additional strategic measures. Many investment programmes are already under way, which will result in a sharp increase in throughput capacity in a few years' time. This in turn will undoubtedly lead to new mergers, take-overs and partnerships. Moreover, this trend might have an impact on prices, given that the present overcapacity at (partly) idle terminals (e.g. Ceres Terminal in Amsterdam, Cagliari, etc.) has a downward effect on transshipment charges at more successful terminals. There is also a considerable likelihood that new forms of integrated partnerships and mergers will occur between shipping companies and terminal operating companies that go beyond the level of dedicated terminals.

The scenarios for future developments are more or less fixed, but the speed at which evolutions will unfold the degree to which one will opt for certain subscenarios is an area of considerable uncertainty. This will depend upon a battery of exogenous and endogenous variables, such as the occurrence of opportunities for acquisitions, the availability of capital, cash flow, etc. The decision framework is inevitably preoccupied by a concern for profit maximisation, increasing market share and/or expansion of market power.

The scholar can, however, provide tools for acquiring insight into the consequences of specific decisions. Indeed, we urgently need to establish, specify and empirically apply a tool that is able to test the consequences of each of the scenarios described in the present article in terms of cost structure, demand developments, pricing, market share, etc. Perhaps this is the research agenda for the coming years, preferably in a partnership between scientists and industry.

NOTES

1. Containerised liner shipping is a cyclical business. The cycles are, however, increasingly short. As a result, shipping companies are no longer merely sticking to their own operational and strategic requirements, but they appear to be increasingly inclined to take risks.

2. These declining rates also occurred outside the container business. Bocimar, a shipping company that transports iron ore for steel producers, was confronted during that period with extremely low freight rates. Up to August 2002, the freight rate for a Capesize vessel with a capacity of 150,000 DWT amounted to US$10,725/day, compared to US$24,400 in 2000 and US$15,234 in 2001.

3. In 2002, one could commission a super post-Panamax vessels of 6,300 TEU for just US$62 million, which is about US$10 million less than the price charged 1 year before.

4. Wijnolst et al. (1999) assert that 'with a twin propulsion the ship can be powered by existing diesel engines. The hull can be constructed with commercially available steel qualities, while plate thickness can be kept at acceptable levels'.

5. Other plans have been formulated in this sense, including a concept based on 15,000 TEU containerships. This would involve round-the-world services on

the equator, and portal cranes with a capacity of over 500 tonnes that can handle containers in blocks of 4–48 TEU and with a performance of 1,000 TEU per hour. Dream or delusion? Clearly, the answer to this question depends on what is technically feasible and economically sensible and optimal.

6. Towards the end of 2002, the names NOL and P&ONL were mentioned quite frequently in the context of possible mergers. Indeed, both these shipping companies reported very bad results for this financial year, and they both faced overcapacity and low freight rates. Moreover, their networks seem rather complementary, with P&ONL occupying a weak position on the Pacific, but a much stronger one on the other trades. In meantime, P&ONL announced that it would de-merge again.

7. According to Midoro and Pitto (2000, p. 39), the present structure of strategic alliances will not yield the anticipated results, among other things because of the considerable degree of intra-alliance competition between the various partners, whereby the stability of the alliances themselves is undermined. This could be prevented by reducing the number of partners, diversifying the role of those that remain, and co-ordinating marketing and sales. In the short term, the authors predict a period of market instability.

8. Shipping companies were being taken over all the time. In August 2002, for example, CP Ships acquired Italia Di Navigazione for the sum of US$40 million. The new acquisition became CP Ships' seventh brand after Canmar, Cast Lykes, TMM Lines, ANZDL and Contship Containerlines.

9. We shall not deal in the present article with the role that government can play in port operations. We merely refer to Winkelmans (1999, p. 2), who asserts quite categorically that "in order to achieve the necessary port efficiency, governments need not be the provider of services, not even in the so-called public interest. Therefore, the 'state' should not determine individual port strategies, management and operations. In principle, and for sound reasons too, governments are asked to stimulate, to co-ordinate, and to facilitate port development and port investments".

10. For an extensive and systematic survey of the topic of port competition, we refer explicitly to Huybrechts et al. (2002), where it is demonstrated quite unequivocally that competition between ports will increasingly become a matter of overall supply chain logistics. For a disaggregate analysis of factors influencing a port's attractiveness we refer to Malchow and Kanafani (2001).

11. Such stakes can assume various forms, e.g. that of an equity participation or of a contractual agreement.

12. The regulatory authorities are also quite aware of this. The European Commission, for example, wants to see transparency and flexibility in Europe's ports, and to this end issued the so-called Port Package. This package contains a set of general procedural rules that are transparent, non-discriminatory, objective and equitable. Local specifications may be taken into account, but this can never be used as an excuse to restrict or exclude competition. The European Commission was, however, aware that it was unable to impose a harmonisation of ports, including in respect of their structure and management.

13. Cf. European Directive 2000/59/EG.

14. The Belgian ports favour the deposit system. Indeed, it has many advantages. For one thing, it is uniform, so that no distortion of competition occurs. Furthermore, the port authorities are not involved in the collection and processing of waste,

so that they bear no direct financial responsibility for the system. The shipping company, finally, only needs to pay for the real cost of collection and processing.

15. Externalities are costs that the transport user causes to a third party and for which he does not pay. Noise pollution, air pollution and safety risks for local residents are typical examples. External costs are part and parcel of the social cost of transport, but they are not included in the private cost that is charged to the transport user (Blauwens, De Baere, & Van de Voorde, 2002).

ACKNOWLEDGMENTS

The author would like to express his gratitude to T. Vanelslander for his help in the writing of this article. Thanks are also due to Prof. H. Meersman and other colleagues from the Department of Transport and Regional Economics and from the Institute of Transport and Maritime Management Antwerp (ITMMA) at the University of Antwerp for the many interesting conversations about the topics under consideration. It speaks for itself that the author is solely responsible for the content of this paper.

REFERENCES

Blauwens, G., De Baere, P., & Van de Voorde, E. (2002). *Transport economics*. Antwerp: De Boeck Ltd.

Heaver, T., Meersman, H., Moglia, F., & Van de Voorde, E. (2000). Do mergers and alliances influence European Shipping and port competition? *Maritime Policy and Management*, *27*, 363–373.

Heaver, T., Meersman, H., & Van de Voorde, E. (2001). Co-operation and competition in international container transport: Strategies for ports. *Maritime Policy and Management*, *28*, 293–305.

Huybrechts, M., Meersman, H., Van de Voorde, E., Van Hooydonk, E., Verbeke, A., & Winkelmans, W. (2002). *Port competitiveness. An economic and legal analysis of the factors determining the competitiveness of seaports*. Antwerp: De Boeck Ltd.

Lorange, P. (2001). Strategic re-thinking in shipping companies. *Maritime Policy and Management*, *28*, 23–32.

Malchow, M., & Kanafani, A. (2001). A disaggregate analysis of factors influencing a port's attractiveness. *Maritime Policy and Management*, *28*, 265–278.

Meersman, H., Van de Voorde, E., & Vanelslander, T. (2003). Analyzing the maritime sector of the next decades. In: W. Dullaert, B. Jourquin & J. Polak (Eds), *Across the border: Building upon a quarter century of transport research in the benelux*. Antwerp: De Boeck Ltd.

Midoro, R., & Pitto, A. (2000). A critical evaluation of strategic alliances in liner shipping. *Maritime Policy and Management*, *27*, 31–40.

Notteboom, T. E. (2002). Consolidation and contestability in the European container handling industry. *Maritime Policy and Management*, *29*, 257–269.

Ocean Shipping Consultants Ltd. (2002). *Containerisation in North Europe to 2015*. Surrey: OSC.

Slack, B., Comtois, C., & McCalla, R. (2002). Strategic alliances in the container shipping industry: A global perspective. *Maritime Policy and Management*, *29*, 65–76.

Song, D.-W., & Panayides, P. M. (2002). A conceptual application of cooperative game theory to liner shipping strategic alliances. *Maritime Policy and Management*, *29*, 285–302.

Terada, H. (2002). An analysis of the overcapacity problem under the decentralised management system of container ports in Japan. *Maritime Policy and Management*, *29*, 3–16.

Wijnolst, N., Scholtens, M., & Waals, F. (1999). *Malacca-max. The ultimate container carrier. Design innovation in container shipping*. Delft: Delft University Press.

Winkelmans, W. (1999). The issue of shipping and port management in the 21st century. *Proceedings of VIII Congreso de Trafico Maritimo y Manipulacion Portuaria*. Algeciras. Mimeo.

PORT INVESTMENTS IN AN UNCERTAIN ENVIRONMENT

Hilde M. A. Meersman

ABSTRACT

This contribution focuses on some specific aspects of port investments related to the large amounts of capital needed for some types of port investments and the long payback time of the projects. The first part treats the problem of private and/or public involvement and the impact on the competitiveness of a port. Next, the consequences of uncertainty in combination with large sunk costs for the rules which guide the decision to invest are examined. The last part considers the problem of forecasting port traffic and its volatility because they are crucial for an accurate evaluation of the investment projects.

1. INTRODUCTION

The success of a port in attracting investors will depend to a large extent not only on its current, but also its future competitive position. Good and well-planned investments can help to strengthen the competitive position of a port. Investment in a port may create additional capacity, increase productivity, generate added value and income, boost international trade, and

Global Competition in Transportation Markets: Analysis and Policy Making
Research in Transportation Economics, Volume 13, 279–298

ISSN: 0739-8859/doi:10.1016/S0739-8859(05)13013-3

create economic growth within the port and also in other sectors of the economy.

Due to the complex nature of a port, analysing investments in ports has to be done at several levels by looking at: investments in port infrastructure, investments in the superstructure, investments in hinterland connections, investments by logistics companies and investments by companies from the industrial and service sectors. Analysing the factors that affect port investment is a difficult task because investment operates at different levels and involves a large number of decision takers. The analysis is further complicated by the interconnections between the different investment levels. For this reason, this chapter is restricted to some specific aspects of port investment.

The first part treats the problem of private and/or public involvement and the impact on the competitiveness of a port. It is clear that for some types of port investment private capital can easily be attracted. But given the large amounts of capital needed for some other types of port investments and the long payback time of the investment projects, public money is often the only option. However, one needs a good and reliable social cost–benefit analysis to find out whether public investment is worthwhile from a social point of view. One of the dangers of public involvement in economic activity is that it can lead to distortions of competition and efficiency losses, which should be avoided.

The second part examines the consequences of uncertainty for the rules, which guide the decision to invest. Investment in port infrastructure often involves large sunk costs. In combination with uncertainty this requires adjustment of the traditional net present value rule, whether applied to private or public investments. In the presence of uncertainty and irreversibility the decision will not solely be whether to invest, but when to invest.

The last part considers the problem of forecasting port traffic and its volatility. Given the long time horizon of port investments, it is crucial to have good tools to forecast the costs, and especially the future returns of the investment project. Too often forecasts of the returns are based upon some form of trend extrapolation or some simplified scenario. However, one should at least pay attention to two important elements that have an impact on the returns of the port investment, the future economic situation, and the competitive position of the port. These factors will not only affect the investment in port infrastructure, but will also have an impact on the investment decisions about the superstructure and the location of service providers and industrial companies.

2. FINANCING PORT INVESTMENTS: PUBLIC AND/OR PRIVATE?

A port is a complex entity bringing together a variety of activities and many kinds of infrastructure (Fig. 1). Its accessibility by sea is straightforward, but it also needs good, reliable links with the land-side. As a consequence, port investments cover not just the internal infrastructure (berths, docks, storage areas, etc.) and superstructure (cranes, terminals, etc.), but also the infrastructure for maritime access (breakwaters, locks, buoys, etc.) and for hinterland connections (roads, railways, etc.).

The port industry is clearly an industry in transition. In the past, ports were considered as suppliers of services, which were of general (economic)

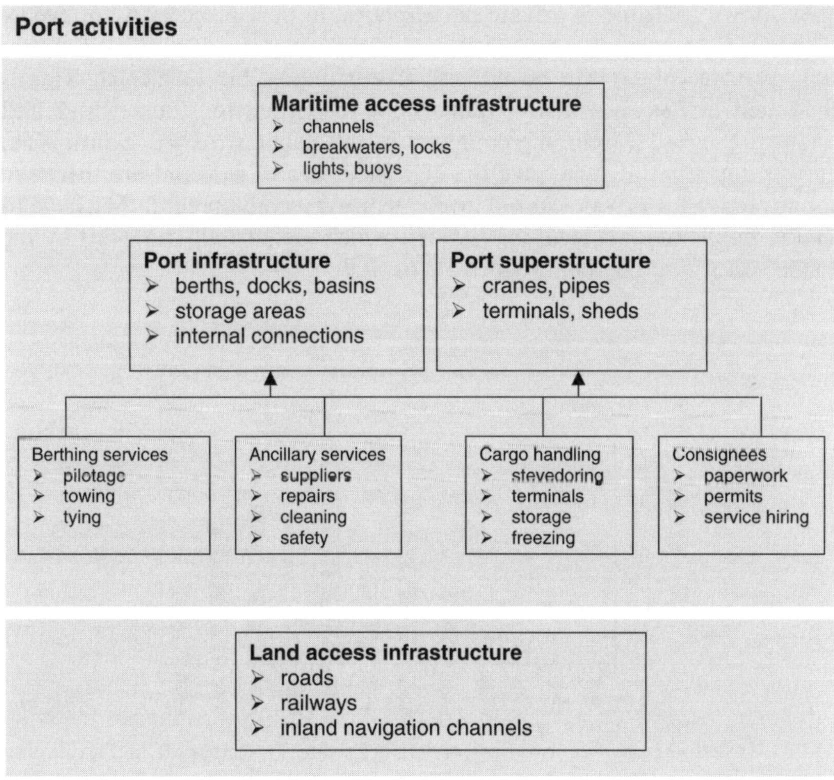

Fig. 1. Port Activities and Infrastructure. *Source:* Estache & de Rus (2000).

interest. For that reason they were in the hands of the public sector and financed by the taxpayer. There was no immediate concern for efficiency and cost minimisation. Modern developments, such as containerisation and the introduction of IT, have put pressures on ports to become competitive and efficient.

Ports are becoming commercial entities providing optimal services at profitable prices in such a way that they recover their costs from port users. As a consequence, since 1990 the private sector has become more and more involved in port activities. Fig. 2 is based on the World Bank's Private Participation in Infrastructure (PPI) Project Database that records details of all projects owned or managed by private companies in developing countries. It gives the evolution of the annual investment in seaport projects with private participation in developing countries in billions of US dollars.

As shown in Table 1, private participation in port projects is not evenly distributed. In regions where risk and uncertainty are experienced as too high, private capital can be difficult if not impossible to attract. This is illustrated in Fig. 3, which contains three indicators for country risk and creditworthiness. Developing countries in sub-Saharan Africa, South Asia, Europe and Central Asia have the highest degree of risk and are therefore not attractive for private capital investment. As a consequence, they have to rely on public money, resulting in ports which are partially financed by the public sector and partially by private capital.

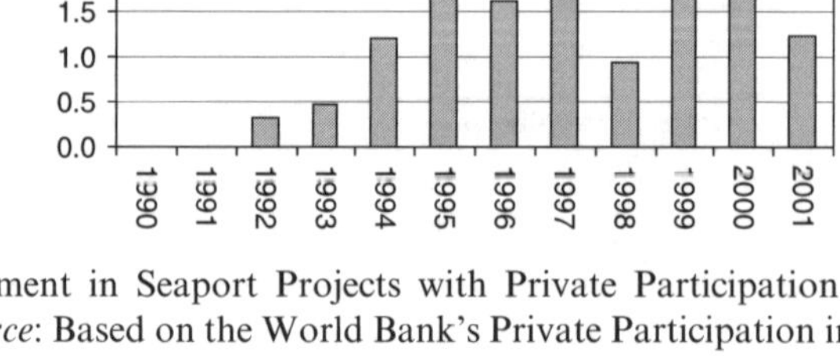

Fig. 2. Investment in Seaport Projects with Private Participation in Developing Countries. *Source*: Based on the World Bank's Private Participation in Infrastructure (PPI) Project Database.

Table 1. Port Projects with Private Participation in Developing Countries by Region (1990–1998).

Region	No. of projects	Total Investment (1998 US$ millions)
East Asia and the Pacific	38	5,410.5
Europe and Central Asia	8	23.4
Latin America and the Caribbean	48	2,497.7
Middle East and North Africa	5	376.5
South Asia	9	942.6
Sub-Saharan Africa	4	32.0
Total	112	9,282.7

Source: Sommer (1999).

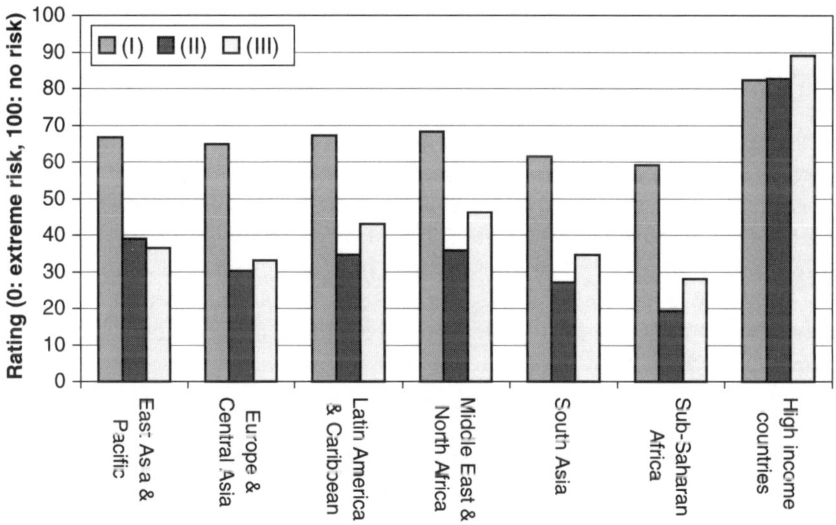

Fig. 3. Country Risk and Creditworthiness Indicators: (I) Composite ICRG risk rating; (II) Institutional Investor credit rating; (III) Euromoney country creditworthiness rating. *Source*: World Bank, World Development Indicators 2000, Washington DC (Table 5.3).

Even in developed countries with less risk, ports are not fully privatised and there is often still an important role for the public sector. According to Gwilliam (1997) the reasons for this are twofold:

- on the one hand, governments may resist full privatisation because it is believed that
    - strategic issues are at stake,
    - public ownership is necessary to control social impacts,
    - and/or a private monopoly would exploit users; and
- on the other, private companies do not want to enter the sector because
    - there is no apparent revenue flow,
    - they think that there is a high probability of damaging government interventions,
    - and/or sunk capital will not be recoverable.

As a consequence, there are different types of ports varying from fully publicly owned to fully privately financed. Table 2 gives an overview of four types of port organisation in the European Community resulting from an investigation of the European Commission (Commission of the European Communities (2001)). According to this investigation, 90% of European maritime traffic is estimated to be handled in ports where decisions on funding for infrastructure and charging of expenditure are, to varying degrees, influenced by public authorities (Types I–III).

Table 3 gives an overview of public financing per investment category in ports of the European Community. Investment is mainly in port infrastructure (32%) and may reflect construction in existing port areas by spending on such infrastructure as docks, quay walls, or internal docks. Although Juhel (1998) states that all superstructure and equipment tends to be privately owned, there was still in the EC a considerable amount of public money going into port superstructure and port services (663.6 million Euro or 41% of the total amount of public investments in EC ports).

Privatisation of port infrastructure and services can take different forms: full privatisation, joint ventures, concessions, leasing, licensing, management contracts, etc. The type chosen depends on a number of aspects such as types of traffic, traffic levels, port facilities, etc.

One of the big issues in the wave of privatisation, is the guaranteeing and safeguarding of competition. Developments on a worldwide scale indicate a tendency towards more private initiative and more transparency when concessions are given. Fair and efficient pricing, the problem of infrastructure charging, and the access to port services are issues which are at the heart of the European Commission's policy for ports and its competition policy. The major aim is to create transparency and fair competition to stimulate port investment in modern, innovative and cost-efficient infrastructure and

Table 2. Organisational Structure of European Ports.

	Type I	Public Type II	Type III	Private Type IV
Ownership	Public	Public	Mixed	Private
Autonomy of port management	Very restricted	Limited	High	Complete
Responsibility of port management	State operated/ 'Tool 'port'[a]/ 'Landlord port'[b]	'Landlord port'[b] (predominant) 'Tool port'[a]	'Full service port'[c]	'Full service port'[c]
External public funding	Extensive	Important	Very limited	No public aid
Cost recovery practices	Not principal objective	Partial recovery predominant	Full services, Some infrastructure investments	Full cost recovery
Access to provide services	Open tender/ direct agreement	Direct agreement predominant	Direct agreement	Normally closed
Relative importance in traffic terms[d]	Limited 8%	Very important 75%	Limited 7%	Limited 10%
EU states employing organisation types I–IV	Dk, Gr, F, P, D, I	B, Dk, Fin, F, D, Gr, NL, P, E, S, I	Dk, Ir, S, U.K.	Mostly U.K., but also in other member states

Source: Commission of the European Communities (2001).
[a]A port where the public authority is not only providing basic infrastructure but also (some) facilities to port operators.
[b]A port where the public authority is co-ordinating port development and manages only basic infrastructure.
[c]A port operating company runs the port entirely. This company is very often established in a mixed holding between public and private operators.
[d]Traffic estimates based on EU member states replies and best evidence available.

equipment. A lack of transparency in the structure of ports, their ownership, the involvement of central, regional and local governments, pricing rules, administrative practices, etc. will only increase uncertainty and hamper investments.

Table 3. Public Financing per Investment Category.

Investment Category	1997 Euro (million)	Split per Investment Category (1997) (%)	Evolution 1995–1997 (%)
1. Land purchase	69.4	4	139
2. Maritime access	77.1	5	−28
3. Port infrastructure	507.6	32	55
4. Port superstructure	358.4	22	6
5. Infrastructure links	24.1	2	−47
6. Port maintenance works	219.1	14	30
7. Port services	305.2	19	31
8. Other port activities	35.7	2	−7
Total public financing	1,596.6	100	24

Source: Commission of the European Communities (2001).

3. PORT INVESTMENTS, IRREVERSIBILITY AND UNCERTAINTY

A correct evaluation of the investment project, its costs and its returns, is always a necessity whether the project will be financed with public or with private capital. The major characteristics of investment projects in port infrastructure are that they often are of a large scale, are highly irreversible and, due to the long time horizon, the returns of the project have a high degree of uncertainty.

The presence of uncertainty in combination with the high degree of irreversibility of many of the investment projects in port infrastructure has consequences for the decision to invest. The traditional net present value (NPV) rule for evaluating investment projects is no longer valid and needs to be adjusted. Recent developments in investment theory borrow from real option theory to analyse this kind of investment decision. The result is that the timing of the investment becomes important and it can become important to design the project as a number of consecutive stages.

Discussing the importance of uncertainty of the economic environment on the decision to invest is not new. Traditionally, taking into account uncertainty in investment analysis was done by introducing expectations on a set of variables in the investment function, varying from simple adaptive expectations to more complex models with rational expectations.

During the past two decades, a new strand of theory has emerged, linking uncertainty and investments. This theory is based on the simultaneous existence of three phenomena: uncertainty, irreversibility of investment and some freedom of choice on the timing of investment. It starts from the fact that investment decisions are to a large extent irreversible, i.e. it cannot be reversed except at a high cost (the cost is largely 'sunk'). Combining irreversibility with the existence of uncertainty over the future behaviour of variables that affect the value of the investment (such as future output prices) leads to the following intuitive reasoning. Suppose there is some leeway in delaying investment until more information about the uncertain future becomes available, it may then be optimal to wait for some time before investing. It is clear that waiting to invest implies risks (e.g. entry of competitors) and foregone profits, but it may prevent one from being trapped in an irreversible investment project, which turns out to be very costly when an adverse future materialises.

The theory states that an investment project, which satisfies these three characteristics is best treated as analogous to holding a (American-type) financial call option. For some specific time period, an investor (a firm) has the right, but not the obligation to pay a certain price (the investment cost) in return for an asset (an investment project) that has some value. When the investment decision is made the option is exercised, which is an irreversible decision. Like a financial option, the option itself has some (non-negative) value a/o. because of uncertainty over the future value of the investment project. As a consequence, option-pricing theory can be used to 'price' investment decisions and decide on the optimal timing of the exercise. This theory has given rise to a large body of new literature and a new class of models usually referred to as 'real options' models.[1]

The basic model was originally developed by McDonald and Siegel (1986). For a project whose present value is V and for which a sunk cost I has to be made, the NPV rule will lead to investing actually in the project if $V \geqslant I$. Now, it is assumed that the value V varies over time and follows a geometric Brownian motion with drift

$$dV = \alpha V\, dt + \sigma V\, dz \qquad (1)$$

where α is the mean growth rate, σ^2 the variance of dV, and dz the increment of a Wiener process such that

$$dz = \varepsilon_t \sqrt{dt} \quad \text{where } \varepsilon_t \sim N(0,1); \qquad E(\varepsilon_i \varepsilon_j) = 0 \quad \forall i,j \;\; i \neq j$$

This implies that the future values of the project are log-normally distributed with a variance that grows exponentially with time (which is not always the case in reality).

One has to find the point in time T at which it is optimal to invest an amount I. This point will be reached if the expected present value of the option to invest, $F(V)$, is maximised:

$$F(V) = \max E[(V_T - I)e^{-\rho T}] \tag{2}$$

where ρ is the discount rate. Holding the option and delaying the investment is similar as holding an asset, which yields no cash but which may appreciate over time. The optimality condition is given by the Bellman equation

$$\rho F \, dt = E[dF] \tag{3}$$

which shows that the total expected return of the investment opportunity is equal to its expected rate of appreciation.

Using Ito's lemma,

$$dF = F'(V)dV + \tfrac{1}{2}F''(V)(dV)^2 \quad \text{with } F' = dF/dV \text{ and } F'' = d^2F/dV^2 \tag{4}$$

Substituting (1) for dV into (4) gives

$$E[dF] = \alpha VF'(V)\,dt + \tfrac{1}{2}\sigma^2 V^2 F''(V)\,dt \tag{5}$$

Substituting this expression into (3) after dividing throughout by dt, gives for the Bellman equation,

$$\rho F = \alpha VF'(V) + \tfrac{1}{2}\sigma^2 V^2 F''(V) \tag{6}$$

Apart from this condition, the optimal investment rule requires that the value of the option to wait must also satisfy three boundary conditions

$$F(0) = 0 \tag{7}$$

$$F(V^*) = V^* - I \tag{8}$$

$$F'(V^*) = 1 \tag{9}$$

The first condition states that the option to invest will be of no value when $V = 0$. The second condition determines the net payoff at the value V^* at which it is optimal to invest. When Eq. (11) is rewritten as $V^* = I + F(V^*)$, the value of the project is compared to the full cost of making the investment (direct cost I plus opportunity cost $F(V^*)$). The third condition is needed to guarantee that the function $F(\cdot)$ is continuous and smooth around the optimal investment timing point.

The solution to (6) subject to conditions (7)–(9) is

$$F(V) = AV^b \tag{10}$$

with $b > 1$.

Substituting (10) into (8) and (9) gives

$$V^* = \frac{b}{b-1} I \tag{11}$$

and

$$A = \frac{V^* - I}{(V^*)^b}$$

where

$$b = \frac{1}{2} - \frac{\alpha}{\sigma^2} + \sqrt{\left(\frac{\alpha}{\sigma^2} - \frac{1}{2}\right)^2 + \frac{2\rho}{\sigma^2}} \tag{12}$$

Since $b > 1$, Eq. (11) states that the traditional NPV rule is incorrect because investment only becomes worthwhile for values which are bigger than the investment cost ($V^* > I$). Uncertainty and irreversibility drive a wedge between the critical value V^* and the investment cost. The exact magnitude of this wedge will depend upon the growth rate α of the value of the project, the variance σ^2 of the value of the project, and the expected rate of capital appreciation ρ. An important consequence is that the value of b will increase with higher values of σ^2, and hence with higher volatility of the value of the project. This is illustrated in Table 4. For an expected growth rate of 2%, an expected rate of capital appreciation of 4% and a variance of 0.0001, the present value of the investment project must be twice as high as the investment cost to make the investment worthwhile. The higher the rate of capital appreciation, the smaller will be the wedge. Due to the fact that the opportunity cost of waiting goes up when ρ increases, it becomes worthwhile to kill the option sooner. A higher expected growth rate of the value of the project increases the wedge, because it is worthwhile to postpone the killing of the option.

The most important consequence resulting from this model solution is that uncertainty (and irreversibility) introduces a difference between the minimum-acceptable value of the project in order to invest (V^*) and the cost of investment (I). More importantly, the greater the level of uncertainty, the more the value of the investment opportunity must exceed its cost before

Table 4. Critical Value V^* as a Function of α, ρ and σ.

α	ρ	σ	σ^2	I	b	Wedge	V^*
0.02	0.04	0.01	0.0001	100.00	2.00	2.00	200.50
0.02	0.04	0.02	0.0004	100.00	1.98	2.02	201.98
0.02	0.04	0.03	0.0009	100.00	1.96	2.04	204.41
0.02	0.04	0.04	0.0016	100.00	1.93	2.08	207.71
0.02	0.04	0.05	0.0025	100.00	1.89	2.12	211.84
0.02	0.04	0.10	0.0100	100.00	1.70	2.43	242.54
0.02	0.04	0.20	0.0400	100.00	1.41	3.41	341.42
0.02	0.04	0.30	0.0900	100.00	1.26	4.84	483.65
0.02	0.04	0.40	0.1600	100.00	1.18	6.70	670.16
0.02	0.04	0.50	0.2500	100.00	1.12	9.03	902.85
0.02	0.06	0.01	0.0001	100.00	2.99	1.50	150.37
0.02	0.06	0.02	0.0004	100.00	2.94	1.51	151.47
0.02	0.06	0.03	0.0009	100.00	2.88	1.53	153.24
0.02	0.06	0.04	0.0016	100.00	2.80	1.56	155.60
0.02	0.06	0.05	0.0025	100.00	2.71	1.58	158.47
0.02	0.06	0.10	0.0100	100.00	2.27	1.78	178.44
0.02	0.06	0.20	0.0400	100.00	1.73	2.37	236.60
0.02	0.06	0.30	0.0900	100.00	1.47	3.15	314.86
0.02	0.06	0.40	0.1600	100.00	1.32	4.14	413.75
0.02	0.06	0.50	0.2500	100.00	1.23	5.34	534.43
0.01	0.06	0.01	0.0001	100.00	5.86	1.21	120.59
0.01	0.06	0.02	0.0004	100.00	5.50	1.22	122.20
0.01	0.06	0.03	0.0009	100.00	5.07	1.25	124.56
0.01	0.06	0.04	0.0016	100.00	4.65	1.27	127.43
0.01	0.06	0.05	0.0025	100.00	4.26	1.31	130.66
0.01	0.06	0.10	0.0100	100.00	3.00	1.50	150.00
0.01	0.06	0.20	0.0400	100.00	2.00	2.00	200.00
0.01	0.06	0.30	0.0900	100.00	1.61	2.65	264.66
0.01	0.06	0.40	0.1600	100.00	1.41	3.45	345.24
0.01	0.06	0.50	0.2500	100.00	1.29	4.43	442.91

investment takes place. As such, the individual investment decision is, in theory, very dependent on the level of uncertainty.

Fig. 4 illustrates that for a project with a return value V between points 1 and 2, the NPV rule would lead to the decision to invest, whereas the Real Options rule suggests not to invest. Similar analyses can be used to determine the most profitable timing of investments in infrastructure that can be constructed in different, consecutive stages. The general conclusion is that the traditional NPV rule will not necessarily lead to the best decision for

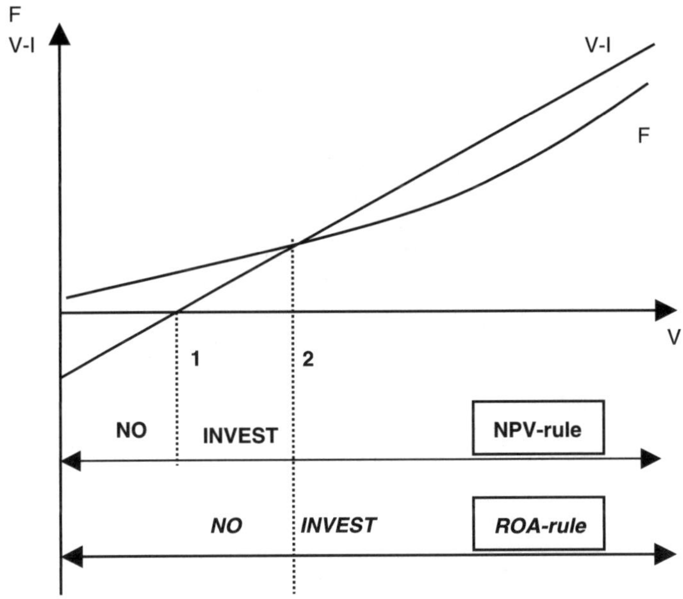

Fig. 4. Real Option Analysis (ROA) versus the NPV Rule. *Source*: Cassimon, Engelen, Meersman, and Van Wouwe (2003), p. 197.

infrastructure projects with high, irrecoverable sunk costs and which face a lot of uncertainty.

4. FORECASTING THE GROWTII AND VARIANCE OF THE EXPECTED VALUE OF PORT INVESTMENTS

From the previous discussion, it is clear that knowledge of the expected growth rate and the variance of the present value of an investment project will be of great help in the decision whether to invest or not. On the one hand, the success of port investments is closely related to the evolution of maritime traffic flows, and on the other, the competitive position of a port.

Maritime traffic is a very general concept that hides the complexity of what really goes on in the sector. It is clear that short sea shipping is quite different from deep-sea traffic and that container traffic and liner shipping are totally different industries to bulk traffic. Forecasting traffic flows to and

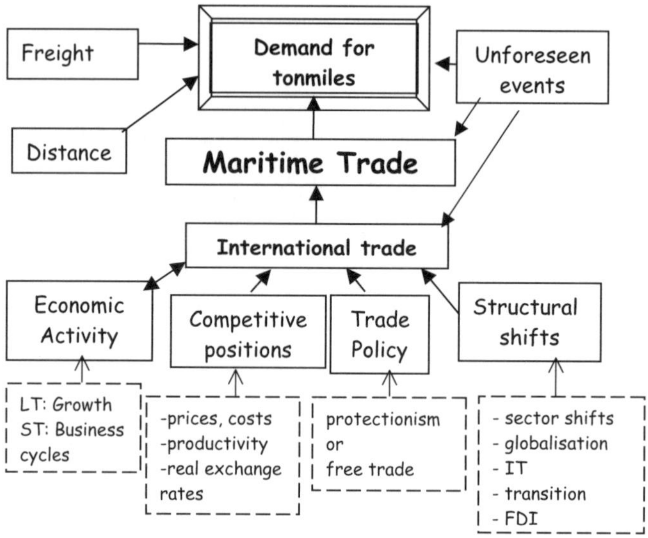

Fig. 5. Factors Influencing the Demand for Maritime Transport.

from a specific port is different from forecasting the overall trend in maritime traffic. The volume of maritime traffic (in general) is the result of a very complex process, which is summarised by the scheme in Fig. 5.

From this scheme it is clear that uncertainties can penetrate maritime trade, the demand for sea transport and, hence, the returns on port investments in a lot of different ways. A good example is the East-Asian crisis, which disturbed economic growth and trade patterns. It also had consequences on the financial markets, which affected investment decisions.

The problem is that long-term evolutions in maritime trade are not under the control of port authorities and port operators. They do not decide on trade policy, on business cycle evolutions, on exchange rates, on economic policies, etc. These are exogenous factors for which one can try to find, buy or produce good and reliable forecasts based on alternative scenarios.

The competitive position of a port will determine which part of maritime traffic it can attract at present as well as in the future. Huybrechts et al. (2002) give a good account of the complexity of determining the competitiveness of seaports. It is clear that it is impossible to give a static definition of a port and of port competitiveness. The role and significance of seaports have changed considerably and the modern seaport has to position itself as a critical node in the logistics chain. Whether investments in port infrastructure will be profitable depends to a large extent on the ability of the port to be one of the key players in the process to minimise the generalised costs of the entire chain.

Modelling and forecasting the competitive position of a port is a complex and not always transparent process. Until now, there has not existed a scientifically sound instrument to simulate the consequences of all kinds of policy measures and strategies on the continuous fight for commodity flows. To some extent this has to do with the lack of reliable information and data, but it is also due to the complexity of the behaviour on which shifts among ports are based. One needs strategic market share models and simulation tools, which can capture the game-theoretic behaviour of all the market

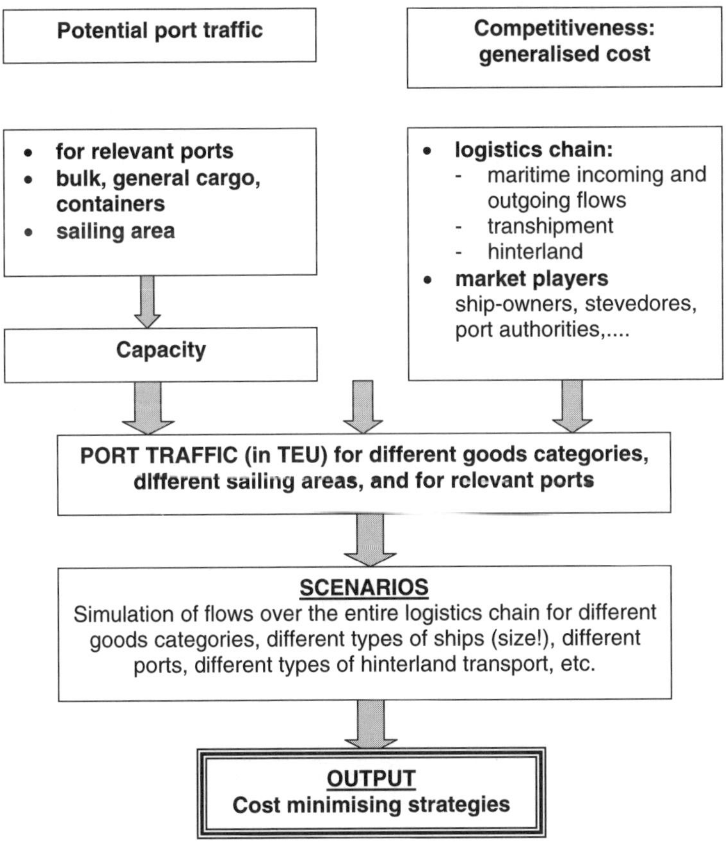

Fig. 6. General Structure of a Forecasting Model for Port Traffic Flows.

players involved in the port choice. It is self-evident that in the long run all parties involved want to maximise profits, but the complexity is mainly due to the different strategies the market players will adopt in the short term to realise their long-term objectives.

Fig. 6 gives an overview of the structure of a model for forecasting port traffic flows, which tries, as much as possible, to incorporate all the relevant decision variables and market players (Winkelmans, Van de Voorde, Huybrechts, Meersman, & Verbeke (1998)).

This model contains two major blocks, the first of which defines potential traffic flows for relevant ports (actual or potential competitors), for different goods categories and for different sailing areas. In this first stage, the potential flows can already be compared with the existing port capacity. The second block analyses in detail all the factors influencing the competitiveness of the port with special attention given to the generalised costs. It is in this part that modelling the strategic behaviour of market players is crucial. Bringing together both blocks results in forecasts of the market shares and realised traffic. Finally, scenarios can be designed to simulate freight flows over the entire logistics chain starting from the port of origin to the final destination of the commodities.

From the previous discussion, it is clear that it is still a very hard job to forecast port traffic. If one cannot model in a reliable way the factors

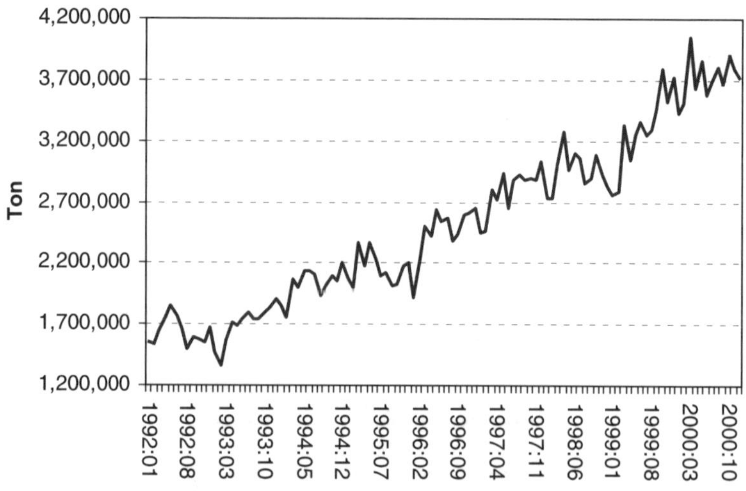

Fig. 7. Container Traffic in the Port of Antwerp (1992–2000). *Source*: Antwerp Port Authorities.

determining the competitive position of a port, it will be difficult to take full account of uncertainty in the decision to invest in port infrastructure.

One way to bypass the difficulties faced when trying to model in detail the competitive position of a port is to use forecasting methods based on time-series analysis. This is illustrated for the case of the port of Antwerp. Fig. 7 gives the evolution of container traffic in the port of Antwerp from 1992 to 2000 on a monthly basis. There is clearly some volatility present in the traffic over this time period.

The evolution of the container traffic can be modelled in logarithms by means of an ARIMA(12,1,24)–GARCH(1,1) specification:

$$\text{dln } y_t = \sum_{i=1}^{12} \alpha_i \text{dln } y_{t-i} + \sum_{j=1}^{24} \beta_j u_{t-j} + u_t$$

and

$$\sigma_t^2 = c + \gamma\sigma_{t-1}^2 + \rho u_{t-1}^2$$

where

dln y is the rate of change in container traffic in Antwerp (monthly changes), u a stochastic error term, σ^2 the variance of the error term and of dln y and α_i, β_j, c, γ, ρ are the coefficients to be estimated.

Table 5. ARIMA–GARCH Estimation of Container Traffic in the Port of Antwerp.

	Coefficient	S.E.	*z*-Statistic	Probability
AR(1)	−0.350411	0.096609	−3.627099	0.0003
AR(12)	0.336994	0.090640	3.717929	0.0002
MA(1)	−0.287928	0.056208	−5.122509	0.0000
MA(12)	0.055680	0.045186	1.232249	0.2179
MA(24)	0.557891	0.058584	9.522980	0.0000
	Variance Equation			
C	0.001251	0.000530	2.361890	0.0182
ARCH(1)	−0.107506	0.019511	−5.509923	0.0000
GARCH(1)	0.648493	0.193055	3.359112	0.0008
R^2	0.464332	Mean-dependent variable		0.009741
Adjusted R^2	0.421232	S.D.-dependent variable		0.069094
S.E. of regression	0.052564	Akaike info criterion		−2.959009
Sum-squared residue	0.240382	Schwarz criterion		−2.743945
Log likelihood	148.5529	Durbin–Watson statistics		1.949687

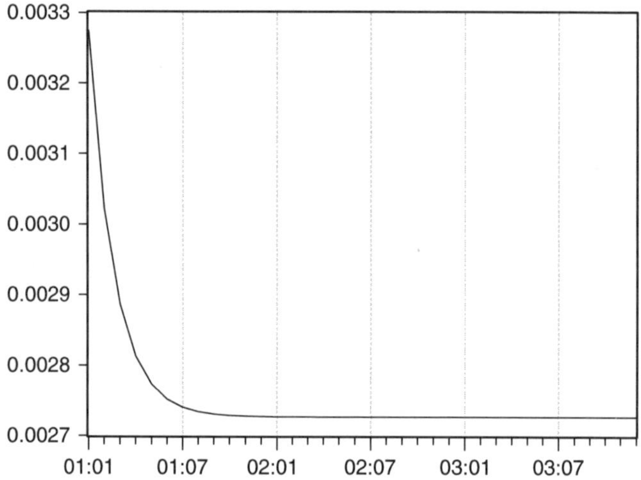

Fig. 8. Forecasts of the Variance of the Rate of Change of Container Traffic in Antwerp Based on an ARIMA(12,1,24)–GARCH(1,1) model.

Table 5 gives the results of the ARIMA–GARCH estimation.

The advantage of this method is that it allows one to forecast the growth rate of the traffic and also the variance (Fig. 8). This information can be very helpful in the evaluation of investment projects in ports.

This methodology has the advantage that it can be used with a limited number of variables, in contrast to a model that tries to capture all the factors influencing the competitive position of a port.

5. CONCLUSION

Uncertainty is a crucial aspect of the future. It is impossible to make waterborne transport more attractive without the necessary investments in infrastructure, in ports, in intermodal nodes, etc. It is quite obvious that those investment projects need to generate enough benefits above costs (private or social) to attract the necessary capital, whether private or public, and should be evaluated in the correct way.

For the traditional NPV rule one needs forecasts of all the future cash flows. Traditionally, the NPV is calculated with those forecasts and a sensitivity analysis gives an idea of the NPV for alternative scenarios. The uncertainty is not explicitly introduced in the decision.

Investments in port infrastructure are by their nature often irreversible and the amounts invested can be to a large extent sunk costs. The consequence is that, for this type of project, the presence of uncertainty will have a strong impact on the validity of the traditional NPV decision rule. Real option theory, applied to irreversible investments, shows that in the presence of uncertainty it can be better to wait to invest even when the project has a positive NPV. The higher the degree of uncertainty, the greater will be the probability of postponement.

One of the problems is the difficulty in obtaining reliable forecasts of future returns of the investment project and of their volatility. Good forecasts of port traffic can be of great help, but they are also hard to get. On the one hand, port and maritime traffic are derived from international trade. General economic conditions, trade policy, international competitive positions and structural changes will all have an impact on trade and make it difficult to forecast accurately future port traffic. On the other hand, the competitive position of a port will determine its attractiveness and its market share. Modelling port competitiveness is complex and difficult because of the need for good information and insight into the decision processes and strategies of all the relevant market players involved. In the meantime, one can rely on time-series methods using ARIMA–GARCH models (or extensions such as EGARCH and GARCH in M models) to generate forecasts not only of the growth rate of port traffic, but also of its volatility.

Potential investors should be well aware of the fact that increasing uncertainty, reflected in a higher volatility, requires a higher critical present value to make it worthwhile. This may be one of the reasons why it is difficult to attract private capital for large, irreversible port infrastructure projects. During the past few years, however, we have seen an evolution towards a larger involvement of the private sector, which has resulted in more efficiency, higher productivity and better service. However, there is still a long way to go. In order to continue this process of privatisation, governments on a national and international level should create and safeguard a climate of fair competition, which might increase returns on investment, but, even more importantly, could also reduce the volatility of future returns. It is especially this last element that is needed to make investments in large, irreversible infrastructure projects more attractive.

NOTES

1. See e.g. McDonald and Siegel (1986), Caballero and Pindyck (1992), Chirinko (1993), Pindyck and Solimano (1993), and Bertola and Caballero (1994).

A comprehensive theoretical treatment is given in Dixit and Pindyck (1994). A recent extensive review of theory and empirics is presented in Carruth et al. (2000).

REFERENCES

Bertola, G., & Caballero, R. J. (1994). Irreversibility and aggregate investment. *Review of Economic Studies*, *61*(2), 223–246.

Caballero, R. J., & Pindyck, R. (1992). Uncertainty, investment, and industry evolution. Paper presented at the *economic policy conference*, Gotheburg, Sweden, August.

Carruth, A., Dickerson, A., & Henlay, A. (2000). *What do we know about investment under uncertainty? Journal of Economic surveys*, *14*(2), 119–153.

Cassimon, D., Engelen, P. J., Meersman, H., & Van Wouwe, M. (2003). Investment, uncertainty and irreversibility: Evidence from Belgian accounting data. In: P. Butzen & C. Fuss (Eds), *Firms' investment and finance decisions* (pp. 194–219). Cheltanham: Edward Elgar.

Chirinko, R. S. (1993). Business fixed investment spending: Model strategies, empirical results, and policy implications. *Journal of Economic Literature*, *31*, 1875–1911.

Commission of the European Communities. (2001). *On public financing and charging practices in the Community sea port sector (on the basis of information provided by the member states)*. Commission staff working document, SEC 234, Brussels.

Dixit, A., & Pindyck, R. (1994). *Investment under uncertainty*. Princeton, NJ: Princeton University Press.

Estache, A., & de Rus, G. (Eds) (2000). *Privatisation and regulation of transport infrastructure: Guidelines for policymakers and regulators*. Washington, DC: World Bank Institute.

Gwilliam, K. (1997). *Contract design for concessions and franchising*. Washington, DC: World Bank Institute.

Huybrechts, M., Meersman, H., Van de Voorde, E., Van Hooydonk, E., Verbeke, A., & Winkelmans, W. (2002). *Port competitiveness. An economic and legal analysis of the factors determining the competitiveness of seaports*. Antwerp: Editions De Boeck Ltd.

Juhel, M. (1998). Globalisation, privatisation and restructuring of ports. Paper presented at the 10th Annual Australasian Summit on Ports, Shipping and Waterfront Reform.

McDonald, R., & Siegel, D. (1986). The value of waiting. *Quarterly Journal of Economics*, *101*, 707–728.

Sommer, D. (1999). Private participation in port facilities – recent trends. *Public policy for the private sector*. Note 193, Washington, DC: The World Bank Group.

Winkelmans, W., Van de Voorde, E., Huybrechts, M., Meersman, H., & Verbeke, A. (1998). *Aspecten van havencompetitiviteit*. Final research report of the concerted action on Port Competitiveness (1995–1998). University of Antwerp.

AN INTERNATIONAL CONTAINER SHIPPING MODEL IN EAST ASIA AND ITS TRANSFERABILITY

Ryuichi Shibasaki, Hitoshi Ieda and Tomihiro Watanabe

ABSTRACT

Under the rapid expansion of container shipping in Eastern Asia with the increased size of vessels seeking economy of scale, hub-spoke operations with efficient transshipment, and global alliances of shipping companies, an Asia-range simulation model to estimate the effects of policy improvement of port facilities and operations upon maritime container flow patterns is proposed. In this chapter, the transferability of the original model, which was based on data from 1993, is ascertained by data from 1998. The model's reproducibility is similar to the original model, although the newer model leads to more extensive hub-spoke transportation than in actual use.

1. INTRODUCTION

Container shipping has rapidly expanded in Eastern Asia because of the region's impressive economic growth. Large ports such as Hong Kong, Singapore, Kaoshiung (Taiwan), and Busan (Korea) occupy major positions

Global Competition in Transportation Markets: Analysis and Policy Making
Research in Transportation Economics, Volume 13, 299–336

ISSN: 0739-8859/doi:10.1016/S0739-8859(05)13014-5

among the world's top container ports every year, replacing European, American, and Japanese ports. Vessels increased in size in the 1990s, seeking economy of scale, which encouraged the concentrated use of large ports equipped with deep berths and efficient logistic systems, increasing the hierarchical movement of freight, namely hub-spoke operations with efficient transshipment. The growing freight transport market and stronger competition sometimes brought frequent and direct operation using medium-sized vessels as well as so-called "global alliances" of shipping companies. On the other hand, the significant shift of calls of large vessels from Kobe to Busan, observed after the 1995 Hanshin-Awaji Earthquake in Japan, aroused concern for improving the efficiency of port operations to keep up with other Asian ports. Disputes over whether investment policies should focus on local or major ports remain unresolved.

Against this dynamic and critical background, we propose an Asia-range simulation model (Ieda, Shibasaki, & Naito, 1999; Ieda, Shibasaki, Naito, & Mishima, 2000) to estimate the effects of such policies as improvement of port facilities and operations upon sea traffic flow patterns, container flow patterns, and the use of ports. In this chapter the transferability of the model, which was developed based on data from 1993, is tested with data from 1998. The reproducibility of the model and the future direction of its improvement are summarized.

2. STUDIES ON A MODEL SIMULATING CONTAINER MOVEMENT

The Freight Network Equilibrium Model (FNEM) (Friesz, Viton, & Tobin, 1985; Harker & Friesz, 1986; Harker, 1987) is a typical freight transportation model that considers the behavior of both shippers and carriers and their interaction with the network. It is argued in many papers (e.g. Matsumoto & Watanabe, 1994; Crainic & Laporte, 1997; Japan Society of Civil Engineering 1998; Pendyala, Shankar, & McCullough, 2000). However, Crainic and Laporte (1997) and Pendyala et al. (2000) pointed out that "this line of research has not yet yielded practical planning models and tools" because it is "more complex to implement than other modeling method and often has more intensive data requirements." Harker and Friesz (1985) and Friesz, Gottfried, and Morlok (1986) are exceptions who studied a domestic freight transportation network in the U.S.A. and mainly concentrated on railways. Recent research in this field has analyzed the detailed or theoretically strict behavior of shippers and carriers with numerical experiments on a very simplified network (see Harker, 1988; Miller, Tobin, & Freisz 1991; Hurley & Pettersen, 1994; Brotcorne, Labbe, Marcotte, & Savard, 2000). On the other hand, for applications of actual freight transportation, a simple disaggregated model, or shortest path model, see Crainic, Florian, and Leal (1990), Southworth and Peterson (2000), and Malchow and Kanafani (2001). Even though the number of carriers is limited, the international container transportation market is considered competitive because it is relatively difficult to differentiate level of service except price. For a market with such a characteristic, a network model that considers interaction between shippers and carriers is an effective and reasonable tool to describe actual transportation: the calculations are relatively easy, and equilibrium can be assumed.

Inamura, Nakamura, and Gu (1997) and Takebayashi and Kuroda (2000) applied this type of network model to the expression of an actual international container movement. However, the former research didn't consider the equilibrium between shippers and carriers, and the latter research assumed a perfectly competitive market with innumerous shippers and carriers. Therefore, Ieda *et al.* (1999) and (2000) proposed a model where container cargoes are assigned to a freight transportation network, assuming that market equilibrium is achieved as a result of the behavior of innumerous shippers who minimize transportation costs and several shipping alliance groups who minimize their total costs because of strong price competition.

3. DESCRIPTIVE ANALYSES OF INTERNATIONAL CONTAINER SHIPPING IN EAST ASIA

Before the introduction of our simulation model, in this section the actual conditions and changes in international container shipping in East Asia are analyzed mainly using data from 1993, which was employed in the original model (Ieda et al., 2000), and data from 1998, which will be employed to certify the model's transferability.

3.1. Amount of Container Handling

Fig. 2 shows the amount in twenty-foot equivalent unit (TEU) containers handled at eight major Japanese ports in 1993 and 1998: Tokyo, Yokohama, Shimizu, Nagoya, Osaka, Kobe, Kitakyushu, and Hakata, as shown in Fig. 1. A comparison between 1993 (left side of the bar charts for each port)

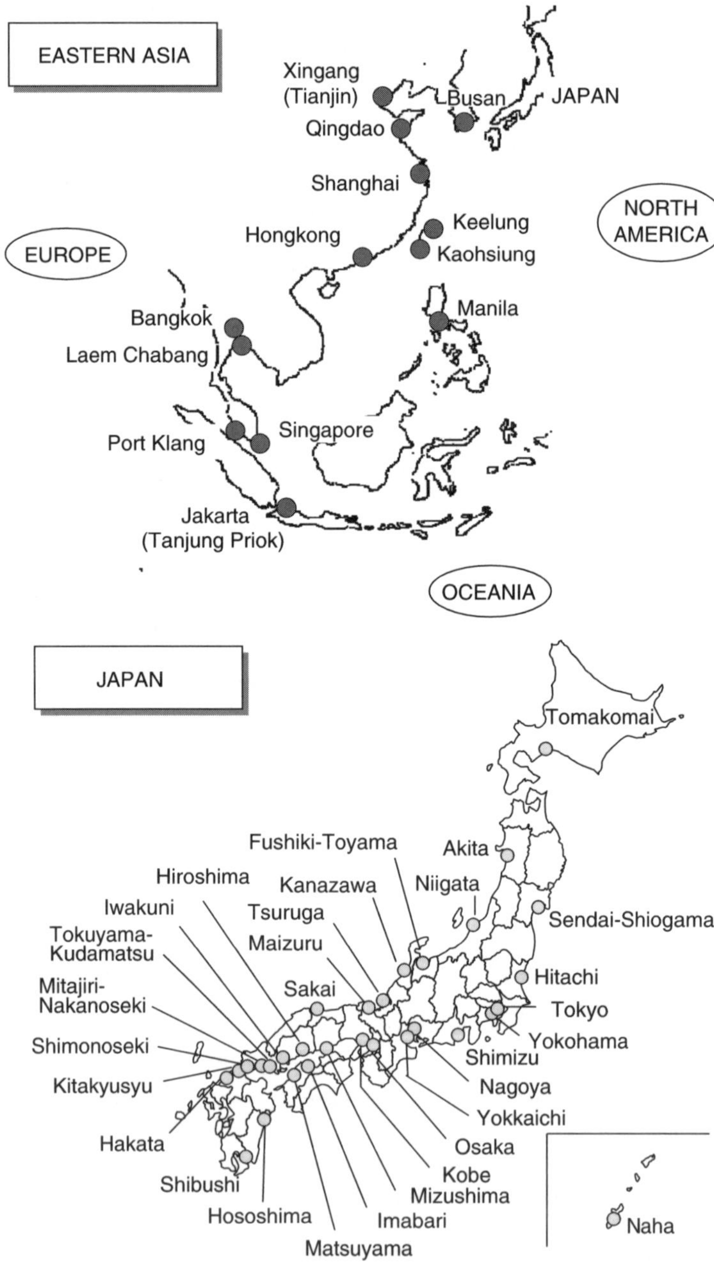

Fig. 1. Ports Mentioned in This Chapter.

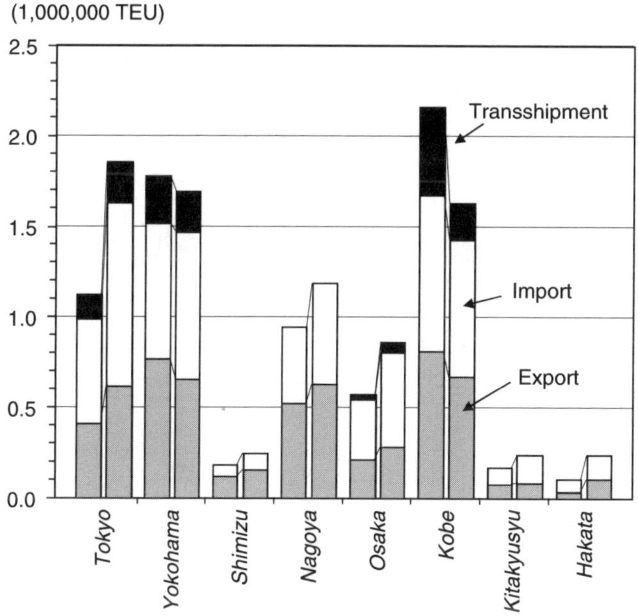

Fig. 2. Amount of Container Handling at Eight Major Japanese Ports: Left Side, 1993; Right Side, 1998. Source: Materials of Ports and Harbors Bureau, Ministry of Transport (each year).

and 1998 (right side) shows a decrease in container handling at the ports of Yokohama and Kobe and an increase in the other ports, as shown in Fig. 2. The decrease for the port of Kobe was caused by the Hanshin–Awaji Earthquake in 1995. Fig. 3 shows the amount of container handling at 22 other Japanese container ports (see Figs. 1 and 3). In almost all the ports the amount of container handling increased, except for Shimonoseki and Mitajiri–Nakanoseki, both located in Yamaguchi Prefecture of the Chugoku area. Fig. 4 shows the amount of container handling in 13 major East Asian ports (see Figs. 1 and 4). In all the ports, both the amount of container handling and transshipment steadily increased, except for Keelung (Taiwan) and Bangkok (Thailand) where container cargoes strategically shifted to the nearby larger ports of Kaohsiung and Laem Chabang by each government. In particular, the amount of containers handled in Chinese ports increased two to three times over the past 5 years, although transshipped cargoes were still not observed.

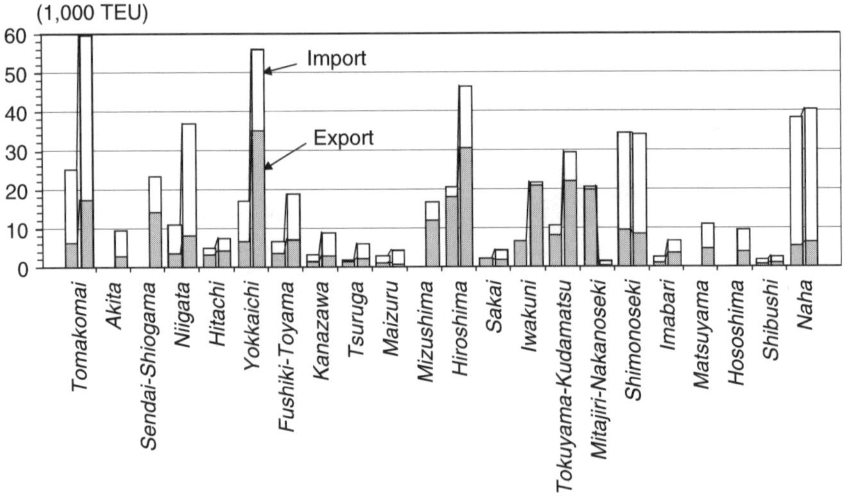

Fig. 3. Amount of Container Handling at 22 Other Japanese Ports. Source: Same as above.

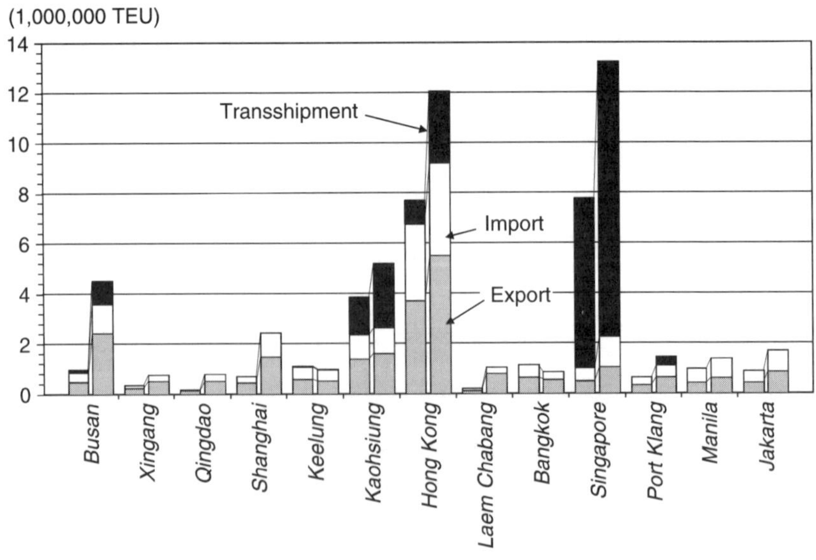

Fig. 4. Amount of Container Handling at 13 Major East Asian Ports (made by authors from Informa Group and Materials of Ports and Harbors Bureau, Ministry of Transport, each year).

3.2. Distribution of Vessel Use by Size Category

Vessel size in operations is one of the key points of our proposed model. International container ships are classified into four size categories by the loading capacity of the containers: over 4,000 TEUs (C-1); 2,250–4,000 TEUs (C-2); 1,000–2,250 TEUs (C-3); and less than 1,000 TEUs (C-4). We determined category boundaries according to the share of each category in the entire capacity (TEU base) and in vessel number. Roughly summarizing, vessels under 2,250 TEUs are used for feeder and intra-Asia container transportation, while vessels over 2,250 TEUs are used for container transportation on trunk routes between East Asia and other regions such as North America and Europe.

Fig. 5 shows the distribution of the amount of arriving and departing container cargo in vessels by size category in major Japanese and other Asian ports in 1995 and 1998. Here, the amount of arriving and departing container cargo by size category is calculated as follows: (i) the capacities of calling vessels for each port are summed up by size category from port call records in the Informa Group and Ocean Commerce (1996) and (1999) and then (ii) multiplied by the loading factor. The loading factor is assumed to be 0.4 when vessels that connect East Asia and other regions navigate in East Asian regions, and 0.8 for other cases, in reference to the survey. In this figure, the

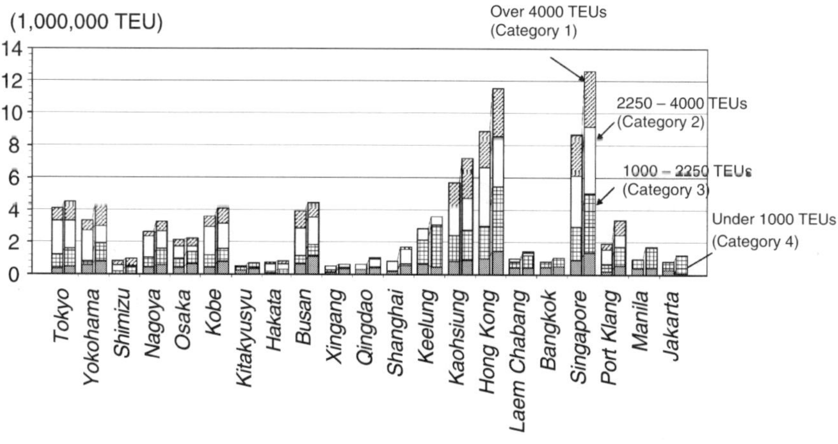

Fig. 5. Distribution of Amount of Arriving/Departing Container Cargoes in Vessels by Size Category in Major Japanese and Other Asian Ports: Left Side, 1995; Right Side, 1998 (estimated by authors).

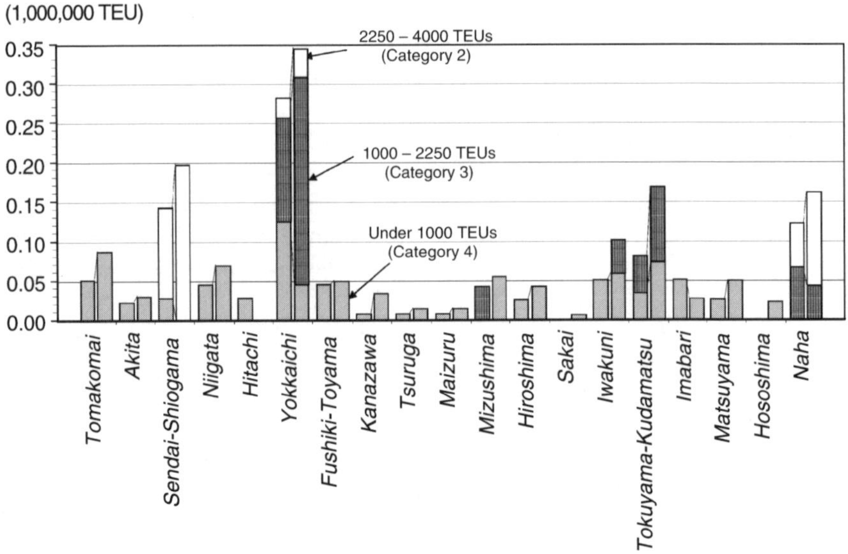

Fig. 6. Distribution of Amount of Arriving/Departing Container Cargoes in Other Japanese Ports (estimated by authors).

amount of arriving and departing container cargo in vessels increased in all ports during the 3 years, mainly due to the increased share of major world shipping companies and their affiliates. (In this chapter, the amounts of arriving and departing container cargoes are only totaled for these companies). The actual load factors are expected to decrease over the 3 years due to intense worldwide competition. Another tendency observed in the figure is the increase of the share of vessels over 4,000 TEUs and between 1,000 and 2,250 TEUs, contrary to the decrease of the share of vessels between 2,250 and 4,000 TEUs and under 1,000 TEUs. This indicates that vessel size increased on both trunk and feeder routes. Fig. 6 shows the distribution of the amount of arriving and departing container cargoes in vessels by size category in other Japanese ports. In these ports, with the exceptions of Sendai Yokkaichi and Naha, only the smallest vessels, under 1,000 TEUs of capacity were calling.

3.3. Global Alliances and Share of Each Alliance Group

In the model discussed in this chapter, container cargo flows are assigned by a group of shipping companies called a "global alliance," which is the operational federation of container transportation. Fig. 7 shows the composition of the major world alliance groups in international container transportation. For simplicity, major operation groups are assumed to be integrated over routes throughout the world. The total vessel capacities of these seven groups is close to 90% of the entire capacity of the trunk routes connecting East Asia with the other regions of the world, and around 80% on intra-Asia routes.

Fig. 8 shows the share of the amount of arriving and departing container cargoes for each alliance group in Japanese ports. For major Japanese and other Asian ports, even with the realignment of shipping companies and the increase in total amount as shown in Fig. 5, no special growth or decline was observed by amount share. On the other hand, the shares changed for the other Japanese ports. In particular, the share of Group A where included Nippon Yusen Kaisha, the leading shipping company in Japan, showed growth. Fig. 9 shows the share of the amount of arriving and departing container cargoes for each alliance group in several major Japanese and Asian ports. Although the total amount of shares of the eight major Japanese ports was balanced, as shown in Fig. 8, an alliance group generally uses either port in a bay, such as Tokyo, Yokohama, Osaka, or Kobe. For other Asian countries, ports are divided into two types according to the share's pattern. In the first type of port, such as the Busan or Shanghai alliance groups, national carriers occupy a tremendous share. In the second type of port, such as Hong Kong or Singapore, all alliance groups call equally.

3.4. Inter-region/Inter-port Container Cargo OD Flows

In the model proposed in the next section, inter-region (inter-port for countries other than Japan) container cargo Origin – Destination (OD) flows are required. Fig. 10 shows the total TEUs of the top 20 OD pairs by volume of change for 5 years from Japan by prefecture to major East Asian ports and vice versa. Here, OD flows – from/to Japanese prefectures and to/from major East Asian ports – were estimated according to the following procedure: (i) acquiring country-based (for countries other than Japan) OD flows from Ports and Harbors Bureau, Ministry of Transport (1993) and (1998) and (ii) dividing them into port-based OD flows by shares of the amount of container handling by port out of the entire country, obtained from Informa Group (1994) and (1999). From these figures, for exports

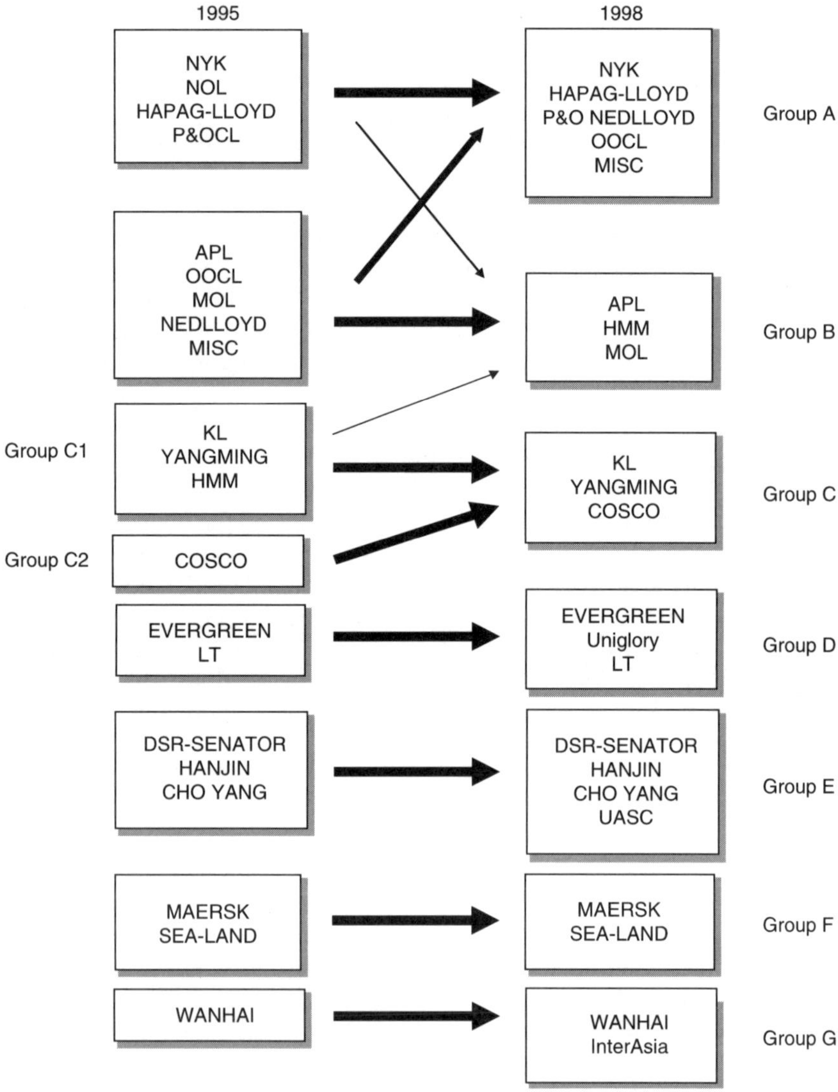

Fig. 7. Global Alliances of Major World Shipping Companies; Global Alliances Started in 1995 (summarized by authors from Ocean Commerce, each year and interviews).

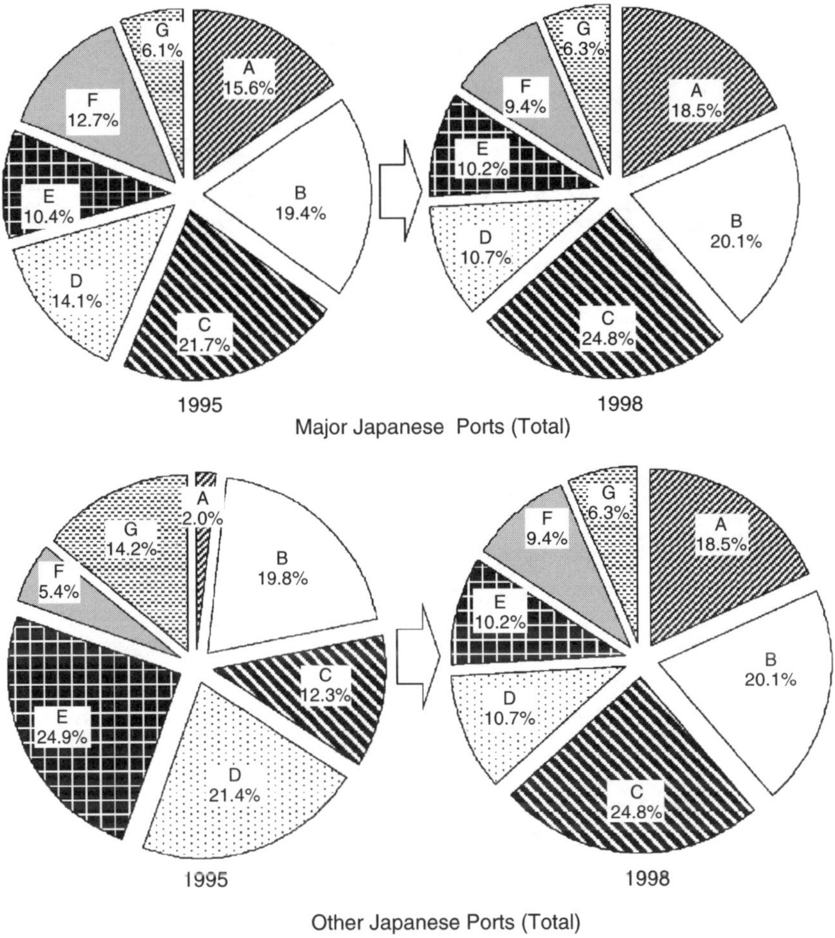

Fig. 8. Share Changes by Alliance Group in Japanese Ports (estimated by authors from Ocean Commerce, each year).

from Japan, the amount of container cargoes to the ports of Shanghai and Laem Chabang increased dramatically, especially from prefectures in the three major urban area in Japan, such as around the city of Tokyo, Osaka, and Nagoya. For imports into Japan, the number of container cargoes from the Hong Kong port increased, also particularly into prefectures in the three major urban area.

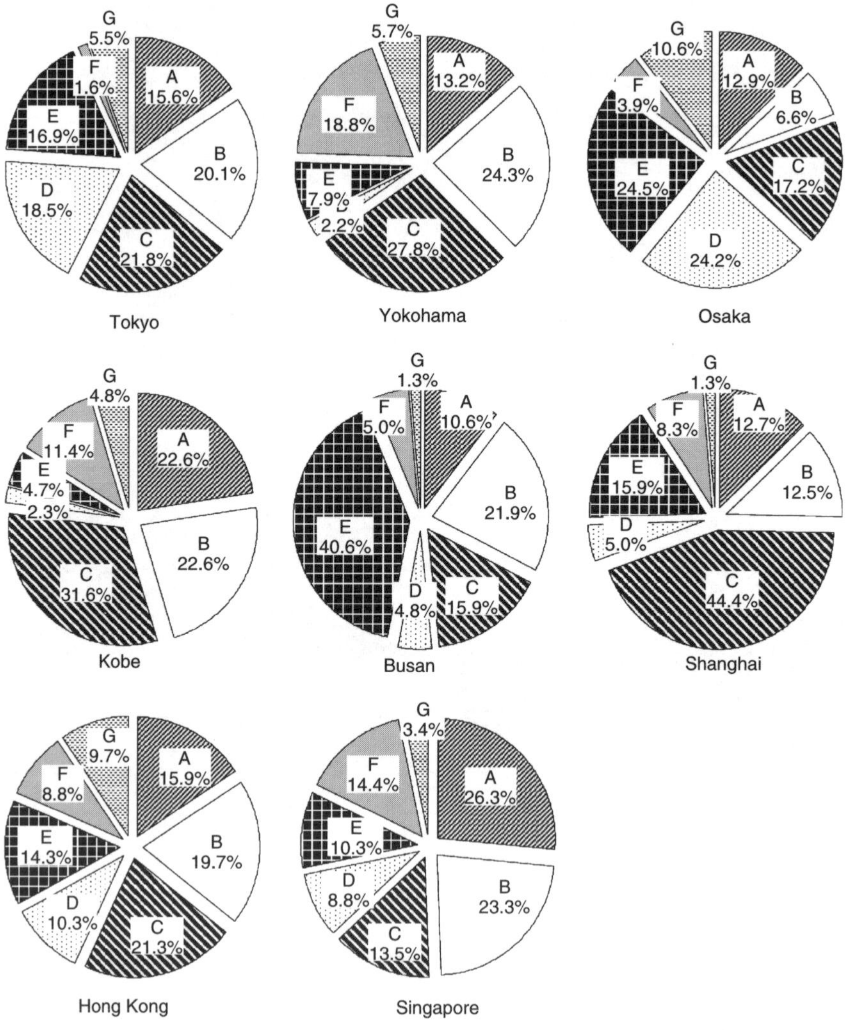

Fig. 9. Shares for Each Group in Major Ports in 1998. Source: As in Fig. 8.

Fig. 11 shows total TEUs for the top 20 OD pairs defined in the same manner as between East Asian ports. Here, OD flows were estimated according to the following procedure: (i) acquiring container cargo OD flows between East Asian countries and from/to East Asian countries to/

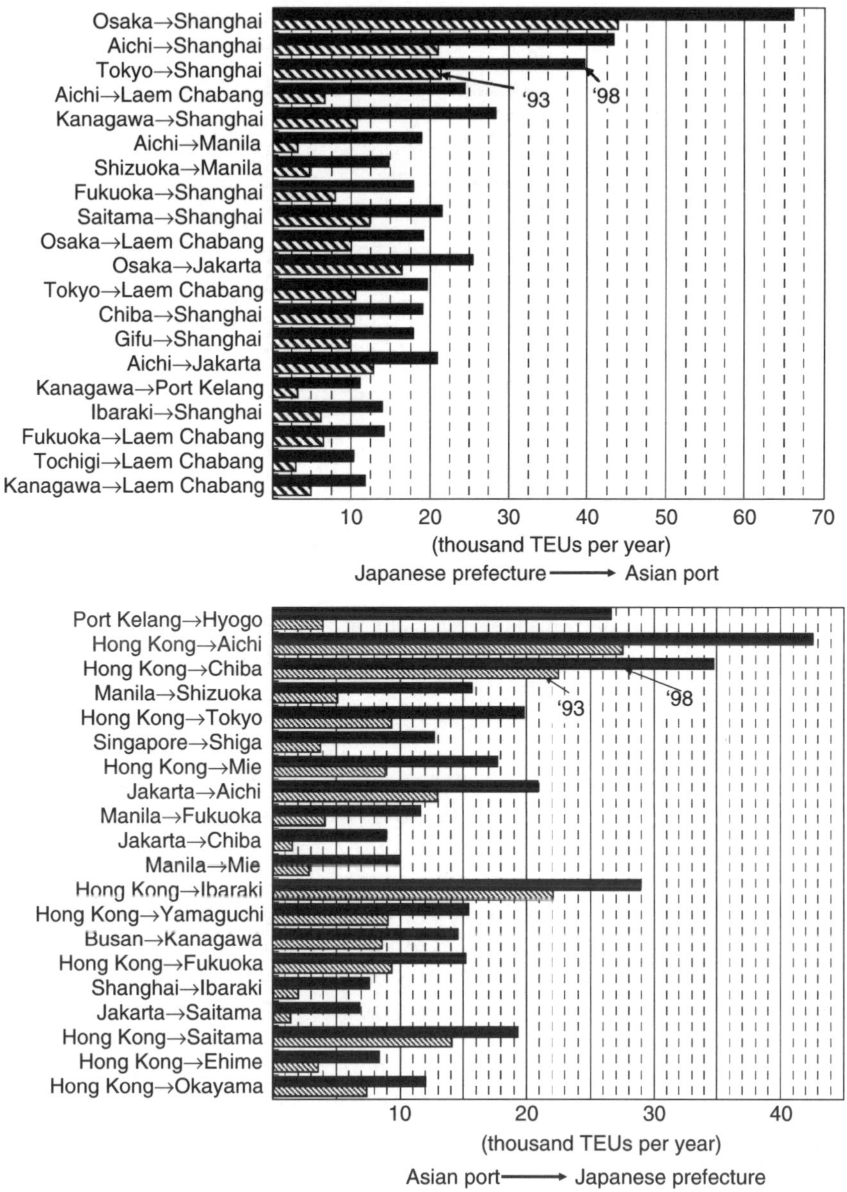

Fig. 10. Total TEUs of Top 20 OD Pairs of Changed Volume for 5 Years; from/to Japan by Prefecture to/from Major East Asian Ports (estimated by authors).

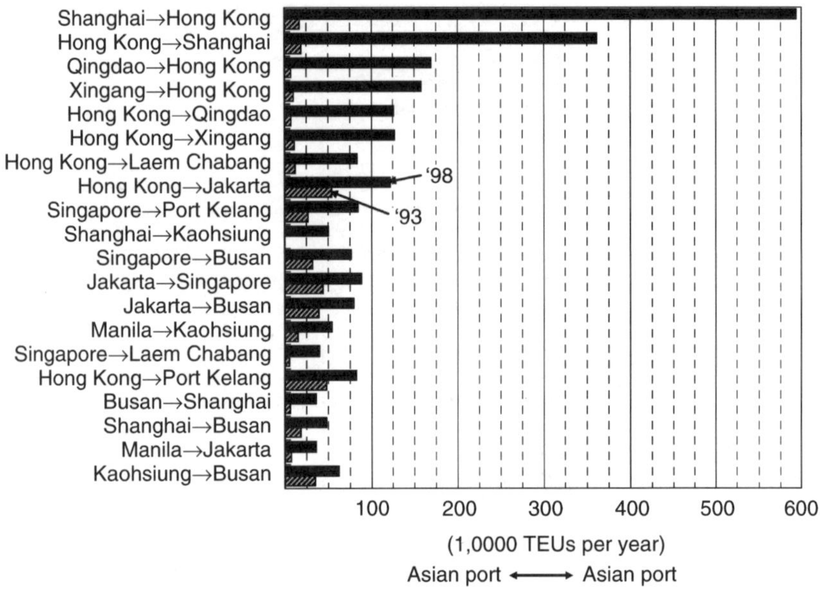

Fig. 11. Total TEUs of Top 20 OD Pairs of Changed Volume for 5 Years between East Asian Ports (estimated by authors).

from other regions of the world, from a roughly estimated container cargo OD matrix (by Ocean Commerce and Research Co-operation Office, Mitsui O.S.K. Lines, 1994 and 1999), and complementing the matrix with a trade conversion matrix between countries (US $ basis) on statistics from the Department of Economic and Social Affairs and the United Nations (1994) and (1999), into a TEU base. Then, (ii) estimating inter-port OD flows by the Flater Method, utilizing acquired container cargo OD flows from (i) as the initial matrix, and the amount of imported/exported container handling for each port as a control total. From this figure, remarkable increases in the volume of container cargoes are observed from/to Chinese ports or the port of Laem Chabang, with a tremendous increase in the rate between Chinese ports and Hong Kong. On the other hand, the volume of container cargoes in some OD pairs, such as between Taiwanese ports and Hong Kong decreased. Fig. 12 shows the distribution of the amount of container cargoes by category of vessel size between Chinese ports and Hong Kong. In this figure, the amount of container

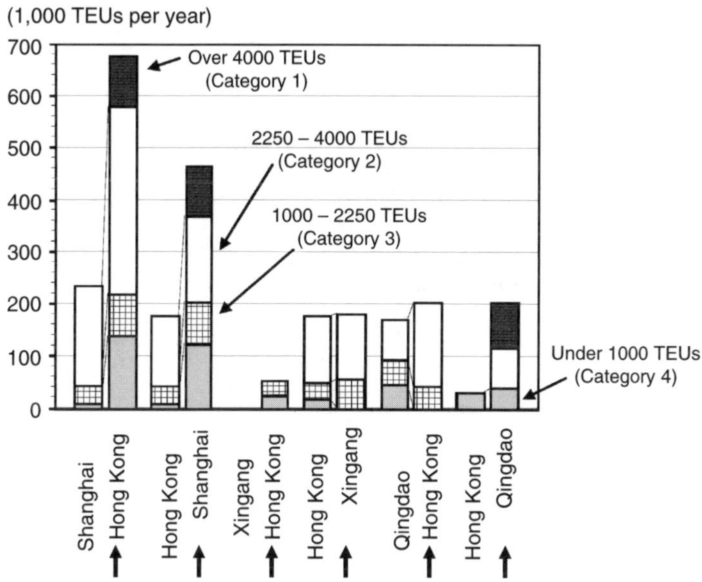

Fig. 12. Share by Vessel Size Category of Amount of Container Cargo Flow between Chinese Ports and Hong Kong Port; Left Side, 1995; Right Side, 1998 (estimated by authors).

cargoes from/to Shanghai to/from Hong Kong increased in all categories of vessel size, while the amount of container cargoes from/to Xingang (New Tianjin Port) and Qingdao to/from Hong Kong increased, especially in larger vessels.

3.5. Investment for Port Facilities

Fig. 13 shows the number of berths for each port in 1993 and 1998. For this 5-year period, for example, operations began at 4 berths on Port Island in Kobe and in the ports of Tokyo (3 new berths), Yokohama (1), Nagoya (4), and Hakata (2). For other Japanese ports, a berth deeper than 12 m was constructed in the ports of Tomakomai (Hokkaido Pref.), Akita, Yokkaichi, and Naha. On the other hand, in most other Asian ports numerous berths, many quite deep, began operations: in the ports of Busan (9), Shanghai (3), Kaohsiung (8), Hong Kong (4), Singapore (6), Klang, Malaysia (6), and so on. The rate of investment was higher than in Japanese ports.

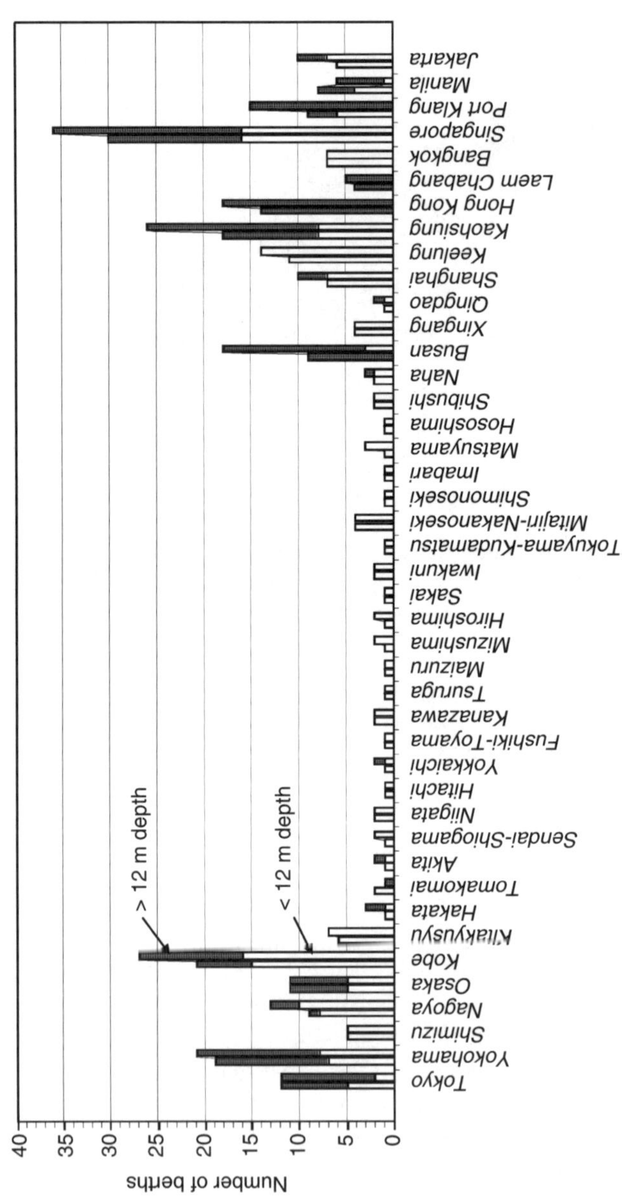

Fig. 13. Number of Berths By Port; Left Side, 1993; Right Side, 1998. Source: Ocean Commerce (each year).

4. FRAMEWORK OF THE MODEL

4.1. Definition of the Problem in a Network Context and its Basic Configurations

Our problem is defined as the reproduction and forecast of the movements of containers in international markets. It includes transshipment between vessels and the multifarious use of vessels of different sizes, when inter-region OD flows (cf. Section 3.4) and various characteristics of container transportation are given. We express the complicated movement of containers as traffic flow on a comprehensive network that covers all Japanese and all major East Asian containers ports (cf. Section 3.1), all transport routes, and all major shipping companies (cf. Section 3.3). It consists of the following nodes and links (cf. Figs. 14 and 15):

Nodes:	(1) origin nodes	by prefecture (in Japan); by port (in other countries)
	(2) destination nodes	same as above
	(3) international arrival nodes	by port by vessel size category
	(4) international departure nodes	same as above
	(5) international dwelling nodes	same as above
	(6) domestic arrival nodes	by Japanese port
	(7) domestic departure nodes	same as above
	(8) domestic dwelling nodes	same as above
	(9) container yard nodes	by Japanese port (international yard and domestic yard)
	(10) prefecture nodes	for each Japanese prefecture
Links:	(1) international cruise links	between each port by vessel size category
	(2) international loading links	by port by vessel size category
	(3) international unloading links	same as above
	(4) international arrival links	same as above
	(5) international passing links	same as above
	(6) transshipment links	between each vessel category by port
	(7) domestic cruise links	between each Japanese port
	(8) domestic loading links	by Japanese port

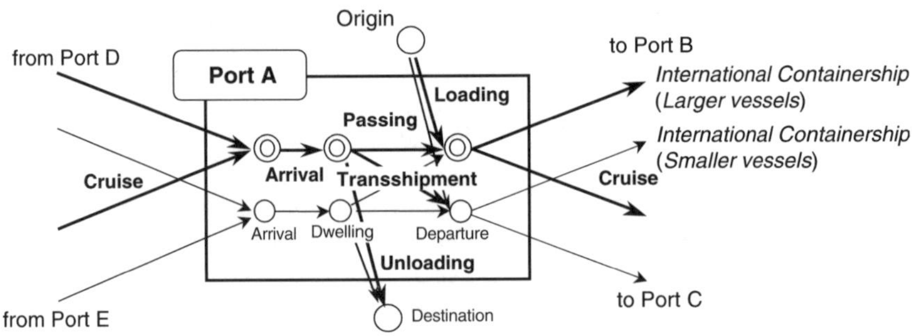

Fig. 14. Conceptual Structure of Network for Foreign Ports.

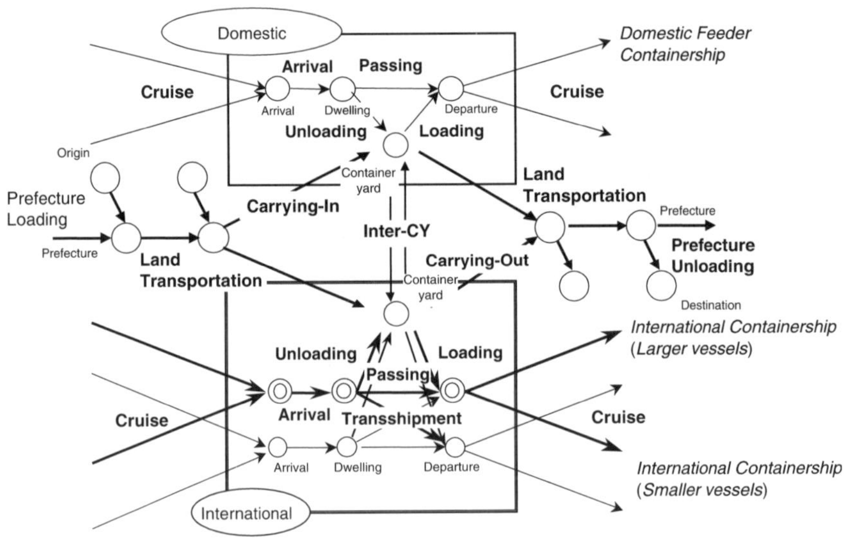

Fig. 15. Conceptual Structure of Network For Japanese Ports.

(9) domestic unloading links	same as above
(10) domestic arrival links	same as above
(11) domestic passing links	same as above
(12) inter-container yard links	same as above
(13) land transportation links	between each Japanese prefecture
(14) carrying-in links	from Japanese prefecture to port
(15) carrying-out links	from Japanese port to prefecture

(16) prefecture loading links by Japanese prefecture
(17) prefecture unloading links same as above

Nodes (6)–(10) and Links (7)–(17) are added to express Japanese domestic container transportation.

Each link has a corresponding flow-dependent or independent link cost. Containers will be carried from: prefecture of origin, or in the case of foreign countries, a port of origin, to a destination port, and in Japan to a destination prefecture through paths on the network according to path costs defined as the sum of link costs on the paths. Therefore, our problem is defined as a traffic assignment problem of OD flows on a network under a particular set of link costs.

The basic configurations of the model are as follows:

(1) Amounts of container cargoes are calculated in TEUs.
(2) Thirty Japanese and 13 large Asian ports that handled container cargoes in 1995 comprised the network (cf. Fig. 1). Three composite ports were prepared to represent North America, Europe, and Oceania.
(3) Berths in ports were classified into two categories: those serving only the smaller two categories of vessels, that is, not deeper than 12.0 m berth depth, and those serving all vessels, berths deeper than 12.0 m, (cf. Section 3.5).
(4) The nodes and links of a group's network reached approximately 842 and 10,390, respectively.
(5) Link cost functions reflecting various transport costs were prepared for each link.
(6) Since it became a non-convex numerical optimization problem, we applied a current flow pattern (cf. Section 3.2) as initial values.

4.2. Five Hypothetical Assignment Methodologies of OD Flow

It is necessary to assume traffic assignment rules of OD flows on the network upon which the behavior principles of the shipping companies, freight forwarders, and shippers must be either reflected explicitly or implicitly. First, consider two simple cases; the first is a system optimum assignment rule for the entire container transportation network: (i) SOALL that is intended to find a flow pattern that minimizes the total of all link costs. The second is a user equilibrium assignment rule for the entire container transportation network, (ii) UEALL that minimizes costs along the path of each container transport. Both can be calculated by Frank–Wolfe's optimization algorithm.

In addition to these two simple assignment rules, three additional assignment methodologies were proposed to reflect the interaction between the behavior of the shipping alliance groups and shippers. As already mentioned, the international container transportation market is approximated to be very competitive. Therefore, when equilibrium on the international container transportation market is observed, innumerous shippers minimize transportation costs, and oligopolistic shipping alliance groups minimize their total costs. All shippers choose both a port for loading or unloading their cargoes and an alliance group that carries them, comparing service levels provided by international shipping alliance groups and domestic transportation conditions. Each shipping alliance group decides its own allocation pattern of vessels on the container transportation network to minimize their total costs.

According to the behavioral structure of both objects, in the following three methodologies, assignments based on two different rules are alternately repeated until all container flows have converged. On the assignment based on shippers' behavior, the user equilibrium assignment rule is adopted in all three methodologies. Note that at this stage, container cargoes are assigned on the entire network, including Japanese domestic transportation. In addition, the maritime transportation fares, which are a key factor in the decision of transportation routes for shippers, are normally set by each OD pair. However, for simplicity, they are assumed to be uniquely defined for each link on the network. For assignments based on the shipping alliance groups' behavior, two different rules are prepared: one is that assignment is based on the assumption that only one group is operating all international container transportation to minimize the total cost of international container shipping. The other is that several oligopolistic alliance groups are operating independently of each other and minimizing their costs.

The methodology, which is composed of the user equilibrium assignment for shippers' behavior and system optimum assignment on the entire international shipping network for the monopolistic shipping company, is named, (iii) UE–SO assignment. On the other hand, the methodologies, which include a "by-group system optimum" (GSO) assignment for the shipping alliance groups' behavior, are further divided according to an estimation method of the fares for shippers. It is based on information from the link costs calculated from the allocation patterns of vessels decided upon by the shipping alliance groups. The first methodology, (iv) UE–GSO assignment, assumes that shippers regard the fare of the link for each alliance group as being equal to each alliance group's cost. In the second methodology, (v) UE–GSO–MINCOST assignment, shippers are assumed to regard the fare of the link for each alliance group as being equal to the cheapest of all alliance groups on the link. The assumption in this methodology corresponds to a more competitive situation in the international container transportation market. Thus, five assignment methodologies are proposed.

4.3. Characteristics and Formulation of Link Costs

Concerning the identification of link costs, we emphasize two points. One concerns the berth capacity of each port. When the arrival rate of vessels increases, the waiting time for mooring grows since it is dependent on the capacity of berths of vessel size. Thus, the costs on arrival links increase as flow increases. The second point concerns the economy of scale use of vessels. Larger vessels are potentially more efficient. Also, in a given vessel size, transporting a larger volume would naturally be advantageous because the load improves; even if it does not change, the improved operation frequency resulting from denser flow could lower transport costs for shippers. This feature makes the cost for all cruise links a decreasing function of link flows.

Due to these two elements of increasing and decreasing link costs, our problem resembles a flow-dependent network assignment problem to be solved by Frank–Wolfe's optimization algorithm (cf. Sheffi, 1985), for example. Moreover, our problem has become non-convex since some of the links have a decreasing flow function. This means that instead of having a unique solution to our problem it is rather dependent on the initial values of the solving process. The practical interpretation of this context is that the determination of the actual transport situation appears strongly dependent upon historical background, if the container market is somewhat ruled by an economy of scale and agglomeration. This seems reasonable because the current predominance among east Asian ports of large ports such as Hong Kong and Singapore cannot only be explained by their geopolitically and economically advantageous locations; it also results from their previous port and economic development. Since every future forecast must be based on the present, we applied the current situation of container flows to the model as an initial solution.

We formulated the following link cost functions based on interviews and surveys with several shipping companies. Cruise-link and arrival-link costs are flow dependent, while loading and unloading costs, transshipment, and passing links are independent of flow. A cruise-link cost function can be used for both the entire flow-based and the group-based flow calculations. On the other hand, the arrival-link cost function is applied only for the entire flow. Thus, for group-based assignments an iterative procedure is required: (1) group-based assignment, (2) summation of group-based flow on arrival links and convergence check, (3) revision of arrival-link flow, then back to group-based assignment.

(1) Cruise link costs

(i) International cruise link. Let i denote origin port; j the destination port; k the vessel category by size of international container ships ($k = 1-4$); and n the shipping alliance group. When link a is a cruise link between ports i and j by vessel category k of company group n, we formulated the cost for one TEU on this cruise link as follows:

$$C_a = \frac{a_{1k} + a_{2k}}{f_a cap_k} l_{ij} + vt \left\{ \frac{l_{ij}}{v_k} + b_1 \left(\frac{T f_a cap_k}{2q_a} \right)^{b_2} \right\} \tag{1}$$

C_a	cost of one TEU on link a (1,000 JPY/TEU),
q_a	annual flow at cruise link a (TEU),
l_{ij}	distance between ports i and j (nautical miles),
v_k	velocity of vessels for vessel category k (knots),
cap_k	loading capacity of vessel category k (TEU/vessel, average in each category),
f_a	load factor on link a,
T	time period (1 year = 8,760 h),
vt	value of time for container transport (1,000 JPY/h/TEU),
a_{1k}	unit operation cost of vessel category k (1,000 JPY/NM/vessel),
a_{2k}	unit fixed cost of vessel category k (1,000 JPY/NM/vessel), and
b_1, b_2	parameters related to operation interval costs (b_1, $b_2 > 0$).

The first term of Eq. (1) denotes the monetary cost of the cruise for one TEU, which consists of the following operation costs of the vessels: fuel, personnel, and capital. The first term in the parentheses refers to the time consumed during the cruise, and the second term represents the cost to shippers related to the operation interval on link a. We introduced parameters b_1 and b_2, considering that substantial time-costs for shippers may not always be proportional to the operation interval.

(ii) Domestic cruise link. Let i denote origin port; j the destination port; and k the vessel category for domestic container ships ($k = 5$). When link a is a cruise link between ports i and j by vessel category k, we formulated the cost for one TEU on this cruise link as follows:

$$C_a = \frac{a_{1k} + a_{2k}}{f_a cap_k} l_{ij} + vt \left\{ \frac{l_{ij}}{v_k} + b_1 \left(\frac{Tf_a cap_k}{2q_a} \right)^{b_2} \right\} \quad (2)$$

The definitions of the variables in Eq. (2) are the same as Eq. (1).

(2) Arrival link costs

(i) International arrival link. When link a is the arrival link at port i of vessel category k of alliance group n, the cost for one TEU at link a is formulated as follows:

$$C_a = \frac{f\_p_{ik}}{f_a cap_k} + vtw_{ik} \frac{24}{h_i} \quad (3)$$

where f_p_{ik} port charge at port i for vessel category k (1,000 JPY/vessel),

w_{ik} expected waiting time for mooring at port i for vessel category k, and

h_i operation hours at port i (h/day).

Term $24/h_i$ represents the inconvenience caused by restricted operation hours at some ports. The expectation of waiting time, which is derived from a queuing theory, is approximated by the power function of the ratio of the arrival rate and the product of the number of windows and the service rate. By approximating the service rate by the reciprocal of the mean time required for loading/unloading/transshipment, waiting time w_{ik} can be formulated as follows:

$$w_{ik} \begin{cases} b_3 \left(\frac{t\_anc_k r_m}{n\_b_{im}} \right)^{b_4} & \text{if } \frac{r_1}{n\_b_{i1}} < \frac{r_2}{n\_b_{i2}} \\ b_3 \left(\frac{t\_anc_k \sum_m r_m}{\sum_m n\_b_{im}} \right)^{b_4} & \text{if } \frac{r_1}{n\_b_{i1}} \geqslant \frac{r_2}{n\_b_{i2}} \end{cases} \quad (4)$$

where n_b_{im} number of container berths of category m at port i, berth category $m = 1$ can only serve smaller vessels in international shipping (category $k = 1$–2),

$m = 2$ can serve all vessels in international shipping (category $k = 1$–4),
- t_anc_k time required for loading/unloading/transshipment at ports for vessel category k (h/vessel),
- r_m arrival rate of vessels classified by berth category m (vessels/h), and
- b_3, b_4 parameters relating to waiting time for mooring (b_3, $b_4 > 0$).

Arrival rates of vessels r_m, can be given by the following expression:

$$r_1 = \frac{\sum_{k=1}^{2} \sum_n q_a}{f_a cap_k T}, \quad r_2 = \frac{\sum_{k=3}^{4} \sum_n q_a}{f_a cap_k T} \tag{5}$$

(ii) Domestic arrival link. When link a is an arrival link at port i of vessel category k, the cost for one TEU at link a is formulated as follows:

$$C_a = \frac{f\_p_{ik}}{f_a cap_k} + vtw_{ik} \frac{24}{h_i} \tag{6}$$

when

$$w_{ik} = b_3 \left(\frac{t\_anc_k r_m}{nb_{im}} \right)^{b_4} \tag{7}$$

and when berth category $m = 3$ can only serve vessels of domestic shipping (category $k = 5$). Arrival rates of vessels can be given as

$$r_3 = \frac{q_a}{f_a cap_k T} \tag{8}$$

(3) Loading, unloading, transshipment, and passing link costs

In both international and domestic shipping (except transshipment between domestic container ships, which is not considered in this model), when link a is loading, unloading, "transshipmenting," or passing links at port i of vessel category k, the cost of loading, unloading, and transshipment links consists of time and monetary elements. However, the cost for container cargoes that are merely passing is only the waiting time until departure. Therefore,

$$C_a = df\_l_i + vtt\_anc_k \tag{9}$$

where f_l_i loading or unloading charge for one TEU at port i (1,000 JPY/TEU),

d dummy variable $= 0$ for passing links, $= 1$ for loading and unloading links, and $= 2$ for transshipment links.

(4) Japanese land transportation link costs

The cost of container trailers on land transportation and carrying-in and carrying-out links also consists of time and monetary elements. Moreover, the monetary cost is divided into the term in proportion to the link distance and fixed term. A fixed term of monetary cost is included only in carrying-in or carrying-out links because it must be summed up just once during container cargo transportation from origin to destination.

$$C_a = a_3 l_{ij} + d a_4 + vt \frac{l_{ij}}{v_{\mathrm{car}}} \tag{10}$$

where v_{car} velocity of container trailers (km/h),

a_3 unit flow cost of container trailers (1,000 JPY/km/TEU),

a_4 unit fixed cost of container trailers (1,000 JPY/TEU),

d dummy variable $= 0$ for land transportation links, and $= 1$ for carrying-in and carrying-out links.

(5) Inter container yard link costs

For transshipment between international and domestic container yards, the transfer time and charge are normal costs because international and domestic container yards are normally separated.

$$C_a = f\_tc_i + vtt\_tc_i \tag{11}$$

where f_tc_i a transfer charge between international and domestic container yards for one TEU at port i (1,000 JPY/TEU), and

t_tc_i transfer time between international and domestic container yards at port i (h).

5. REPRODUCIBILITY OF THE PROPOSED MODEL USING 1993 DATA

5.1. Choice of Assignment Methodologies and Estimation Results of Unknown Parameters

The link cost functions mentioned above have numerous parameters, most of which were estimated through interviews with shipping companies and examining statistics. Those parameters are not shown in this paper due to space limitations (see Ieda et al., 1999).

Such parameters as b_1, b_2, b_3, b_4 and vt, which relate to the perceived cost for shippers or shipping companies, were estimated through model calibration procedures to minimize the sum of the squares of the errors at cruise link flows (between the output of the model and the estimated values described in Section 3.2) under each of the five assignment hypotheses. Calibrated parameters are shown in Table 1. On the whole, the results are logical. For example, time value vt should be around 1,000 JPY/h, which corresponds reasonably with past studies (e.g. Watanabe, Higuchi, & Morikwa, 2000).

5.2. Reproducibility of the Model

After comparing such aspects as amount of container handling at Japanese ports and amount of container transshipping at foreign ports, the UE–GSO assignment is the most reproducible of the current patterns of container transportation of the five prepared assignment methodologies (Ieda et al., 1999). This section introduces the reproducibility of the UE–GSO assignment. Fig. 16 compares the actual and model estimated amounts of container handling in 1993 at Japanese ports. For the major Japanese ports of Tokyo and Shimizu, the model considerably overestimated, while at the port

Table 1. Estimated Parameters of the Model.

Assignment Methodologies	vt (1,000 JPY/TEU/h)	b_1	b_2	b_3	b_4
SOALL	0.9	13.0	0.9	120	5
UEALL	0.6	12.5	0.7	120	5
UE–SO	1.1	13.0	0.7	120	5
UE–GSO	1.0	13.5	0.4	120	5
UE–GSO–MINCOST	0.9	13.0	0.3	120	5

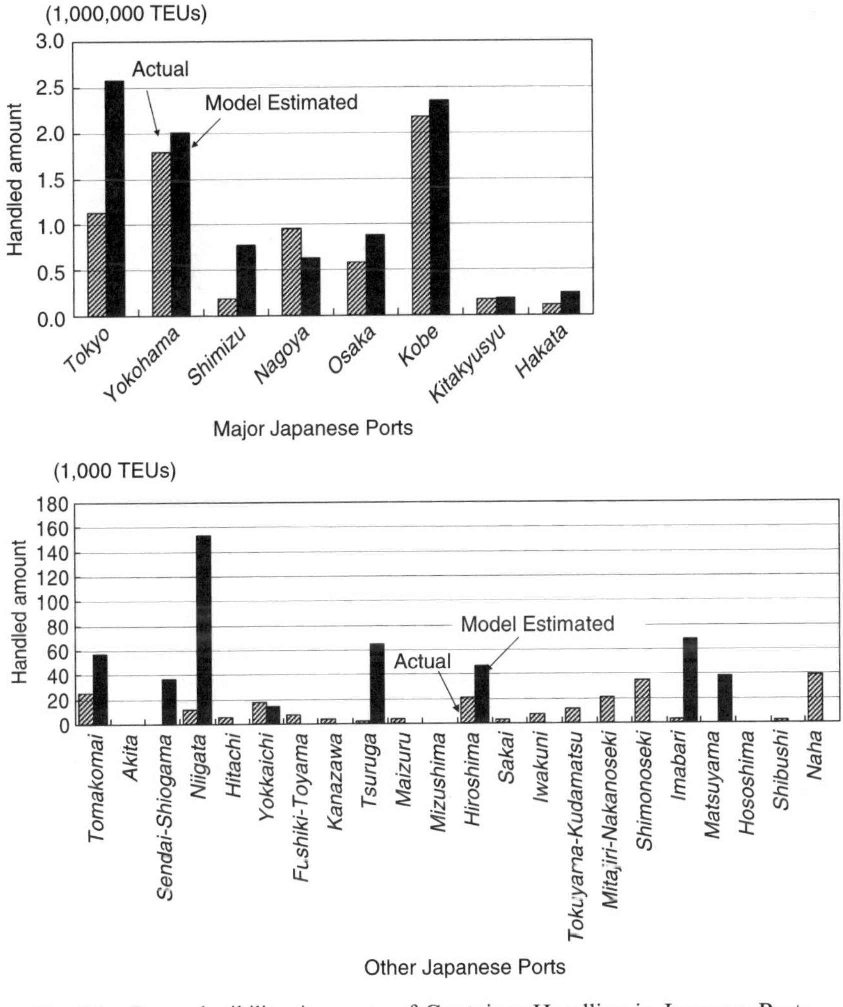

Fig. 16. Reproducibility Amounts of Container Handling in Japanese Ports.

of Nagoya it underestimated. Most overestimation was attributed to overestimating the amount of transshipped container cargoes. For other Japanese ports, the model was limited to reproducing handling at ports such as Tomakomai, Sendai, Niigata, Yokkaichi, Tsuruga (Fukui Pref.), Hiroshima, Imabari and Matsuyama (Ehime Pref.). These locations have the potential to become main container handling ports for each region in

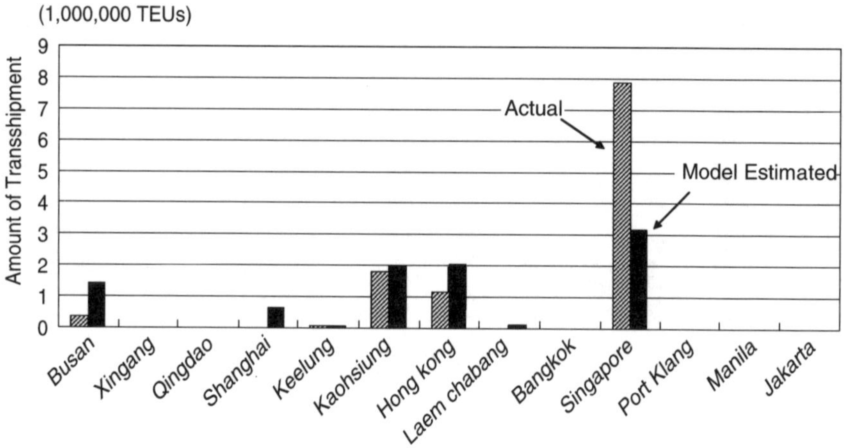

Fig. 17. Reproducibility Amount of Transshipment in Major Asian Ports.

Japan, except for the eight major ports. Fig. 17 compares the actual and model estimated amounts of transshipment in major Asian ports because comparing the amounts of transshipment by the amounts of export and import container cargoes in these ports were given. In the port of Singapore, the estimated amount was less than half of the actual amount, while in the port of Busan the estimated amount was more than three times the actual amount. In addition, some transshipment amount was observed in the port of Shanghai in the estimated results.

6. TRANSFERABILITY OF THE MODEL USING 1998 DATA

6.1. Transferability of the Model Considering Different Initial Flow Patterns

To investigate the model's transferability, the UE–GSO assignment with the estimated parameters in Table 1 is calculated using 1998 data. The following two patterns of initial flow were input into the model to calculate the initial cost for each link: (i) current flow pattern in 1998 (cf. Section 3.2), and (ii) flow pattern estimated by the model using 1993 data. The second flow pattern is based on the expectation that the behavior of shippers and shipping alliance groups are determined at the starting point of the previous flow patterns of a few years earlier.

Fig. 18 compares the actual and model estimated amounts of container handling at Japanese ports. For the major Japanese ports, in either case of initial flow patterns, the amount of container handling was overestimated at the ports of Tokyo, Shimizu, Kobe, and Kitakyushu and underestimated at Osaka. In addition, in type (ii) initial flow patterns, it was overestimated at the port of Yokohama and underestimated at Hakata. For type (i), the

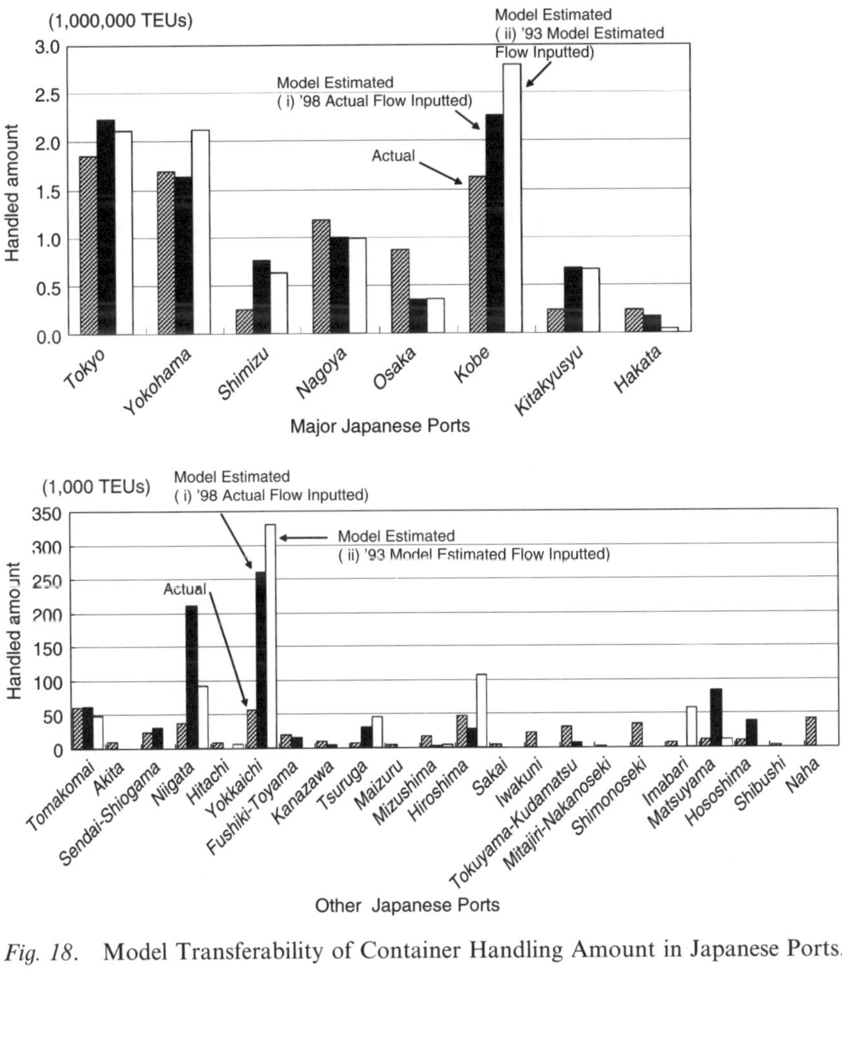

Fig. 18. Model Transferability of Container Handling Amount in Japanese Ports.

reproducibility of the model seemed to result in a similar pattern to the original model using 1993 data, except in Osaka. The different estimation results at Osaka Bay are because a fair portion of container cargoes was forced to switch to the port of Osaka from the port of Kobe, which was severely damaged in the 1995 Hanshin-Awaji Earthquake. For other Japanese ports, in type (i) initial flow patterns, ports where container cargoes were assumed to be handled to some degree were almost the same as in the original model, except container cargoes handled at the port of Hososhima (Miyazaki Pref.) instead of Imabari. A few container cargoes were handled in several other ports such as Fushiki-Toyama and Kanazawa (Ishikawa Pref.). In type (ii), ports where container cargoes were assumed to be handled are the same as in the original model; however, the amount of container handling was different. Fig. 19 compares the actual and model estimated amounts of transshipment at major Asian ports. In either case of initial flow pattern, the model's reproducibility also resembles the level of the original model using 1993 data.

Fig. 20 shows the total distribution of vessel use by size category of all international cruise links. As in the original model (Ieda et al., 2000), the use of larger vessels is somewhat exaggerated and emphasized by the model, especially in type (i) initial flow patterns. Fig. 21 shows the distribution of the amount of arriving and departing container cargoes in vessels by size category in representative ports. In type (i), a similar tendency is observed that exaggerates the use of larger vessels, as shown in Fig. 20. However, in

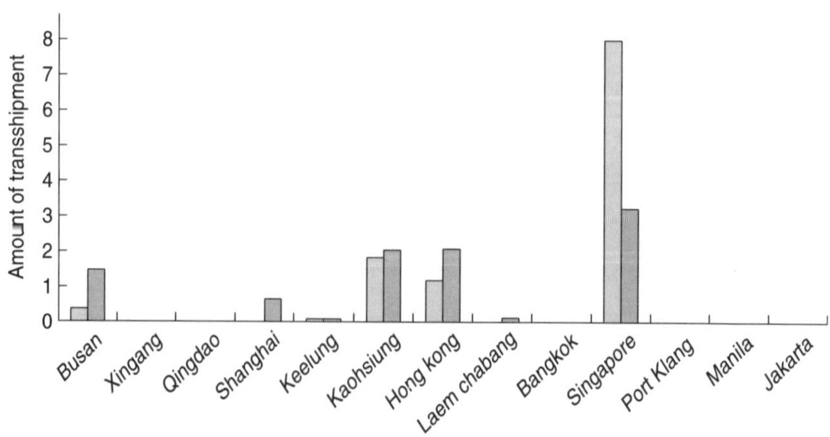

Fig. 19. Model Transferability of Transshipment Amount of Major Asian Ports.

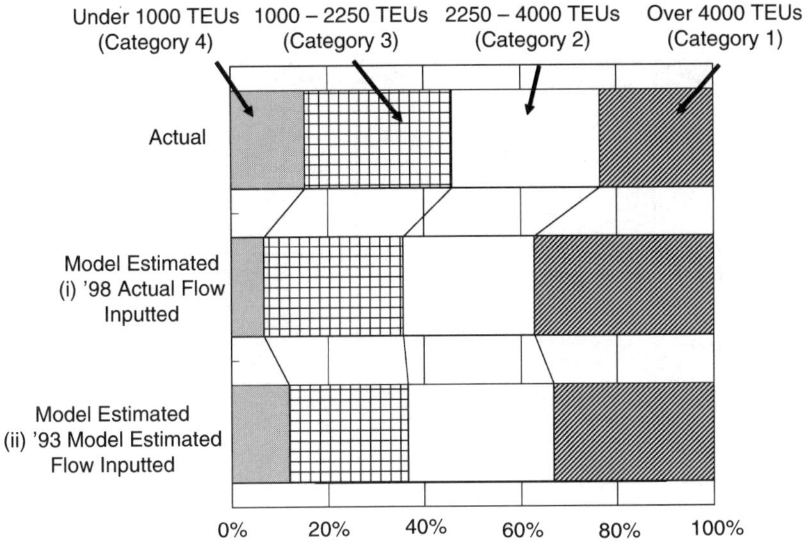

Fig. 20. Distribution of Vessel Use by Size Category for All International Cruise Links.

type (ii), the use of smaller vessels is greatly emphasized in some ports, such as Hakata and Shanghai, because they were both previously characterized as feeder ports in the original model (Ieda et al., 2000). Although during the 5 years from 1993 to 1998 these ports actually developed into ports where larger vessels could call, they could not break out of that position as feeder ports in the model estimations. This indicates the model's inflexibility that creates more extreme hub-spoke transportation than in the actual situation. This particular problem of the model can also be seen in the greatly over-estimated amount of transshipped container cargoes in almost all of the major Japanese and other Asian ports.

6.2. Flexibility of Parameters Estimated by the Original Model

To verify the flexibility of the parameters estimated by the original model, time value parameter vt, which has the most flexibility of the unknown parameters (Ieda et al., 2000), was changed to (i) $vt = 50$ (JPY/TEU/h); (ii) $vt = 200$; (iii) $vt = 1,000$ (original estimation); and (iv) $vt = 5,000$. For each time value parameter, the UE–GSO assignment is calculated using 1998 data and the type (ii) initial flow pattern.

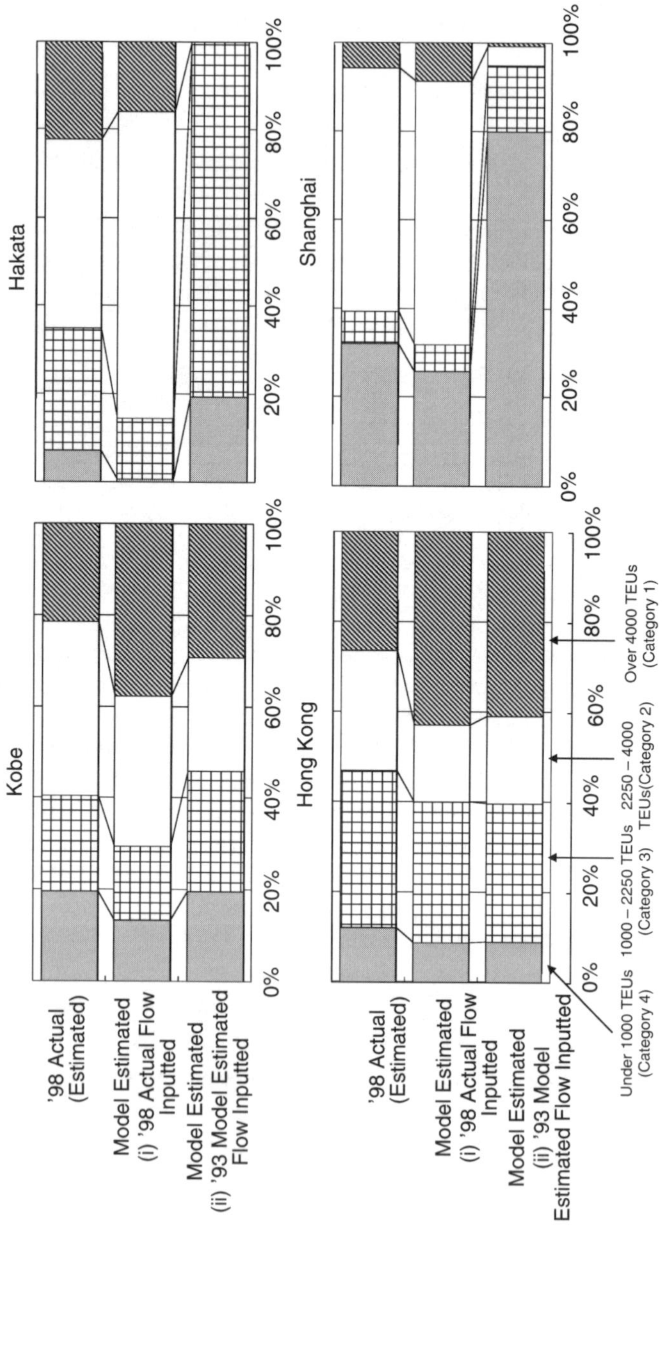

Fig. 21. Distribution of Amount of Arriving/Departing Container Cargoes in Vessels by Size Category.

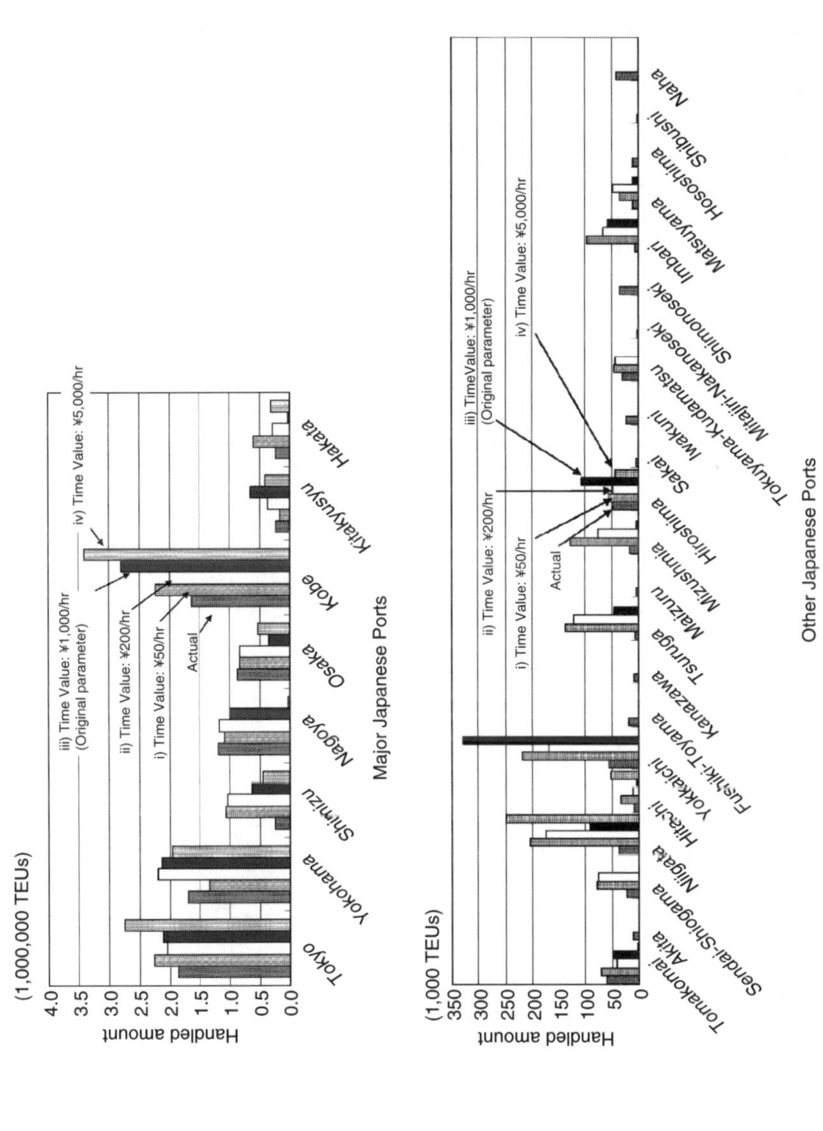

Fig. 22. Amount of Container Handling for Time Value Parameters in Japanese Ports.

Fig. 22 shows the amount of container handling for each time value parameter in Japanese ports. For major ports, case (ii) ($vt = 200$) is the most reproducible except in the ports of Yokohama and Shimizu. For other Japanese ports, case (ii) also seems the most reproducible but no significant differences between reproducibility of four cases. Fig. 23 shows the amount of transshipment of container cargoes for each time value parameter in other Asian ports. Although the rough outlook for reproducibility is almost the same among the cases, case (iii) ($vt = 1,000$; the actual estimated values in the original model) is considered the most reproducible.

Fig. 24 shows the distribution of vessel use by size category for all international cruise links by time value parameters. In this figure, as the time value decreases, the use of larger vessels increases because when the time value is low, shippers can wait for a departure time as long as a larger vessel call. Fig. 25 shows the number of links with any flow by size category from all cruise links. For vessels in Categories 1–3 (i.e. vessel capacity is more than 1,000 TEUs), as time value increases, the number of links with any flow decreases, and the usage of larger vessels are confined. On the contrary, for vessels in Category 4 (less than 1,000 TEUs of capacity), as time value increases, the number of links with any flow increases, and the usage of smaller vessels is mainstreamed.

Based on the above results, the time value parameter, which was estimated in the original model, is assumed to have some degree of transfer-

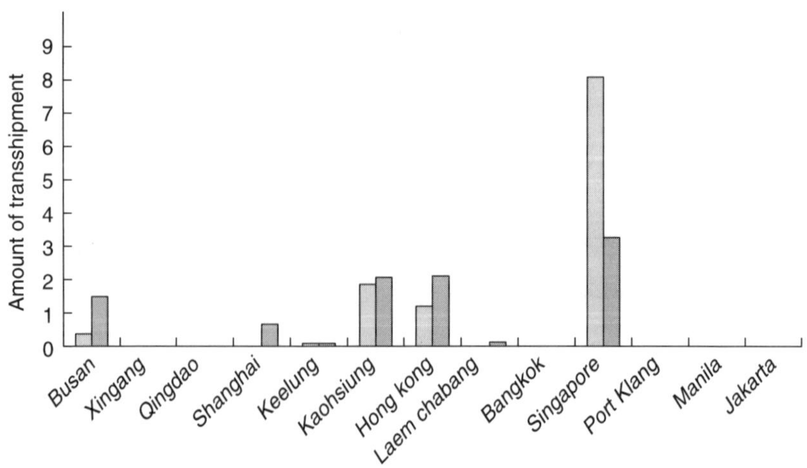

Fig. 23. Amount of Transshipment for Time Value Parameters in Other Asian Ports.

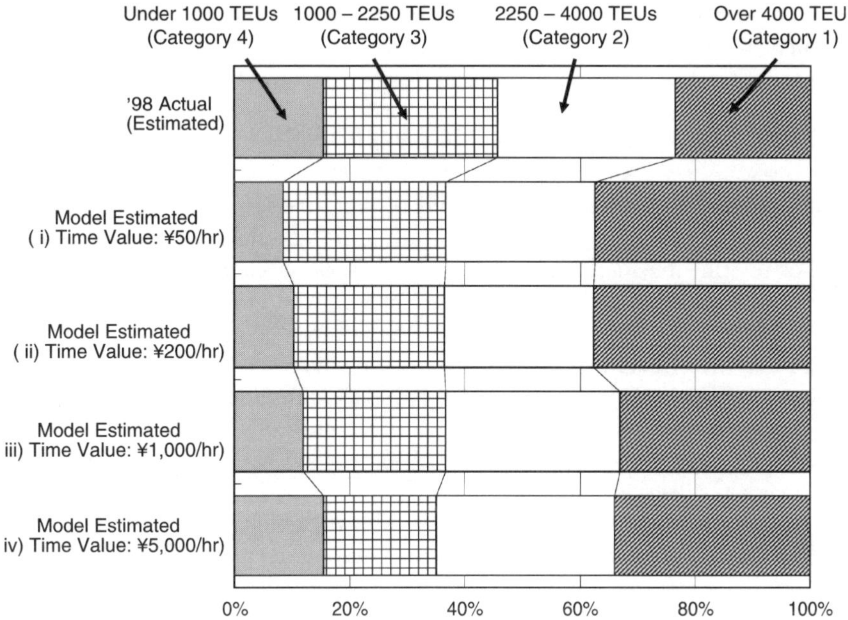

Fig. 24. Distribution of Vessel Use by Size Category for All International Cruise Links for Time Value Parameters.

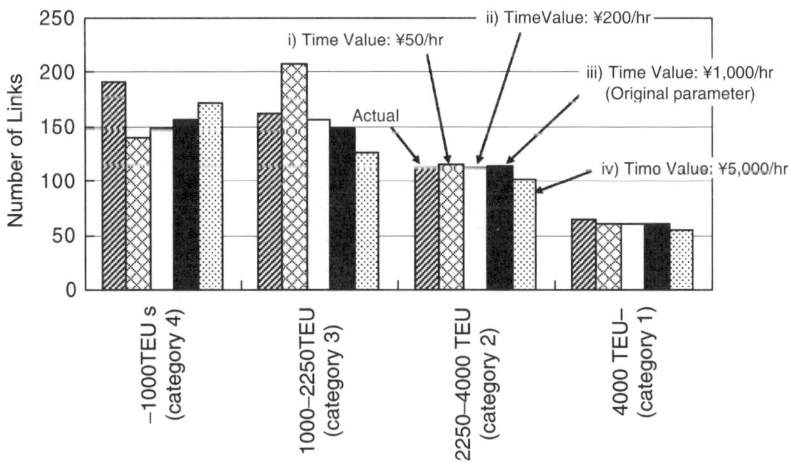

Fig. 25. Number of Links With Any Flow by Size Category.

ability, although it is a little high. In addition, time value amount also influences vessel use.

7. CONCLUSIONS

The dynamic and critical background of container transportation includes such features as the rapid expansion of container shipping in the Eastern Asian region, the increased size of vessels seeking economy of scale, hub-spoke operations with efficient transshipment, and global alliances of shipping companies. We proposed an Asia-range simulation model that estimates the effects of such policies as improvement of port facilities and operations upon the sea traffic flow patterns, container flow patterns, and the use of ports. In this paper, the transferability of the model, which was developed based on data from 1993, was verified by data from 1998. The following results were obtained:

(i) The results of the UE–GSO assignment, which used 1998 data with unknown parameters estimated in the original model using 1993 data for the amount of container handling by port, indicate that the model's reproducibility is similar to the original model.

(ii) However, for the distribution of vessel use by size category, the use of larger vessels is emphasized in ports categorized as hub ports in the original model, while small vessels are only used in the ports categorized as feeder ports. This means that the model may lead to more extreme hub-spoke transportation than in actual situations.

(iii) To determine the flexibility of the parameters estimated by the original model, a UE–GSO assignment was calculated based on four time value parameters. According to the results, the time value parameters estimated in the original model are assumed to have some degree of transferability, although it is considered a little high.

As for the model's reproducibility, there still remain two significant problems: (i) overestimation of transshipped container cargoes, or in other words, extreme progress toward hub-spoke transportation, and (ii) reproducibility, especially in local Japanese ports where the amount of container handling is relatively small. To solve the first problem, a feedback mechanism is needed in the model to restrict the progress of hub-spoke transportation. In this model, the expression of the faults of hub-spoke transportation is considered insufficient. For example, there are no restrictions between arriving and departing vessels (i.e. no correlation between the number of arriving and departing vessels by size category), no additional time needed for transshipment as with container cargoes that are merely passing (i.e. no special waiting time for transfer of vessels), and so on. To solve the second problem, further research is required on the port choice behavior of shippers. We also need to ascertain whether shippers behave logically in real situations.

REFERENCES

Brotcorne, L., Labbe, M., Marcotte, P., & Savard, G. (2000). A bilevel model and solution algorithm for a freight tariff-setting problem. *Transportation Science*, *34*(3), 289–302.

Crainic, T. G., Florian, M., & Leal, J. E. (1990). A model for the strategic planning of national freight transportation by rail. *Transportation Science*, *24*(1), 1–24.

Crainic, T. G., & Laporte, G. (1997). Planning models for freight transportation. *European Journal of Operational Research*, *97*, 409–438.

Department of Economic and Social Affairs, United Nations (annually). Yearbook of international trade statistics. New York: United Nations.

Friesz, T. L., Gottfried, J. A., & Morlok, E. K. (1986). A sequential shipper-carrier network model for predicting freight flows. *Transportation Science*, *20*(2), 80–91.

Friesz, T. L., Viton, P. A., & Tobin, R. L. (1985). Economic and computational aspects of freight network equilibrium models: A synthesis. *Journal of Regional Science*, *25*(1), 29–49.

Harker, P. T. (1987). *Predicting intercity freight flows*. Utrecht, The Netherlands: VNU Science Press.

Harker, P. T. (1988). Multiple equilibrium behaviors on networks. *Transportation Science*, *22*(1), 39–46.

Harker, P. T., & Friesz, T. L. (1985). The use of equilibrium network models in logistics management: With application to the U.S. coal industry. *Transportation Research B*, *19*(5), 457–470.

Harker, P. T., & Friesz, T. L. (1986). Prediction of intercity freight flows, I: Theory & II: Mathematical formulations. *Transportation Research B*, *20*(2), 139–174.

Hurley, W. J., & Pettersen, E. R. (1994). Nonlinear tariffs and freight network equilibrium. *Transportation Science*, *28*(3), 236–243.

Ieda, H., Shibasaki, R., & Naito, S. (1999). An Asian-range model of international container shipping including domestic transport in Japan. *Infrastructure Planning Review*, *16*, 731–741 (in Japanese).

Ieda, H., Shibasaki, R., Naito, S., & Mishima, D. (2000). Model development for east Asian container shipping considering multifarious use of vessels and ports. *World Conference on Transportation Research Proceedings*, *8*(3), 213–226.

Inamura, H., Nakamura, M., & Gu, J. Y. (1997). An international container cargo demand modeling taking account of coastal feeder transport. *Journal of Infrastructure Planning and Management*, *562/IV-35*, 133–140 (in Japanese).

Informa Group (annually). Containerisation international yearbook. London: Informa Group plc.

Japan Society of Civil Engineering (1998). *Equilibrium analysis of transportation networks* (Chap. 11) (in Japanese).

Malchow, M., & Kanafani, A. (2001). A disaggregate analysis of factors influencing port selection. *Maritime Policy and Management*, *28*(3), 265–277.

Matsumoto, S., & Watanabe, Y. (1994). Predictive intercity freight transportation models: The state of the art of econometric and network models. *Journal of Infrastructure Planning and Management*, *494/IV-24*, 117–125 (in Japanese).

Miller, T. C., Tobin, R. L., & Freisz, T. L. (1991). Stackelberg games on a network with Cournot–Nash oligopolistic competitors. *Journal of Regional Science*, *31*(4), 435–454.

Ocean Commerce (annually). International Transportation Handbook. Tokyo: Ocean Commerce Ltd. (in Japanese).

Pendyala, R. M., Shankar, V. N., & McCullough, R. G. (2000). Freight travel demand modeling – synthesis of approaches and development of a framework. *Transportation Research Record*, *1725*, 9–16.

Ports and Harbors Bureau, Ministry of Transport (every 5 years). Survey Report of International Container Cargo Flow. Tokyo: Ministry of Transport.

Research Co-operation Office, Mitsui O.S.K. Lines (annually). Present Condition of Regular Marine Transportation. Mitsui O.S.K. Lines, Ltd., Osaka (in Japanese).

Sheffi, Y. (1985). *Urban transportation networks*. Englewood Cliffs: Prentice-Hall.

Southworth, F., & Peterson, B. E. (2000). Intermodal and international freight network modeling. *Transportation Research Part C*, *8*, 147–166.

Takebayashi, M., Kuroda, K., 2000. Equilibrium analysis of Asian-Pacific container transportation market – discussion of the relation of the Chinese and Japanese ports. *Proceedings of 2nd APTE*, (*Vol.1*, pp. 80–88) (in Beijing).

Watanabe, T., Higuchi, N., & Morikawa, M. (2000). A route choice model for domestic shippers in container cargo distribution between Japan and West Coast of North America. *Infrastructure Planning Review*, *17*, 677–685 (in Japanese).

IDENTIFICATION AND ELIMINATION OF BARRIERS IN THE OPERATIONS AND MANAGEMENT OF MARITIME TRANSPORTATION

Kazuhiko Ishiguro and Hajime Inamura

ABSTRACT

Transportation-related costs both at ports and for inland transportation are very high, which is essentially caused by the high cost of labor and construction of infrastructure. This condition, however, is also caused by protectionist policies in the domestic transportation field. This chapter discusses and identifies barriers in the operation and management of maritime and multimodal transportation. The effect of the removal of barriers is calculated by applying a Spatial Computable General Equilibrium (SCGE) model that includes behavior of the transportation sector. The results of case studies show that a 10% reduction of import or export costs produces a 1% increase in total trade.

Global Competition in Transportation Markets: Analysis and Policy Making
Research in Transportation Economics, Volume 13, 337–368

ISSN: 0739-8859/doi:10.1016/S0739-8859(05)13015-7

1. INTRODUCTION

When compared with Korea and China, Japan does not have many maritime and multimodal transportation barriers. Many foreign shipping lines, agents, couriers, container lease companies, and insurance companies are operating in Japan. Imported foreign commodities are very popular, and direct foreign investment is also increasing every year. A foreign freight forwarder, on the other hand, cannot enter Japanese markets because of Japanese traditions and customs. The most prominent and biggest traditional obstacle that must be overcome is the prior consultation system managed by the Japanese Harbor Transportation Association (JHTA).

Transportation-related costs, which are high both at ports and for inland transportation, are essentially caused by high labor costs and expensive infrastructure. This is, however, the result of protectionist policies in domestic transportation fields as well. The prior consultation system mentioned above is also one of the biggest reasons for high costs in Japan. Since 1988, the Japanese government has made a great effort to deregulate transportation industries. However, Japan still has many regulations that protect agriculture, manufacturing, and transportation industries.

Japan has one of the lowest custom duty rates in the world. In particular, manufacturing commodities have the lowest tariff rates among industrialized countries. However, a significant number of agriculture products, such as rice and oranges, still have high tariff rates. This special duty system sometimes causes international friction. Japan's special duty system includes the following kinds: countervailing, antidumping, emergency, retaliatory, etc. In December 2000, Japan's first safeguard investigation commenced on three items: leeks, shiitake mushrooms, and tatami reed-mat facing. Provisional emergency tariffs were imposed on these products in April 2001. In February 2001, Teijin and four other companies filed an application for antidumping duties to be imposed on certain polyester staple fibers from the Republic of Korea and Taiwan; investigations began in April of the same year. An investigation into allegations of the dumping of Chinese ferrosilicon–manganese was the first such survey in Japan. Eventually, an antidumping duty was imposed on most Chinese ferrosilicon–manganese.

Barriers in operation and management of maritime and multimodal transportation can be classified into five types:

(1) Legal and institutional
(2) Financial and traditional
(3) Technological

(4) Physical
(5) Discrimination/restrictions

However, so far, nonlegal discrimination does not exist in Japan, and restrictions are directly related to legal and institutional barriers. Issue 5 will be combined into other sections when the other four subjects are discussed in this chapter.

2. LEGAL AND INSTITUTIONAL BARRIERS

2.1. Operation and Management of Maritime Transport

2.1.1. Identification of Barriers and Historical Changes of FMC Pressures
In June 1853, the U.S. East India Fleet commanded by Commodore Matthew C. Perry entered Uraga Harbor near Yokosuka and Yokohama. His four, well-built, black ships undoubtedly left a deep impression on the Japanese people. In March 1854, following Commodore Perry's demands backed by threats of force, Japan signed a peace treaty with the United States. This was Japan's very first step toward joining the international community. The historic treaty, which opened the ports of Shimoda and Hakodate to the United States to supply American ships with fuel, food, etc. was soon followed by a series of similar treaties with other countries.

Since then, Japan's maritime policies have always operated under American pressure. Worldwide, the Federal Maritime Commission (FMC) unofficially controls international trade rules. However, its status has been changing based on changes in the situation of the American shipping carriers and in world transportation market.

Container transport was started in 1960 by Sealand Co. Ltd, an American company. Matson, another U.S. shipping company, joined the container market in 1961. This era might be the peak of FMC's power, proved by the following two facts. The maritime Shipping Act of 1953 was called the international constitution of world trade and greatly influenced the container trade. Shipping conferences and alliances received exceptions in the Anti-monopoly Law.

In 1999 the last major American shipping line, the American President Line, was sold to Neptune Orient Lines of Singapore, who changed the company's name to APL. The Maritime department at Washington University in Seattle was abolished, marking the setting of the sun on the FMC in the United States. Most Americans had lost interest in the maritime transportation field. The FMC was scheduled to be abolished in 1995. Suddenly, it started to attack Asian countries to show its importance to the American economy. They have succeeded in erasing the article of abolishment in 1998 by the outcome.

In August 2001 the FMC issued "Section 15 order," which said that certain restrictions and requirements on the use of ports and terminal facilities in Japan would continue, aggravating unfavorable shipping conditions between United States and Japan. The FMC was seeking additional information concerning these conditions. Such restrictions and requirements for the use of ports and terminal facilities can be easily identified as the barriers in the operation and management of maritime transport. They appear not only in Japan–U.S. trade relations but also in Japan's trade relations with Korea and China.

These are the major problems:

(1) A prior consultation system for entering a port on related business.
(2) A licensing system for getting licenses to enter a port on related business.
(3) Other practices that contribute to unfair Japanese port practices.

The FMC issued a final decision on February 26, 1997 that imposed a $100,000 per voyage fee on Japanese carriers entering U.S. ports. It was suspended on November 13, 1997 after comprehensive government-to-government and industry-to-government accords were reached that substantially reformed Japanese port practices.

The FMC expected to periodically collect information to remain apprised of changes in port conditions resulting from implementation of the Agreements. Port Restrictions and Requirements in the United States/Japan Trade, 27 S.R.R. 1301, 1302 (1997), 62 FR 61648 (November 19, 1997).

In May 1998, the FMC ordered three Japanese and two U.S. carriers to file information and documents in response to an Information Demand Order served pursuant to section 19(6) of the Merchant Marine Act, 1920, 46 U.S.C. app. § 876(6). The FMC was again canvassing the conditions affecting service at Japanese ports in May 1999, concluding that:

The pace of progress and reform in Japan's port transportation sector has been slow. It has been reported that no foreign shipping lines have applied for or received licenses to operate their own terminals. No carrier appears to have invoked or tested the prior consultation dispute settlement procedures or other procedural safeguards that were agreed to, and no alternative to JHTA's prior consultation system for reconciling carrier service issues with the concerns of Japanese labor has been developed. Moreover, proposals for broader reform under consideration by Japan's Government fall well short of full deregulation.

The FMC noted that the following factors had contributed to the lack of progress in Port Restrictions and Requirements in the United States/Japan Trade, 28 S.R.R. 822 (1999), 64 FR 30245 (June 7, 1999):

(1) The strong opposition, including threats of work stoppages or other labor actions, communicated by Japanese labor unions to foreign lines hoping to establish terminal operations; and
(2) Government of Japan regulatory requirements for "close ties" (through equity exchange or long-term contracts) with subcontractors, minimum manning standards, and proof by license applicants that issuance of a license would not result in an excess of supply over demand.

The FMC concluded at that time, however, that the record of the proceeding was stale, and therefore it withdrew its existing final ruling and ordered the Japanese and U.S. carriers to continue to file periodic progress reports addressing port reforms.

Deregulatory measures proposed by Japan's Transport Policy Council Harbor Transport Subcommittee in 1998 were incorporated in amendments to the Port Transportation Business Law and enacted on May 10, 2000, taking effect in November 2000.

These amendments eliminated supply/demand tests, changed licensing to "permission," and ended regulatory approval of fees and charges for port services. Additional changes have also occurred regarding access on Sundays and extended working hours at Japanese ports as a result of labor agreements concluded earlier this year.

2.1.2. Prior Consultation System

The JHTA, an association of companies providing harbor transportation services, includes terminal operators, stevedores, and sworn measurers. Under this system, carriers serving Japan must consult with JHTA about operational matters involving Japanese ports or harbor labor. After JHTA consults with a carrier, it may conduct consultations with labor interests and then approve or deny the line's request.

Virtually all operational plans and changes made by carriers serving Japan must be submitted for prior consultation. These include:

(1) any changes in berth, route, or port calls;
(2) inauguration of new services or new vessels;

(3) the addition of extra port calls (either permanent or temporary) or calls by noncontainer ships at container berths;
(4) increasing the size of vessels or making changes in vessel technology that affect stevedoring or terminal operations;
(5) temporary assignment of vessels as substitutes (even if only for one voyage) or the renaming of vessels;
(6) rationalization agreements between carriers that involve vessel sharing or berthing changes;
(7) assignments of a stevedoring contractor or terminal operator to a carrier and any subsequent change in assignment;
(8) requests for Sunday work;
(9) changes in mandatory weighing and measuring arrangements; and
(10) any other changes that affect stevedoring or terminal operations.

To adopt any of the above actions, a carrier must submit a written request to JHTA and outline its proposal.

2.1.3. Recently Removed Problems

Mandatory Weighing and Measuring. Mandatory weight and measurement data for all cargo is not required for administrative functions or documentary procedures in Japan, nor do carriers require measurement of export box load cargo. Some carriers stated that they have attempted unsuccessfully to refuse sworn measurement services and charges; however, JHTA and union representatives threatened work delays, stoppages, and other retaliation.

The majority of carriers have not attempted unilaterally to stop weighing and measuring. Estimates of per-container weighing and measuring costs ranged from $41 to $85 per twenty-foot equivalent unit (TEU), and the majority of responses were in the $60–$68 range.

Between December 1995 and January 1996, agreements were reached involving JHTA, the sworn measurement companies, JSPC, and JFSA (the Japanese and foreign carrier groups) to phase out mandatory weighing and measuring over a 5-year period. Under the plan, carriers were required to pay a lump sum to the sworn measurers each year from 1996 to 2000 based on the amount paid for weighing and measuring in 1994. The lump sum payments decreased each year by about 17%: from 83.3% to 66.6% to 49.9% to 33.2% and finally to 16.5% of the 1994 total.

Sunday Work. In January 1995, a huge earthquake struck the Kobe region and disabled most of the port facilities, substantially increasing the volume of cargo moving through other Japanese ports. Harbor workers immediately began working on Sundays on an emergency basis to accommodate the additional loads. In May 1995, a 1-year agreement was reached between JHTA and the unions to continue working on Sundays at Japan's six major ports: Tokyo, Yokohama, Nagoya, Osaka, Kobe, and Kanmon.

The 1-year agreement, which included several requirements and restrictions, limited Sunday work to moving containers between vessels and the carriers' container yards. Therefore, cargo arriving at the gate on Sunday could not be loaded that day, and cargo discharged on Sunday could not be released the same day to the consignee. Also, the agreement provided that receiving cargo on Saturdays should be minimized because Saturday was a day off for most harbor workers. Vessels may be loaded and unloaded on Sundays only between 8:30 a.m. and 4:30 p.m.

According to the agreement, a shipping company that wishes to work on Sunday must apply by the previous Friday. An additional charge is imposed for Sunday work. Sunday work is limited to shipping companies that have fully implemented the Ministry of Land, Infrastructure, and Transportation (MLIT) approved rates and charges and to carriers that have observed the harbor industrial labor/management agreement concerning numbers of hours and days that union laborers may work, and the availability of overtime.

The restrictions on Sunday work apparently had a number of negative effects on respondent carriers, causing inefficiency and creating gate congestion on Saturday and Monday. The Sunday work surcharges resulted in extra costs. Applying in advance and the shortened working hours were also a burden and posed planning problems.

Then the situation dramatically changed. In April 2001, JHTA and the unions agreed to perpetuate Sunday work by extending hours of gate operations and reducing the year end and the New Year holidays. In November, they also extended operating hours to 24 hours a day everyday except New Year's Day. The gates are even open on Sundays.

2.2. Import/Export Procedures through Seaports

2.2.1. Necessary Applications

Various applications must be submitted to government agencies when cargo is imported or exported. Even though such regulations are common institutional barriers, they are appropriate because the current situation of the Electronic Data Interchange (EDI) system, which is widely used for these applications, is classified into technological barriers. We explain the current situation and the future prospects of EDI systems in the field of international trade.

The necessary applications that must be made by a customs agent company or a consignee in a typical case of importation are as follows:

Customs:
    Import declaration
Ministry of Economy, Trade, and Industry:
    Import approval/permission
Quarantine Office:
    Quarantine inspection
    Declaration of food imports.

These are required in typical cases, but more or less in other cases. Ocean carriers are required to declare at Customs, the Quarantine Office, the Immigration Bureau, the Port Authority, and to the Port Commander. The number of declarations often exceeds 30, which can cause a lot of trouble. Before the introduction of the current EDI systems, all applicants were required to submit directly the necessary documents to each department, which used to take all day long. Over the last decade EDI systems have been improved and expanded; therefore, most applications can be processed through EDI systems. The following problems remain, however:

- Each department developed its own EDI systems, resulting in a current situation far from "one-stop service."
- Even though EDI systems can accept ordinary cases, there are many exceptions.
- The high initial cost of EDI systems is a deterrent to their introduction, especially at small and medium companies.

2.2.2. Time Required

The Japanese Customs Bureau surveyed the time required at each process of importation. Table 1 shows the results of the most recent surveys. The rapid diffusion of EDI systems has dramatically reduced the time required to declare and wait for clearance through customs. Yet the current average time required – 67.1 h – which is still much longer than other countries. According to Table 1, which includes the time required for air cargo, there is a great time difference between maritime and air cargo.

In conjunction with the above survey, Japan Customs conducted a questionnaire to discover the reasons for the inordinately long time required for

Table 1. Average Time Required at Each Process of Importation.

	Carrying in Bonded Area	Declaration (h)	Permission (h)	Total (h)
1993	39.4	63.7	13.1	116.2
1996	35.4	49.5	10.2	95.1
1998	35.1	46.0	5.6	86.7
2001	31.1	37.8	4.9	73.8
2003	26.0	36.8	4.3	67.1
Air 2003	2.6	14.0	0.4	17.0

Source: Customs and Tariff Bureau, Ministry of Finance.

Table 2. Reasons for Excessive Waiting from Arrival in Port to Carrying to Bonded Area.

Weekend or holiday	30%
Large volume	14%
Particular reason of container freight station	16%
Congestion from container yard to bonding area	11%
Sorting consolidated shipment	9%
From bonded area to declaration area	
Weekend or holiday	44%
Delay in acceptance of documents from shipper	13%
Not required to declare earlier	10%
Other permission required	8%
No indication from shipper	4%

Source: Customs and Tariff Bureau, Ministry of Finance.

maritime cargo imports. The results are shown in Table 2. The most frequent reason was a weekend or a holiday. Cargo handling and necessary procedure are ordinarily suspended on weekends and holidays. Customs officials strive to process operations and procedure 24 h a day, everyday.

3. FINANCIAL AND CUSTOMS BARRIERS

3.1. Organization and Function of Japanese Customs

A discussion of the organization and functions of customs in Japan is important for understanding the current situation of Japanese customs policies and their problems for trade barriers among three Asian countries.

The Customs and Tariff Bureau was separated from the Customs Department of the Tax Bureau in November 1961. The Bureau currently has four divisions: Coordination, Management, Research and Statistics, and Appraisal. It also includes a Chief Inspector of the Custom house and Customs Inspectors. The Director-General is assisted by two Deputy Director-Generals and a counselor who heads the bureau. Under them are seven directors with various responsibilities, the manager of the Office of Inspection of the Custom houses, the manager of the Customs Administration Research Office, and the manager of the ADP Management Office. These directors and managers deal with all matters relating to Customs administration and tariff policy. The Bureau had a staff of 173, as of fiscal year 2001.

The main responsibilities of the Customs and Tariff Bureau are as follows:

(1) Matters related to research, planning and drafting of Customs duties, tonnage dues, special tonnage dues, and other administrative matters including agreements on Customs duties with other countries;
(2) Matters related to the imposition and collection of Customs duties, tonnage dues, special tonnage dues, and local consumption taxes imposed on international freight as set forth in Section 3, Chapter 2 of the Local Tax Law (Law No. 226 of 1950).
(3) Matters related to the surveillance and control of exports, imports, vessels, aircraft, and passengers in accordance with the provisions of Customs-related laws and regulations;
(4) Matters related to the operation of bonded areas;
(5) Matters related to the supervision of Custom house brokers and specialists;
(6) Matters related to processing the International cargo Nippon Automated Cargo Clearance System (NACCS);
(7) Matters related to Customs Statistics;
(8) Matters related to the education and training of customs officers; and
(9) Matters related to the general support of the Tariff Branch of the Council on Customs, Tariff, Foreign Exchange, and other Transactions.

3.2. Tariff System

3.2.1. Basic Tariffs

Ad valorem Duty (Tax according the value). In Japan, the most popular form of tariff is an ad valorem duty. So it is important that the price of imported merchandise be determined since it serves as the basis for the Customs valuation for ad valorem duty. Determining such value is based on a "Duty Evaluation Agreement." The advantages of ad valorem duties are that they are assessed proportionally to the price of imported goods, responding to inflation and price fluctuations. However, a disadvantage is that determining the appropriate import price of goods is difficult, which weakens its effectiveness to protect domestic industries. If the prices of imported goods are lowered, then the duty is also lowered. Generally, an ad valorem duty is applicable to cargo whose qualities differ substantially and whose prices rarely show short-term fluctuations. About 90% of duty articles, except on duty-free goods, are assessed ad valorem. Ad valorem duty rates increased or decreased by 5% in each duty bracket in 1961. The present duty system reflects the results of the Kennedy and Tokyo Rounds of tariff negotiations and other tariff revisions. Since then, rates of 0.1% in each duty bracket have prevailed.

Specific Duty. Since specific duty is charged on the basis of such objective quantities as the number and volume by weight of imported goods, it is unaffected by price fluctuations. It has the following advantages: duty value can be computed easily; the same duty can be applied to cargo of the same type and quality; and it offers more domestic industry protection than ad valorem duty. On the other hand, it must precisely classify duty rates in accordance with the kind, quality, or degree of imported goods. Tax inequalities will also accompany price fluctuations. Specific duty is generally applicable to cargo of similar qualities and products whose prices are difficult to determine.

Compound Duty. Compound duty involves a combination of ad valorem and specific duties. There are two kinds of compound duty: ad valorem and specific selective duty (selective duty), and ad valorem and specific combination duty (compound duty):

(a) Selective duty involves the determination of both ad valorem and specific duties on the same article and usually charging the higher price, but for some articles the lower price. Ad valorem duty is applied on higher and specific duty on lower duty prices, which enables tariffs to adequately protect domestic industries.

(b) Compound duty involves the assessment of both ad valorem and specific duties. Since specific duty rates drop as the price of imported goods increases, a certain amount of ad valorem duty is additionally charged to protect domestic industries.

3.2.2. Special Forms of Tariff

Price Differential Duty. Price differential duty is a tariff that establishes a duty rate that is equal to the difference between the imported price and a specific or intercepting price. The objective of a price differential duty is to strike a balance between the protection of domestic products and price stability. Even if import prices dip below a certain level, the duty prevents imported products from being sold at lower levels in the domestic market. If prices soar, the tariff burden is automatically lightened.

(Note: Although in the past price differential duties were applied to ports etc., the equivalent part of price differential duty was changed to specific duty and tarifficated under the Uruguay Round Agreement. This agreement maintained the conventional function of the price differential duty system.)

Sliding Duty. For articles whose international prices fluctuate dramatically, such as onions, copper, and lead, a mechanism has been established that serves the interests of both domestic producers and users. This is accomplished by charging a certain amount of duty when import prices decrease but not when prices rise.

Seasonal Duty. Seasonal duty is a tariff that applies different duty rates according to the season. Seasonal duty protects domestic seasonal producers by imposing higher duties on competing imports in season as well as meeting the needs of consumers during the off-season by charging lower duty rates. This duty is currently applied to bananas, oranges, grapefruit, and grapes.

3.2.3. Tariff Quota System

A tariff quota system charges a lower duty rate (primary duty rate) on imports of specific goods up to a certain quantity, but a higher duty rate (secondary duty rate) on quantities that exceed the tariff quota volume. In principle, the primary duty rate tariff volume is determined by subtracting the domestic production volume from the domestic demand volume (insufficient volume = import requirement). This system protects the domestic producers of similar goods while simultaneously benefiting consumers, who will presumably experience cheaper imported goods. Tariff quota volume can be applied in one of the two ways: according to the order in which the request was received or according to prior allocations. Japan utilizes the latter method.

Unlike an import quota system, a tariff quota system does not restrict direct imports since imports can be made without an import quota certificate, provided a high duty is paid. However, if the secondary duty rate is raised excessively or if the quota volume is lowered more than the required import volume, it has the same effect as an import quota system.

3.2.4. Special Duty System

Outline of the Special Duty System. Japan's special duty system includes the following duties: countervailing, antidumping, emergency, retaliatory, etc. Countervailing and antidumping are extra duties imposed to protect Japanese industries from activities that undermine the trade order using either subsidized goods from exporting countries or dumping by exporters. Emergency is an extra duty that protects domestic industries from such unforeseen situations as import increases caused by a drop in prices abroad. Retaliatory is an additional duty imposed to protect Japanese interests, achieve WTO Agreement objectives, or react to cases where Japanese ships, airplanes, export products, or transit goods are being unfairly treated.

Trend of Special Duties in Japan. Numerous examples of suits, surveys, and imposition of special duties have been implemented in the United States and Europe, but until recently there were none in Japan. Since 1982, however, requests for imposition of countervailing and antidumping duties in Japan have surfaced. In December 1982, the Japan Spinners Association filed to impose a countervailing duty against Pakistani cotton yarn and an antidumping duty against South Korean cotton yarn.

These were the first filed applications in Japan for the imposition of special duties. In March 1984, the Japan Ferroalloy Association filed for the imposition of an antidumping duty against Norwegian and French ferrosilicon, which is a steel alloy used to deacidify chemicals in making steel, and for a countervailing duty against Brazilian ferrosilicon.

In October 1988, the Japan Knitting Industry Association filed for the imposition of an antidumping duty against South Korean knit sweaters.

In October 1991, the Japan Ferroalloy Association filed for imposition of antidumping duties against Chinese, Norwegian, and South African ferrosilicon–manganese, which is a type of steel alloy used to deacidify or desulfur chemicals in making steel. The Japan Spinners Association filed an additional imposition of an antidumping duty against Pakistan cotton yarn in December 1993.

In December 2000, Japan's first safeguard investigation was held on three items including leeks, shiitake mushrooms, and tatami reed-mat facing. Provisional emergency tariffs were imposed on these items in April 2001.

In February 2001, Teijin and four other companies filed an application for antidumping duties to be imposed on certain polyester staple fibers from the Republic of Korea and Taiwan, and investigations began in April of the same year.

The investigation conducted into a countervailing duty on Pakistani cotton yarn was the first such investigation made in Japan, and the final decision was made not to impose a countervailing duty. For the Chinese ferrosilicon–manganese, the investigation conducted on dumping was the first such survey in Japan. It decided to impose an antidumping duty on most Chinese ferrosilicon–manganese. As for Pakistani cotton yarn, the final decision imposed an antidumping duty on the majority.

4. PHYSICAL AND INSTITUTIONAL BARRIERS FOR DOMESTIC TRANSPORT

4.1. Introduction

Almost 90% of containers are handled at the five major ports in the biggest three metropolitan areas: Tokyo, Yokohama, Nagoya, Osaka, and Kobe. Such heavy truck traffic causes serious problems: traffic congestion and jams, environmental and energy concerns. The Japanese government is trying to shift the mode of domestic container transport away from trailer to coastal shipping and railway transportation. Although the origin and destination of a majority of containers is located within an area less than 100 km, 32%, 39%, 15%, 29%, and 42% of containers are more than 100 km from Tokyo, Yokohama, Nagoya, Osaka, and Kobe, respectively.

Transportation modes of exporting containers between a port and places of vanning and devanning are 90% by trailer, 4% by ship, and less than 1% by rail; importing containers are 93% by trailer, 1% by ship, and less than 1% by rail. Most export/import containers are transported inside Japan by trailer. Although domestic air cargo transportation is very popular, the amount of feeder transport of export/import goods is quite small. The biggest barrier of domestic transport of export/import goods is the extraordinarily high transportation costs caused by many regulations and high labor costs.

4.2. Railway

The Japan Freight Railway Company (JR Freight) is the only nationwide railway company that transports freight in Japan. JR Freight has a long history of transporting small 12-ft containers with a 5 ton capacity, even though the history of import/export container transport, hereafter simply "container," is very short.

Before 1995, there was much temporary container transport by rail. In 1995, malt for beer and natural rubber began to be transported by container between Yokohama and Utsunomiya (approximately 140 km). The transport of other raw materials for beer extended the container transport route between Yokohama and Koriyama (approximately 260 km) in 1996. In April 1998, standard container train service of 10 cars started between the Yokohama Honmoku Container Terminal and the Sendai Port and everyday transported 40 TEUs of electric machinery and parts.

In the autumn of 1998, JR Freight announced plans to strengthen container transport and enrich its network and capacity. According to a study by JR, with small improvements, most rail networks in Japan could accommodate a standard size of container transport (8′ × 8′6″ × 20′). A high cubic container with a height of 9′6″, however, could only be accommodated by a limited number of rail routes: Tokyo to Nagoya to Osaka to Kobe to Hakata; Yokohama to Tokyo to Sendai.

Regarding flatbed wagon container trains, a new type of car called "Koki-104" and its cousin "Koki-106" were introduced to transport one forty-foot container or two twenty-foot containers; the first model "Koki-100" could only transport one twenty-foot container. In 1999 "Koki-72," which has a 74 cm lowered bed for high cubic containers, was introduced. It can also be used for container railway networks and routes with small tunnels.

Freight train capacity is restricted to 1,600 tons because of such factors as the length of siding, substation capacities, locomotive capability, etc. Since a container train is usually formed by a maximum of 20 cars, the capacity of an ordinary container train is 40 TEUs. JR Freight rushed to lengthen sidings on the major lines involving Tokyo, Nagoya, Osaka, Kobe, and Hakata. JR Freight has the privilege of owing and operating freight trains. Passenger railway companies own almost all of the tracks. Since passenger trains take precedence over freight trains, the expansion of capacity of railroad freight transport is difficult to anticipate.

Since JR Freight already owned equipment that could handle 12-ft containers, it prepared marshaling space to endure the weight of 5-ton containers. However, handling equipment for 20–40-ft, containers were required at container terminal stations. Marshaling space with strengthened pavement to accommodate 35-ton containers were required. Container storage yard space was also needed. Such facilities and equipment were insufficient creating big obstacles for container transport.

4.3. Road

Twenty- to forty-feet containers may be transported by semitrailer on any road. Neither double stacks nor full trailers can be adopted for container transportation. Travel regulations prescribe the maximum size and weight of semitrailers. The maximum size is 2.5 m in width, 3.8 m in height, and 12.0 m in length; the maximum weight is 26 tons. Ordinary 20-ft containers with a semitrailer does not exceed the limits. For ordinary 40-ft containers, high cubic containers, and other special purpose containers, a road transportation company is required to operate trailers only on designated roads. Besides, permission is required for each time of transportation.

Before 1993, the maximum vehicle weight on roads was 20 tons. The limit has been increased to 25 tons, and major roads have already been upgraded. However, since a fully loaded, 40-ft container still exceeds the limits, roads available for 40-ft container transportation have been designated. In the case of high cubic containers, when the height exceeds 3.8 m, they are not permitted to pass through some tunnels. Tunnel improvement is difficult and more expensive than pavement. Many tunnels still refuse high cubic containers, even on expressways and major national routes.

4.4. Coastal Shipping

In 1964, two important acts were revised for domestic water transportation: the "Domestic Water Transportation Business Act" and the "Domestic Water Transportation Union Act." Unlike ocean shipping lines, coastal shipping lines in Japan in the 1960s experienced a terrible economic period. Most ship owners were small businesses that had only one or two small ships. Competition was fierce. As a result, tariffs remained at minimum levels, and many small shipping companies turned bankrupt. Although these two acts supported the business of coastal lines and reduced the number of firms by half, the fundamental situation did not improve for more than 30 years.

Supply and demand adjustment regulations are viewed as hindrances that respond to diversified needs and ensure efficient business management.

However, Japan had to experience an economic recession before its entire economic and social systems were fundamentally reformed.

In December 1996, the Ministry of Transportation set a tentative target date to abolish supply and demand adjustment regulations, and to build a transportation system that takes advantage of market forces to improve convenience for people. The Ministry also laid the groundwork for measures to protect people and secure safety by abolishing such regulations. In April 1997, the Ministry asked the Council for Transportation Policy to consider measures that would abolish supply and demand adjustment regulations. In June 1998, the advisory panel submitted recommendations for the basic direction of transportation policies after the abolition of supply and demand adjustment regulations.

4.4.1. Supply and Demand Adjustments

Article 8 of the Domestic Water Transportation Union Act states that the supply and demand adjustment system is classified by the following:

Supply and demand Adjustments:
(1) Volume adjustment of ship ownership
/ Scrap and Build adjustment
/ Cooperative scrapping
(2) Volume adjustment of service supply
/ Service capacity adjustment
/ Cooperative nonusage of ships

Under the "scrap and build" adjustment rule, a shipbuilder needed to scrap an equivalent volume of ships before being allowed to build a new vessel for domestic transport. If the shipbuilder did not have a ship to scrap, he had to purchase one to scrap from other ship owners. The market dealt with the allowance and the treatment of old or useless ships. The allowance was valuated as a latent asset and considered security for a loan, making it easy for ship owners to get loans. This system led to irresponsible management of domestic coastal shipping liners. It also added to the cost for new shipbuilding and proved to be a big obstacle for newcomers into the market.

4.4.2. Abolishment of the Supply and Demand Adjustment Regulations

In 1998, supply and demand adjustment regulations for domestic coastal cargo transportation vessels were abolished.

To escape from the confusion of small domestic coastal shipping liners produced by this deregulation, the government enacted legislation called the "Temporary Domestic Water Transportation Project." The system included the following:

(1) A new shipbuilder must scrap his own ships or pay the charges indicated in Table 3.
(2) A company that plans to scrap a ship will be granted a subsidy as indicated in Table 3.
(3) The Japan Federation of Coastal Shipping Association will operate this project using loans partially guaranteed by the government.

The project was designed to continue for 15 years. However, many people are worried about its financial condition Table 4 shows a number of scrapped vessel and subsidized amount for each four years after the project started. Relatively many vessels are scrapped in the first year due to decline of subsidy rate. On the other hand, a number of new built vessels are suddenly decreased in the year 1998 also due to decline of charge rate (Table 5). Although, ship building charges is expected to balance out subsidies for scrapping, a short of ship building charge causes the project unsustainable. While ship building charges were paid 24 billion yen until 2002 (Table 6), scrap subsidies should be provided 108 billion yen until 2001 (Table 4).

As of 1999, a company with foreign managers can register its vessel as a Japanese ship for domestic transport, if all of its representative directors and more than two-thirds of its directors are Japanese.

Table 3. New Shipbuilding Charges and Subsidies for Scrapping.

	New Shipbuilding Charges		Subsidies for Scrapping	
	Tanker (yen per cubic meter)	General cargo ship (yen per DW tonnage)	Tanker (yen per cubic meter)	General cargo ship (yen per DW tonnage)
1998	75,000	125,000	75,000	125,000
1999	70,000	120,000	70,000	120,000
2000	65,000	115,000	65,000	115,000
2001	60,000	110,000	60,000	110,000
2002	60,000	110,000	60,000	110,000
2003	58,500	106,000	58,500	106,000
2004	57,000	102,000	57,000	102,000
2005	55,500	98,000	55,500	98,000

Source: Japan Federation of Coastal Shipping Association.

Table 4. Scrap Subsidy by Vessel.

Year	Vessel Type	Number	Deadweight	Subsidized (millions of yen)
1998	General cargo ship	288	302,333	33,171
	Specified ship	152	139,211	6,812
	Tugboat	27	24,574	243
	Tanker	200	286,437	17,158
	Total	667	752,555	57,384
1999	General cargo ship	58	51,238	5,301
	Specified ship	43	35,690	1,249
	Tugboat	8	7,680	65
	Tanker	37	53,464	2,937
	Total	146	148,072	9,553
2000	General cargo ship	58	57,050	5,675
	Specified ship	65	59,343	2,465
	Tugboat	13	13,750	105
	Tanker	57	72,595	3,597
	Total	193	202,738	11,843
2001	General cargo ship	173	173,443	16,396
	Specified ship	96	126,132	6,393
	Tugboat	8	9,310	68
	Tanker	90	133,717	5,951
	Total	367	442,602	28,808
Total	General cargo ship	577	584,064	60,544
	Specified ship	356	360,376	16,920
	Tugboat	56	55,314	481
	Tanker	384	546,213	29,642
	Total	1,373	1,545,967	107,587

Source: Japan Federation of Coastal Shipping Association.

In October 2000, the supply and demand adjustment regulations for domestic coastal passenger transport vessels were abolished. The government took necessary measures to protect the people's safety when laying the groundwork for the abolition of the regulation.

The following changes must be reported (changed from approval):

(1) any changes in berth, routes, or port calls;
(2) inauguration of new services or new vessels;
(3) the addition of extra port calls (either permanent or temporary);
(4) increasing the size of vessels or changes in vessels;
(5) changes of tariff or fares;
(6) minor changes of operation/management plans.

Table 5. New Ship Building.

Year	Cargo Ship		Tanker		Total	
	Number	Tonnage	Number	Tonnage	Number	Tonnage
1989	237	152,050	87	63,676	324	215,726
1990	212	158,344	93	64,475	305	222,819
1991	216	155,022	115	97,621	331	252,643
1992	458	266,119	383	298,813	841	564,932
1993	139	84,161	54	36,540	193	120,701
1994	175	112,046	54	19,674	229	131,720
1995	216	150,766	55	44,646	271	195,412
1996	153	104,235	16	14,972	169	119,207
1997	136	121,441	26	35,303	162	156,744
1998	27	72,817	6	4,460	33	77,277

Source: Japan Federation of Coastal Shipping Association.

Table 6. New Ship Building Charges: April 1998 to April 2002.

Ship Type	New Ship Building		Replacement		Paid
	Number	Tonnage	Number	Tonnage	(million yen)
General cargo	93	144,661	61	89,393	8,314
Specified	77	268,529	17	26,754	7,923
Tugboat	79	208,754	64	118,583	7,427
Tanker	16	43,600	9	17,800	335
Total	266	665,544	151	252,530	24,001

Source: Japan Federation of Coastal Shipping Association.

5. TECHNOLOGICAL BARRIERS

5.1. EDI System for Government Procedures

5.1.1. Systems

Sea-NACCS. Japan Customs introduced the Nippon Automated Cargo Clearance System (NACCS), which is an EDI system for the declaration of imports and exports, incoming and outgoing vessels, and most necessary procedures at customs. NACCS, which used to be called the Nippon Air

Cargo Clearance System was introduced in 1987 to import and export air cargo only at Tokyo's Narita airport and in the Atsugi district in Kanagawa prefecture. The system has extended its coverage into maritime cargo since 1991, and the name has been changed to its current name. At present, NACCS consists of Air-NACCS for air cargo and Sea-NACCS for maritime cargo.

Sea-NACCS was altered in 1999 to extend its coverage to most customs procedures, and UN/EDIFACT has become an accepted part of procedure. Table 7 shows the number of user companies. About 90% of all declarations for import/export are processed through Sea-NACCS. Since NACCS is operated on individual networks, security, initial, and operating costs for each company are relatively high.

Port Authority EDI System. Port authority EDI systems were introduced in 1999 by the MLIT, the Japan Coast Guard, and each port authority. Faxed application forms have been accepted since 1996, and the EDI system has been used since 1999. Ocean carriers and ship's agents can declare such ship-related procedures as inward or outward vessels and the use of berths to other port facilities, authorities, and commanders. Although initially the system was introduced at only a small number of ports, as of March 2002 it was available at 23 port authorities and 86 port commanders, and a lot of other ports are planning to introduce it in the near future. The systems are also available on the Internet. Applicants can either send forms by e-mail or from websites, explaining the relatively low initial and operation costs for each company.

JETRAS. The Japan Electronic open network TRAde control System (JETRAS), which was introduced by the Ministry of Economy, Trade, and

Table 7. Number of Sea-NACCS Users.

Industry	Number of Users
Ocean carriers	33
Shipping agents	161
Container yards	248
Bond warehouses	463
Customs agents	630
Banks	51
Total	1,586

Source: Nippon Automated Cargo Clearance System Operations Organization.

Industry (METI) in 2000, was developed to computerize applications for import/export permits and related procedures. The system is operated on the Internet, and users have to install the proper software. A function is also available that prints such application forms as invoices and packing lists that must be submitted to customs. Almost all cargo must be declared, and there are many cases that must obtain permission from METI before being allowed as imports. Although JETRAS cannot interface with NACCS yet, such connections are expected in the near future.

Other EDI Systems. EDI systems of quarantine procedures are also available now. The three systems shown in Table 8 are currently operated and are completely interfaced with NACCS. Applicants can declare most plants, animals, and food at NACCS. For example, 87% of food inspections were applied through FAINS (January 2001).

5.1.2. Current Situation and Future Prospects

Fig. 1 summarizes the current situation and future prospects of EDI systems for government procedures. NACCS (customs) is interfaced with PQ-NETWORK (plant quarantines), ANIPAS (animal quarantines), and FAINS (food inspection). Port Authority EDI systems are partially interfaced with NACCS. JETRAS (import/export permits) is not interfaced with any other systems, except for trade finance systems, which are not included in the scope of this chapter. The national government is arranging and developing a comprehensive EDI system for import/export procedures. All EDI systems introduced in this report will be completely interfaced with each other. "One-stop service" will eventually be implemented.

5.2. EDI System for Private Transaction

5.2.1. POLINET

Port Logistics Information Network (POLINET), which was established in 1986 as Japan's first inter-industry EDI system, connects four maritime

Table 8. Other EDI Systems for Import/Export.

EDI System	Procedure	Commencement
PQ-NETWORK	Plant quarantine	1997
ANIPAS	Animal quarantine	1997
FAINS	Food inspection	1996

```
--------------------------------------------------------------------------
Current Situation
      NACCS     ------------------ Port Authority EDI
       |- PQ-NETWORK
       |- ANIPAS
       |- FAINS

      JETRAS
--------------------------------------------------------------------------
Near Future (in a few years)
      NACCS     ---------------------------- Port Authority EDI
       |- PQ-NETWORK
       |- ANIPAS
       |- FAINS
       |- JETRAS
--------------------------------------------------------------------------
```

Fig. 1. Current Situation and Future Prospects of EDI Systems.

transportation-related industries: ocean carriers, freight forwarders, measuring, and weighing. Various interactions among them can be processed on this system. This system used to be called the Shipping Information Network System (SHIPSNET) before the Port Logistics Information System Association, the corporation aggregate, was established. POLINET has been operating since the system was updated in 1995. Since POLINET can accept and translate data from the three VAN services provided by NTT, IBM, and NEC. The user, which introduced one of these VAN services, can easily join this EDI system. UN/EDIFACT and other standard forms are also available on this system.

5.2.2. SC/SF NET

Shipper/Carrier EDI Network (SCNET) is the EDI system for connections between shippers and ocean carriers. SFNET is for shippers and freight forwarders. These two systems are operated by a voluntary organization called SC/SF Net Center. For instance, shippers and carriers provide freight paid information and bills of lading on SCNET. Shippers and forwarders provide shipping instruction information and bills on Shipper/Forwarder EDI Network (SFNET). UN/EDIFACT has been available since 1996. SC/SF NET adopted two private VAN services provided by NEC and NTTD. However, since these two networks are not connected to each other, users have to connect to both networks as needed. This system proved unpopular with shippers, and the number of participating members has fluctuated around 16 since it was founded in 1988. In 2001 it was dissolved after a better organization, TEDI, was formed in 2000.

5.2.3. TEDI

Trade Electronic Data Interchange (TEDI) is the comprehensive private EDI system established in 2000. Most of Japan's major ocean carriers, freight forwarders, trading companies, banks, and insurance companies are organized on this system. It includes not only SC/SF NET functions but also various trade finance functions. UN/EDIFACT is also available, which explored the possibility of the implementation of an interface with government EDI systems such as NACCS and JETRAS. JETRAS had also been discussed as a form of cooperation with the Pan-Asia e-commerce Alliance (PAA), which it joined in May 2002. PAA comprises the following: Infoshare Technology Development (China), Tradelink Electronic Commerce (Hong Kong), Korea Trade Network (Korea), CrimsonLogic (Singapore), Trade-VAN Information Services (Taiwan), DagangNet (Malaysia), and TEDI.

5.3. Current EDI System Problems

EDI systems have been separately developed and managed by government and private companies, forcing users to master many systems. The current situation, however, should improve when all applicants are no longer forced to compose and submit many kinds of documents.

The MLIT conducted a survey in 2000 of freight forwarders, ocean carriers, ship's agents, and container terminals and learned that Government EDI systems are more popular than private EDI systems. One hundred fifty-six companies (85.2%) are using one or more of the following government EDI systems: Sea-NACCS, Port authority EDI, or JETRAS. Forty seven (25.7%) companies, although, use POLINET and SC/SF NET, the private EDI systems. The problems and barriers of EDI systems are clarified and summarized as follows:

(1) Costs
(2) Institution
(3) System
(4) Relationship with customers
(5) In-house facilities

The high initial and operating costs of EDI systems are widely recognized. For instance, Sea-NACCS is operated on an individual network, which requires a special server machine and software. If all EDI systems were operated on a unified network and accessible from conventional machines, costs could be reduced.

The institutional problems of Sea-NACCS were recognized by a large number of companies. The required server machines for Sea-NACCS have no potential to be diverted to private EDI systems. If a company wants to access both Sea-NACCS and one of the private EDI systems, it needs respective machines and software for two systems.

System problems, relationships with customers, and in-house facilities are not common problems because they are caused by individual situations. Since less useful software for using EDI systems is available, users are sometimes required to customize software or input troublesome data. Some users who present a particular EDI system are designated by customers. Often a company is forced to implement more than a few EDI systems.

6. EFFECTS OF REMOVAL OF BARRIERS

6.1. SCGE Model

6.1.1. Framework

Many previous models could estimate trade amounts, and several could also estimate the impact of reductions in transportation costs. If removal is expressed as a reduction of transportation costs, a model can estimate the effect of the removal of barriers. This section introduces an SCGE trade model. The model developed here includes an ocean carrier, which solely takes charge of international transportation. Transportation costs are expressed as freight rate in the model.

The SCGE trade model developed here is mainly base on Liew and Liew (1984), Whalley (1985), Hertel (1997) and Shoven and Whalley (1992). An ocean carrier sector, which takes charge of inter-regional transportation, is newly introduced. The model in this chapter makes the following assumptions:

- The revenue of inter-regional transportation sectors are composed of freight rates and insurance.
- The input structure of the inter-regional transportation sector is not stated in the input/output table. Assuming that the structure is involved in the amount of exports, the structure is calculated as follows. The total output of inter-regional transportation sectors is distributed according

to the input structure of the transportation sector, which is stated in the table.

- Each region has only one port. All trade goods are transported through that port.
- Production factors are capital and labor. Both are regionally immobile. Capital is not transferable across sectors. Labor can move to another sector in the same region, and it is assumed that wage rates are identical in the region.
- Firms, households, and regional governments exist in each region.
- Regional governments levy direct and indirect taxes, which comprise the funds of government expenditure.
- Goods classified to the same commodity and produced in different regions are regarded as different goods. This is the well-known Armington assumption.
- Only one inter-regional transport firm exists whose demand is derived from the demand of goods. Freight rate is suggested by inter-regional transportation firms.
- Freight rate of each transportation link is calculated based on competition. Inter-regional transportation freight rate plays a role "numeraire."
- Intra-regional transportation freight rates are decided while the profits of inter-regional transportation firms are set at zero.
- Final demand is composed of households' consumption, government consumption, fixed capital formation, and increase in stocks.
- Indirect taxes and increases in stocks are given exogenously.
- The price of products in the rest of the world (ROW) is 1.

6.1.2. Production

CES and Leontief technology are adopted as production functions of industry j in region s, mainly due to data availability. Each industry in each region has a CES value-added function with capital and labor as substitutable primary inputs and fixed coefficient intermediate requirements for composite goods. Fixed requirements of composites can be met by a substitutable mix of comparable domestic and imported goods. CES functions are used at a choice of produced regions for each fixed composite requirement. Value-added and production functions are expressed as follows:

$$V_j^s = v_j^s\left[\varphi_j^s(\mathrm{K}_j^s)^{(\sigma_j^s-1)/\sigma_j^s} + (1-\varphi_j^s)(\mathrm{L}_j^s)^{(\sigma_j^s-1)/\sigma_j^s}\right]^{\sigma_j^s/(\sigma_j^s-1)} \quad (1)$$

$$X_j^s = \min\left(\frac{V_j^s}{a_{VAj}^s}, \frac{x_{1j}^s}{a_{1j}^s}, \ldots, \frac{x_{ij}^s}{a_{ij}^s}, \ldots, \frac{x_{Nj}^s}{a_{Nj}^s}\right) \tag{2}$$

$$X_j^s = \sum_i a_{ij}^s X_j^s + V_j^s \tag{3}$$

$$x_{ij}^s = a_{ij}^s X_j^s \tag{4}$$

$$a_{ij}^s = \left[\sum_r \beta_{ij}^{rs} (a_{ij}^{rs})^{(\theta_j^s - 1)/\theta_j^s}\right]^{\theta_j^s/(\theta_j^s - 1)} \tag{5}$$

$$\varphi_j^s = \left(\frac{K_j^s}{L_j^s}\right)^{1/\sigma_j^s} \Bigg/ \left[1 + \left(\frac{K_j^s}{L_j^s}\right)^{1/\sigma_j^s}\right] \tag{6}$$

$$v_j^s = (K_j^s + L_j^s) \Big/ \left[\varphi_j^s (K_j^s)^{(\sigma_j^s - 1)/\sigma_j^s} + (1 - \varphi_j^s)(L_j^s)^{(\sigma_j^s - 1)/\sigma_j^s}\right]^{\sigma_j^s/(\sigma_j^s - 1)} \tag{7}$$

$$p_j^s = \frac{1}{X_j^s}\left(\sum_i \sum_r q_{ij}^{rs} x_{ij}^{rs} + \rho_j^s K_j^s + \omega^s L_j^s\right) \tag{8}$$

$$x_{ij}^{rs} = a_{ij}^{rs} X_j^s \tag{9}$$

Eq. (6) is derived from the first-order conditions for cost minimization in each industry. If the price of all goods and factors are identical, the profits of each industry are zero; Eq. (7) is obtained, where V_j^s is value added in industry j in region s. v_j^s and φ_j^s are parameters, x_{ij}^{rs} is intermediate input goods i produced in region r, K_j^s is capital, and L_j^s labor. a_{ij}^{rs} is a technical coefficient of industry j in region s. Considering transportation costs, two types of prices are introduced: producer and consumer. The producer price of goods j in region s and the consumer price of goods i produced in regions r and s are expressed as p_j^s and q_i^{rs}, respectively. Consumer price q_i^{rs} is represented as the sum of producer price p_j^s and transportation cost c_{ij}^{rs}, which is introduced below. The behavior of industry j in region s is formulated as a minimization of costs. σ and θ possess substitution elasticity. ρ_j^s is the rent of the capital of industry j in s and ω^s is wages in region s.

6.1.3. Demand

Household behavior is formulated as a utility maximization problem constrained by income. The Cobb–Douglas utility function is adopted:

$$\max \quad U_k^s = \prod_i \left(\prod_r (y_{ik}^{rs})^{\beta_{ik}^{rs}} \right) \tag{10}$$

$$\text{s.t} \quad \sum_i \sum_r q_{ik}^{rs} y_{ik}^{rs} \leqslant W_k^s$$

$$\sum_k \sum_i \sum_r \beta_{ik}^{rs} = 1$$

where y_{ik}^{rs} is consumption of goods i produced in region r at final demand sector k in region s, W_k^s is total consumption of final demand sector k in region s, and β_{ik}^{rs} is a parameter. Final demand sector k indicates: 1, household consumption; 2, government consumption; 3, fixed capital formation; and 4, increase in stocks.

To solve the above maximization problem, y_{ik}^{rs} is derived as follows:

$$y_{ik}^{rs} = \frac{\beta_{ik}^{rs} W_k^s}{q_{ik}^{rs}} \tag{11}$$

Suppose an income transfer of region s from other regions is TR^s, and an indirect tax paid by industry j in region s is IT_j^s; the total final demand G^s is represented as follows:

$$G^s = \sum_j \rho_j^s K_j^s + \omega^s \sum_j L_j^s + \sum_j IT_j^s + TR^s \tag{12}$$

After subtracting taxes from G^s, household disposable income W^s is derived.

$$W^s = (1 - \tau_K^s) \sum_j \rho_j^s K_j^s + (1 - \tau_L^s) \omega^s \sum_j L_j^s + TR^s \tag{13}$$

where τ_K^s and τ_L^s indicate corporate and income taxes, respectively.

Supposing household savings rate in region s is σ^s, then household consumption is

$$W_1^s = (1 - \sigma^s) W^s - \gamma_1^s W_4^s \tag{14}$$

and government consumption is

$$W_2^s = \tau_K^s \sum_j \rho_j^s K_j^s + \tau_L^s \omega^s \sum_j L_j^s + \sum_j IT_j^s - \gamma_2^s W_4^s \tag{15}$$

where W_4^s is an increase in stocks, and γ_1^s and γ_2^s are weight parameters.

6.1.4. *Savings and Investment*

Since it is assumed that firms do not have internal reserves, the savings in region s is $\sigma^s W^s$. These are the funds for investment. Considering an increase in stocks, investment, that is, fixed capital formation, is represented as follows.

$$W_3^s = \sigma^s W^s - \gamma_3^s W_4^s \tag{16}$$

6.1.5. *Inter-regional Transportation*

The relationship between transportation costs and intermediate input from inter-regional transportation firm to industry i in region r is

$$\sum_j \sum_k \sum_s \sum_i \sum_r (c_{ij}^{rs} x_{ij}^{rs} + c_{ik}^{rs} y_{ik}^{rs}) = \sum_i \sum_r \tau T_{ti}^r \tag{17}$$

$$c_{ij}^{rs} = m_i^r d^{rs} \tau \tag{18}$$

where c_{ij}^{rs} is a transportation cost unit of transaction x_{ij}^{rs}, τ a transportation fee unit, m_i^r the tonnage of unit price worth of goods i produced in region r, d^{rs} the distance between region r and region s, and T_{ti}^r transportation service input from inter-regional transportation firms to industry i in region r at period t. Period t indicates 0, the benchmark year, and 1 equals the future. Transportation service inputs in the future are

$$T_{1i}^r = (N_1 - N_0)\delta_i^r + T_{0i}^r \tag{19}$$

where N_t is the total output of inter-regional transportation firms at period t and δ_i^r is a proportion parameter. Besides, exports from industry i in region r to ROW in the future is calculated by

$$E_{1i}^r = (M_1 - M_0)\varepsilon_i^r + E_{0i}^r \tag{20}$$

where M_t is the total output of ROW in period t and ε_i^r is a proportion parameter.

Transportation costs between regions r and s c^{rs} are assumed to be identical, irrespective of the industry, represented as

$$c^{rs} = \text{Tariff}^{rs} - \phi \sum_i \sum_j m_i^r d^{rs} (x_{ij}^{rs} + y_{ik}^{rs}) \quad (r \neq s) \tag{21}$$

$$c^{r'r'} = \frac{(\tau T - \sum_r \sum_{s(\neq r)} c^{rs})}{\sum_r (x_{ij}^{rr} + y_{ik}^{rr})} \left(\tau T - \sum_r \sum_{s(\neq r)} c^{rs} \right) \tag{22}$$

where Tariff^{rs} is a standard freight rate between regions r and s, and ϕ (>0) is a competition parameter derived from the difference between producer and consumer prices in each transaction. Since parameter ϕ cannot be obtained because of data availability, the sensitivity of ϕ is observed in the numerical experiment.

6.1.6. Numerical Specification

The model is applied to four major economic regions: Japan, United States, EU, and Asia. The basic source is the four regional international input/output table of 1990. This table is an accounting matrix that includes domestic transactions and inter-regional trade. The intermediate sectors of the table are reformed to demonstrate inter-regional transportation sectors shown in Fig. 2. The financial statements of the three major ocean carriers in Japan, which are NYK, MO, and KL, are referred to, and then input structures and inter-regional transportation sector sales amounts are estimated.

6.2. Case Study

The model constructed can estimate the effect of the removal of barriers, if the removal is regarded and expressed as a reduction of transportation costs in the model. Supposing that import/export procedure is simplified and streamlined, substantial transportation costs are reduced. The effect is larger

		Region 1 Goods 1 ·· Goods n	·· ·· ·· ··	Region m Goods 1 ·· Goods n	Transport
Region 1	Goods 1 : Goods n				
: :	: :				
Region m	Goods 1 : Goods n				
Transport					

Fig. 2. Input Data Form.

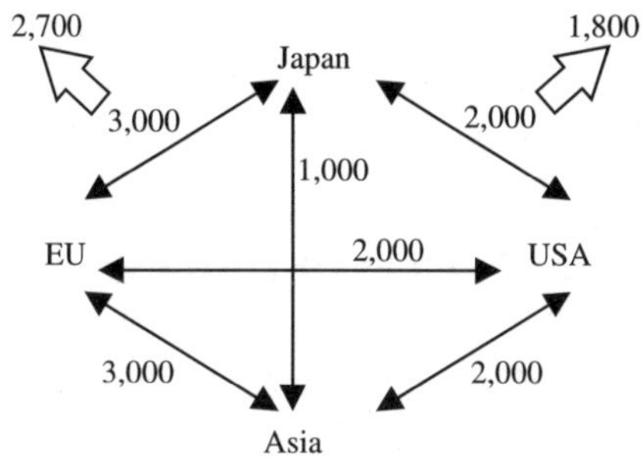

Fig. 3. Transportation Cost Changes.

Table 9. Change of Transaction Amount (%).

	Intermediate Input				Final Demand			
	Japan	U.S.A.	EU	Asia	Japan	U.S.A.	EU	Asia
Japan	0.0410	0.9250	0.8110	−0.1123	0.0200	0.7617	0.9105	0.0808
U.S.A.	1.7210	0.0030	0.0010	0.0019	1.3102	0.0038	0.0119	0.1055
EU	1.1120	0.0090	0.0010	−0.1059	1.6182	0.0183	0.0031	0.0082
Asia	0.2190	0.0700	0.0020	−0.0209	0.2834	0.1166	0.0936	−0.0025
ROW	0.1100	−0.1800	−0.0990	−0.1171	0.1052	−0.0313	−0.0194	−0.0008

in Japan–U.S. and Japan–EU trade where the world's largest vessels operate. For example, Japan Customs starts to accept import declarations only after the unloading of cargo is finished; therefore, the required time is longer at such trade pairs.

We assume that unit transportation costs τ between Japan, EU, and the United States are reduced by 10% as in Fig. 3. A competition parameter ϕ is set to zero. Goods are grouped into three divisions: primary product, secondary product, and tertiary product. Parameter m_i^r is set to 1 for primary and secondary products and at 0 for tertiary products, mainly due to data availability.

Table 9 shows the results of the case study. Transaction amounts are standardized by price in the benchmark year, and therefore each value in the table shows quantity changes. Here, transactions of tertiary products are ignored because they do not involve the movement of freight. As a result, trade amounts increased on the whole, especially between Japan and U.S./EU, where transportation costs were reduced by 10%. On average, trade amounts between Japan and U.S./EU increased by 1.07%. Roughly speaking, a 10% reduction in transportation costs led to about a 1% increase in trade among particular pairs.

7. CONCLUSION

This chapter discussed and identified four types of barriers in the operation and management of maritime and multimodal transportation: legal and institutional, financial and traditional, physical, and technological. The major factors are as follows:

(1) Prior consultation system and licensing system (legal and institutional barriers)
(2) Custom system (financial and customs barriers)
(3) System of inland transport modes and regulation on them (physical barriers)
(4) EDI systems (technological barriers)

Applying an SCGE model that included the behavior of transportation sectors, the effect of the removal of barriers was calculated. The results of the case study showed that in cases where import/export costs are reduced by 10% total trade increased by 1%. The removal of barriers in Japan stimulates not only the national economy but also the world economy. Next we must consider the removal of such barriers.

REFERENCES

Hertel, T. W. (1997). *Global trade analysis*. New York: Cambridge University Press.

Liew, C. K., & Liew, C. J. (1984). Measuring the development impact of a proposed transportation system. *Regional Science and Urban Economics*, *14*, 175–198.

Shoven, J. B., & Whalley, J. (1992). *Applying general equilibrium*. Cambridge: Cambridge University Press.

Whalley, J. (1985). *Trade liberalization among major world trading areas*. Cambridge, MA: MIT Press.

INTERNATIONAL CONTAINER TRANSPORTATION NETWORK ANALYSIS CONSIDERING POST-PANAMAX CLASS CONTAINER SHIPS

Katsuhiko Kuroda, Mikio Takebayashi and Toshiaki Tsuji

ABSTRACT

This paper presents the network competition model concerning the supply–demand interaction in the international maritime transport market and evaluates the impact of the introduction of Post-Panamax vessels. First, the Nash equilibrium-based competition model is formulated by considering the behavior of both carriers and shippers, which works as the equilibrium constraints. Later accuracy of the model is evaluated which resulted with a good accuracy in duplication of container throughput. Findings of the different numerical examples based on application of proposed model suggest that the major use of Post-Panamax vessel leads to enforce the hierarchy between Asian ports.

Global Competition in Transportation Markets: Analysis and Policy Making
Research in Transportation Economics, Volume 13, 369–391

ISSN: 0739-8859/doi:10.1016/S0739-8859(05)13016-9

1. INTRODUCTION

Recent economic development in East Asian countries has remarkably changed the marine transportation market, particularly the container transport market. In the early 1970s, the ports of Kobe and Yokohama were the gate ports of Asian region and until the Hanshin Earthquake in 1995 the port of Kobe ranked at no. 6 while handling the cargo volumes in the world. At present ports of Hong Kong, Singapore, Kaoshueng, and Busan are ranked in world's top 10 where as port of Kobe fell to 23rd. Drastic changes in ranking handling cargo volume came from the change of commodity flows in the world, corresponding to the globalization of economy including Asian NIES and ASEAN countries. Another reason is the hard competition in the container marine transport market, particularly in the Asian–Pacific and Asian–Europe markets, where ship companies are very competitive and interport competition has also intensified.

Under this atmosphere, ship companies are making strategic efforts to survive in the market by reducing operational costs, which include making alliances and introducing huge container ships aiming for an economy of scale. In accordance with these strategic efforts of ship companies, port administrators are rushing to survive in the market by constructing deep-water container berths, reducing port and cargo handling charges, and by providing "one-stop service" by electric data information systems (EDIS). Most of the main ports in East Asia including Japanese ports are rushing to construct large container terminals with deep-water berths.

However, there are few unresolved problems like (i) can all of these ports accommodate such huge ships? (ii) are there any changes of routing by ship companies and in that case what will happen to the market? Solutions to these questions are crucial for port administrators. No research has directly found any solution to these questions. Previous research on container marine transport market (Imai & Papadimitrious, 1996; Osman et al., 1997; Ieda et al., 1998a, 1998b) focused on the carriers' behavior of port choice and failed to model transportation tariffs. Also they did not consider the domestic shippers' behavior or management policies of port administrators. In actual transport market, three kinds of players exist: carriers; domestic shippers; and port administrators. At least, market predictions must give equilibrium solutions of the behaviors of carriers and shippers under various port management policies. From this point of view, Kuroda et al. (1997a, 1999, 2000, 2001) and Kuroda and Yang (1997b) developed different models of container marine transportation.

The present paper proposes an analytical tool, which finds a solution to unresolved problems and few numerical examples are provided to demonstrate the validity of the proposed model.

2. MODEL FORMULATION

In an actual marine transportation market, such cooperation among carriers as alliances and sunken costs is needed to enter the market. Dynamic market changes are observed. However, it is also true that competition among carriers is very intense and alliances among carriers are not necessarily stable. The present paper assumes that the market is "completely competitive." In modeling the behavior of carriers and shippers, it is also assumed that carriers intend to optimize routing networks and vessel size, while shippers optimize inland transport routes by choosing importing and exporting ports and by assigning cargo volumes at each port.

The concept of the model is shown in Fig. 1. In this study, a route that connects with two specific ports is called as "link," and a particular link can only accept one kind of vessel size. For example, if two different sizes of ships are navigating on a route between two specific ports, there will be two different links and each link will accept only one vessel size.

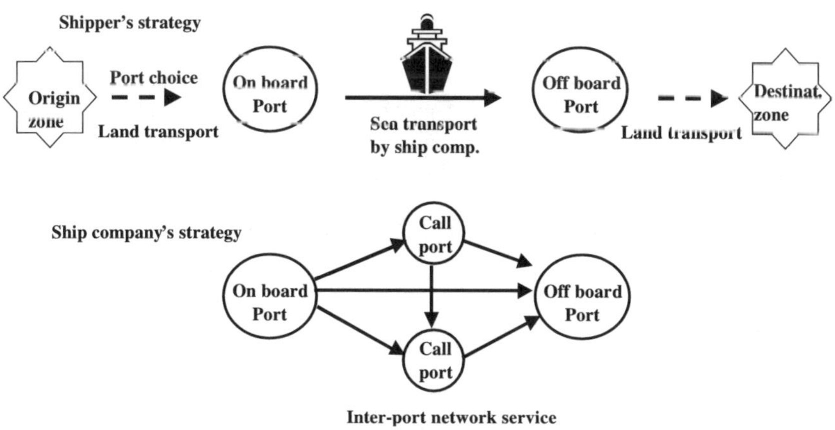

Fig. 1. Concept of the Model.

2.1. Premises and Assumptions

In the present study, the market is assumed to be perfectly competitive, and premises and assumptions include

(1) The market has many carriers that provide the same quality of service and homogeneous shippers.
(2) Carriers and shippers have complete information about the market.
(3) No sunken costs are assumed for participating in and retiring from the market.
(4) Negotiations and special cooperation between carriers and shippers are ignored.
(5) Shippers can freely select ports for importing and exporting.
(6) O.D. cargo volume is given *a priori* and does not influence service provided by carriers.
(7) Carriers have to transport all OD cargo.
(8) Carriers can choose some classes of vessel size.
(9) In numerical examples 1,000, 3,000 and 6,000-twenty foot equivalent unit (TEU) vessels are considered and navigation time on a specific link is assumed to be same for ships of different size.

2.2. Carrier's Behavior

Carriers create strategies to maximize the profit and their surplus equals to be zero based on the assumption of "perfectly competitive market." To minimize the operation costs each carrier will choose a routing network among ports that includes service frequency and vessel size. Thus, denoting vessel sizes on a specific route provided by carriers as m, the carriers' objective function and constraints are expressed as:

$$\underset{f_l}{\text{Min}} \quad Z_c = \sum_l C_l(f_l, T_l) \tag{1}$$

subject to

$$\sum_i \sum_j \sum_r \delta_{ij}^{rl} x_{ij}^r = x_l \quad \forall l \tag{2}$$

$$x_l \leqslant f_l CP_l = \sum_m \delta_l^m f_l CP_l^m \quad \forall l \tag{3}$$

$$\sum_l \delta_l^h f_l \leqslant VC_h \quad \forall h \tag{4}$$

$$\sum_l \delta_h^l \cdot \delta_l^m \cdot f_l \leqslant VC_h^m \quad \forall h \tag{5}$$

$$f_l \geqslant 0 \quad \forall l \tag{6}$$

where

f_l	total service frequency on link l
T_l	navigation time on link l
x_{ij}^r	container volume transported through route r from the origin zone i to the destination zone j
x_l	container volume transported on link l
C_l	total operation cost of a vessel on link l
CP_l^m	maximum loading capacity of vessel of size m on link l
VC_h^m	maximum capacity of a port h for m size vessels
δ_{ij}^{rl}	$= \begin{cases} 1 & \text{when a link } l \text{ is included in route } r \text{ chosen by shippers from } i \text{ to } j \\ 0 & \text{otherwise} \end{cases}$
δ_h^m	$= \begin{cases} 1 & \text{when vessel size is } m \text{ of link } l \text{ connected with port } h \\ 0 & \text{otherwise} \end{cases}$
δ_l^h	$= \begin{cases} 1 & \text{when port } h \text{ is included in link } l \\ 0 & \text{otherwise} \end{cases}$

Eq. (2) means that a container volume from zone i to zone j using link l of route r should equal the total containers flowing on link l. Eq. (3) represents the link capacity constraint. Eqs. (4) and (5) represents the port capacity constraints. Eq. (6) is a non-negative constraint for service frequency.

The total operation cost of a vessel on link l is defined by

$$C_l = F_l(f_l, T_l) + \sum_h \delta_l^{hL} x_l CW_h \tag{7}$$

$$F_l = \{T_l(MFO_l + CA_l) + PC_l\} f_l \Phi(f_l) \tag{8}$$

$$\Phi_h(f_l) = \theta_1 \left[\frac{\sum_l \delta_l^h f_l}{VC_h} \right]^{\theta_2} \tag{9}$$

where

F_l	total navigation cost on link l

CW_H	unit cargo handling charge at port h
MFO_l	unit fuel cost of vessel on link l
CA_l	unit ship cost of vessel on link l
PC_l	port charge at port h for vessel on link l
δ_l^{hl}	$= \begin{cases} 1 & \text{when loading or unloading is done at port } h \text{ included in link } l \\ 0 & \text{otherwise} \end{cases}$
θ_1, θ_2	parameters

In Eq. (8), function $\Phi(f_l)$ means an additional cost coefficient due to congestion as increases of service frequency at port h expressed by Eq. (9).

2.3. Shippers' Behavior

Shippers are assumed to send one container by choosing importing and exporting ports to minimize transport costs. Since "users' equilibrium" is assumed in the present study, their behavior is expressed as

$$\underset{x_{ij}^k}{\text{Min}}\ Z_{ij}^u = SC_k \Gamma(x_{ij}^k) \tag{10}$$

$$SC_k = \delta_k^r \left\{ \sum_l \delta_l^r (TV \times T_l)(1 + \xi(x_l)) + P_r \right\} + \delta_{l'}^k C_{l'} \tag{10'}$$

subject to

$$\sum_k x_{ij}^k = X_{ij} \quad \forall i, j \tag{11}$$

$$\sum_i \sum_j \sum_k \delta_{ij}^{lk} x_{ij}^k = x_l \quad \forall l \tag{12a}$$

$$\sum_i \sum_j \sum_k \delta_{ij}^{l'k} x_{ij}^k = x_{l'} \quad \forall l' \tag{12b}$$

$$x_{ij}^k \geqslant 0 \quad \forall i, j, k \tag{13}$$

where

Z_{ij}^u	total transport cost for shipper
SC_k	unit container transport cost on route k
x_{ij}^k	container volume on route k from a zone i to zone j

$\Gamma(\cdot) \quad = \begin{cases} 1 & \text{when } x_{ij}^k \succ 0 \\ 0 & \text{otherwise} \end{cases}$

TV time value of container

P_r unit tariff per container on navigation route r

C_l land transport cost on land link l'

X_{ij} OD cargo volume from zone i to zone j

$\delta_k^r \quad = \begin{cases} 1 & \text{when route } r \text{ is included in route } k \\ 0 & \text{otherwise} \end{cases}$

L' land transport link connecting shipper's site with port

$\delta_{l'}^k \quad = \begin{cases} 1 & \text{when link } l' \text{ is included in route } k \\ 0 & \text{otherwise} \end{cases}$

x_{ij}^k container volume transported on route k from zone i to zone j

$\delta_{ij}^{jk} \quad = \begin{cases} 0 & \text{when route } k \text{ from zone } i \text{ to zone } k \text{ includes links } l \\ 1 & \text{otherwise} \end{cases}$

In Eq. (10′), function $\xi(\cdot)$ denotes the additional cost coefficient due to congestion at a port connected with link l, which is given by

$$\xi(x_l) = \rho_1 \left(\frac{x_l}{CP_l f_l} \right)^{\rho_2}, \quad \text{where} \quad \rho_1, \rho_2 \text{ are parameters} \tag{14}$$

Eq. (10′) refers to the generalized transport costs for a shipper. Eq. (11) is the constraint for OD cargo volume, and Eqs. (12a) and (12b) refers to the constraints for container volume transported on a link. To avoid confusion, the concept of two kinds of routes defined in the above formulations is explained in Fig. 2.

Marine transportation tariffs are given by the prices determined as market equilibrium resultant from the behavior of carriers and shippers. The present paper assumes a monopolized price market. From Eq. (12a), container volume equilibrium is derived and presented as

$$\sum_i \sum_j \sum_k \delta_{ij}^{lk} x_{ij}^k = \sum_i \sum_j \sum_k \sum_r \delta_{ij}^{rl} x_{ij}^r = x_l \tag{15}$$

Since a navigational route r is given by a combination of import and export ports, the equilibrium tariff is given by

$$P_r = \sum_l \delta_r^l \frac{C_l}{x_l} \quad \forall l \tag{16}$$

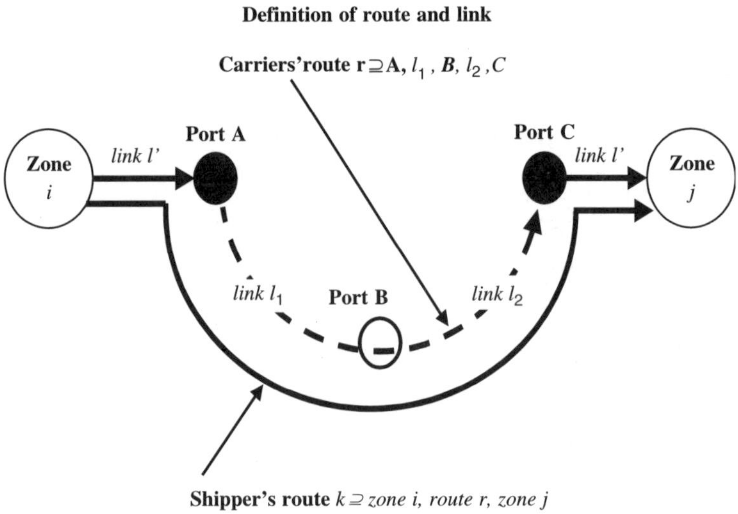

Fig. 2. Definition of Route Equilibrium Tariff.

On the other hand, when "users' equilibrium" is assumed, the following relationship must be satisfied.

$$\begin{array}{ll} \text{if} \quad x_{ij}^k \geqslant 0 \quad \text{then} & SC_k = SC_{ij}^* \\ \quad\quad \text{otherwise} & SC_k \geqslant SC_{ij}^* \end{array} \tag{17}$$

where SC_{ij}^* is the optimal transport cost from zone i to zone j.

3. NUMERICAL EXAMPLES

To answer the unresolved questions raised in Section 1, some numerical examples are presented here to validate the proposed analytical model. Since most deep-water container berths planned in East Asia will be in service in 2010, the numerical examples are computed for the future market in year 2010.

In Japan, to examine the shippers' port choice behavior in detail, the local zone of shippers are divided into 47 prefectures as shown in Fig. 3. Table 1 presents the list of main ports and their hinterlands that are considered in numerical examples. Tables 2 and 3 present the port capacity and port charges.

3.1. Model Behavior

3.1.1. On Japanese Ports in 1993

At first, the model is applied to 1993 market to examine its validity. In computation, 1,000 and 3,000 TEU vessels were employed in the market. Model validity is examined for Japanese ports by comparing observed data with computed results. Fig. 4 shows the computed results from the proposed analytical tool and the observed total cargo handling share for Japanese ports with a total estimation error of 5.1%.

Figs. 5 and 6 show the observed and computed hinterland cargo handling share for each Japanese port for exports and imports. The model explains well in hinterland expect for the port of Nagoya. In case of port of Nagoya, computed hinterland values are overestimated compared to the observed

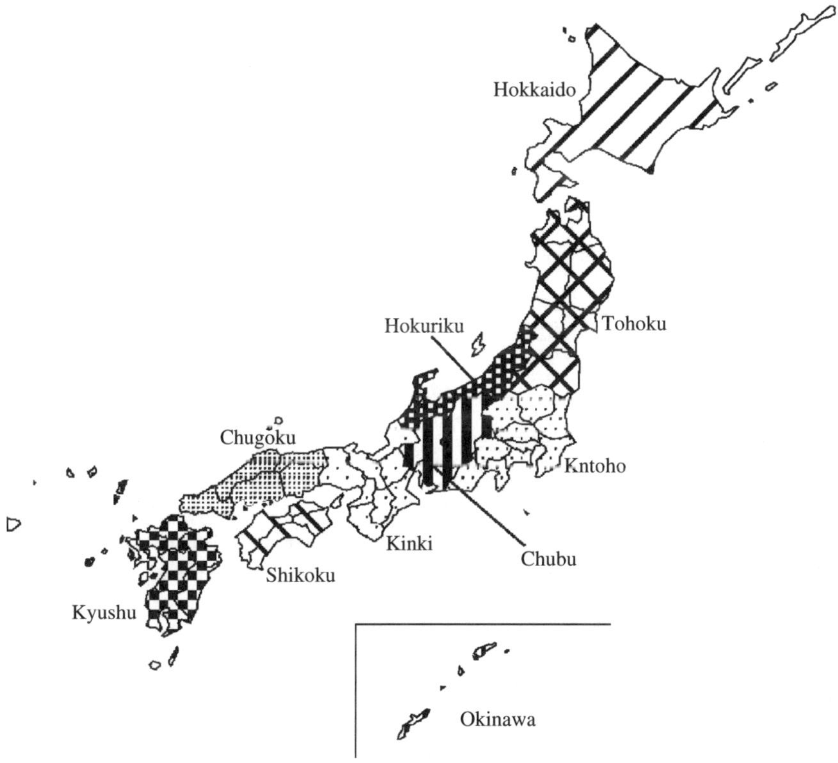

Fig. 3. Local Zone in Japan.

Table 1. Ports and Hinterlands.

No.	Port	Hinterland
1	Keihin	47 prefectures of Japan
2	Nagoya	
3	Hanshin	
4	Kanmon	
5	Busan	Korean Peninsula
6	Shanghai	Middle/South China
7	Hong Kong	Hong Kong
8	Kaoshiung	Taiwan, Philippines
9	Singapore	ASEAN
10	Rotterdam	Europe
11	Los Angeles	N.A./Canada

Table 2. Port Capacity.

Port	Capacity (10,000 TEU/year)		
	1,000 TEU	3,000 TEU	6,000 TEU
Keihin	888	576	200
Nagoya	360	168	—
Hanshin	936	480	320
Kanmon	384	—	—
Busan	440	360	320
Shanghai	256	—	—
Hong Kong	1,200	680	640
Kaoshiung	769	400	1,200
Singapore	1,184	864	520
Rotterdam	2,400	2,400	2,400
Los Angeles	2,400	2,400	2,400

values and this is due to the anchoring effect of the past habits of shippers in Chubu region. Before the development of port of Nagoya, shippers of Chubu region were customers of port of Hanshin.

3.1.2. On Future Market in 2010 Excluding 6,000 TEU

The developed model is applied to the future market in 2010 by considering different size of vessels. OD cargo volume in 2010 is estimated based on 1993 data, considering the GDP growth ratio for each zone. Case 1 is a base

Table 3. Port Charge (10,000 Yen).

Port	Port Charge (per vessel)			Handling Charge (per TEU)
	1,000 TEU	3,000 TEU	6,000 TEU	
Keihin	65.3	184.3	348.3	2.0
Nagoya	71.9	202.7	—	2.0
Hanshin	67.1	189.3	357.8	2.0
Kanmon	47.9	—	—	2.0
Busan	24.5	69.1	130.6	0.6
Shanghai	27.3	—	—	0.2
Hong Kong	32.0	90.2	170.4	1.9
Kaoshiung	32.0	90.2	170.4	1.9
Singapore	25.5	71.9	135.4	0.9
Rotterdam	78.6	221.6	418.8	1.3
Los Angeles	44.9	126.6	239.3	0.6

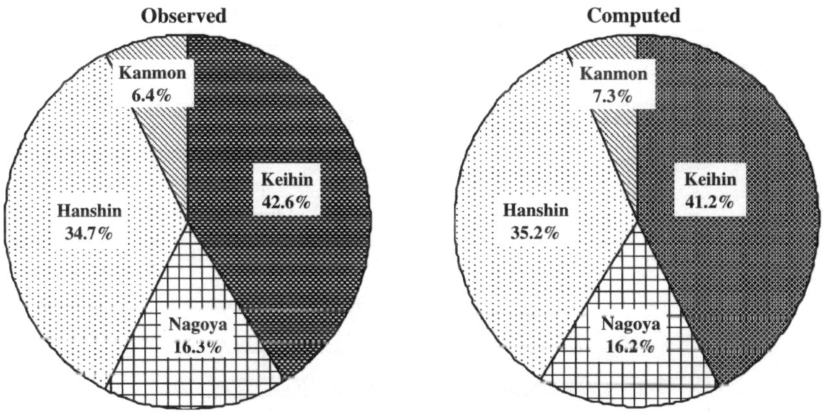

Fig. 4. Comparison of Cargo-handling Share of Japanese Ports.

case, in which only 1,000 and 3,000 TEU are considered by excluding 6,000 TEU vessels in the future market.

Figs. 7 and 8 show the computed navigation route and service frequency per year chosen by carriers for 1,000 and 3,000 TEU vessels. From these figures, 1,000 TEU vessels are restricted in intra-Asian routes, concentrating in Hong Kong and Hanshin. With this Hong Kong will act as a Hub port

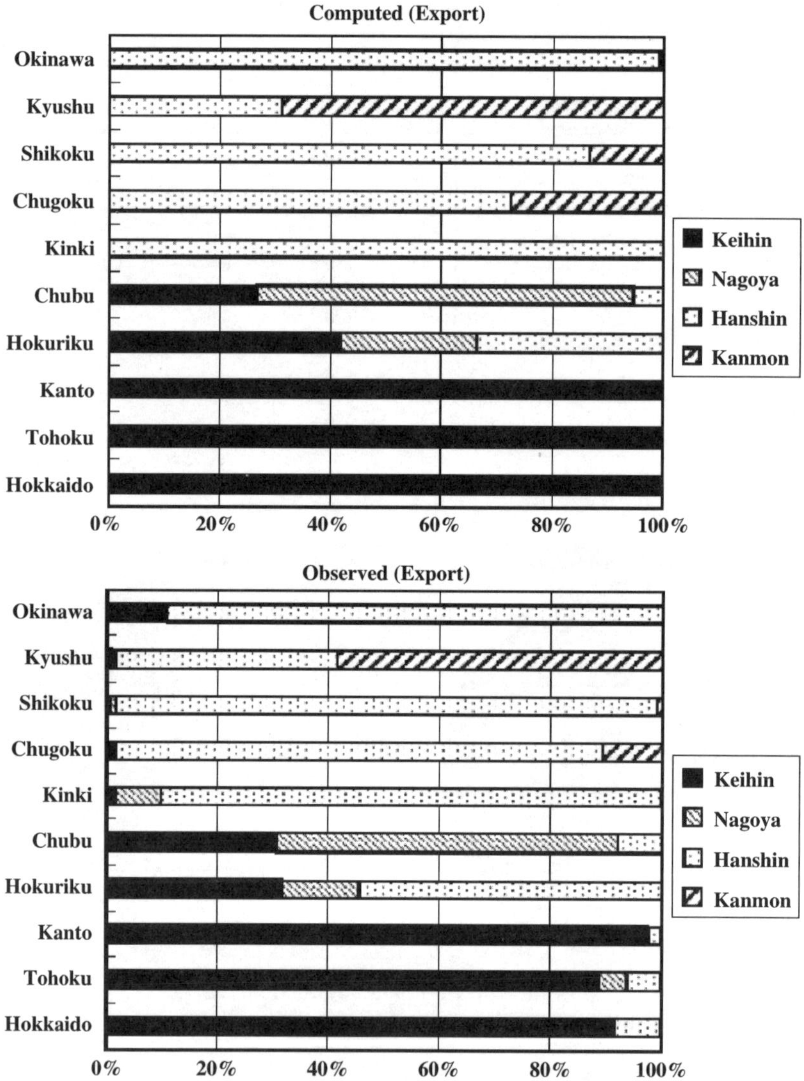

Fig. 5. Hinterland of Japanese Ports (Exports).

in East Asian region. On the contrary, 3,000 TEU vessels are concentrated in between Hong Kong to North America, and Singapore to Keihin. Consequently, Busan, Hanshin, Keihin, and Hong Kong are the gate ports

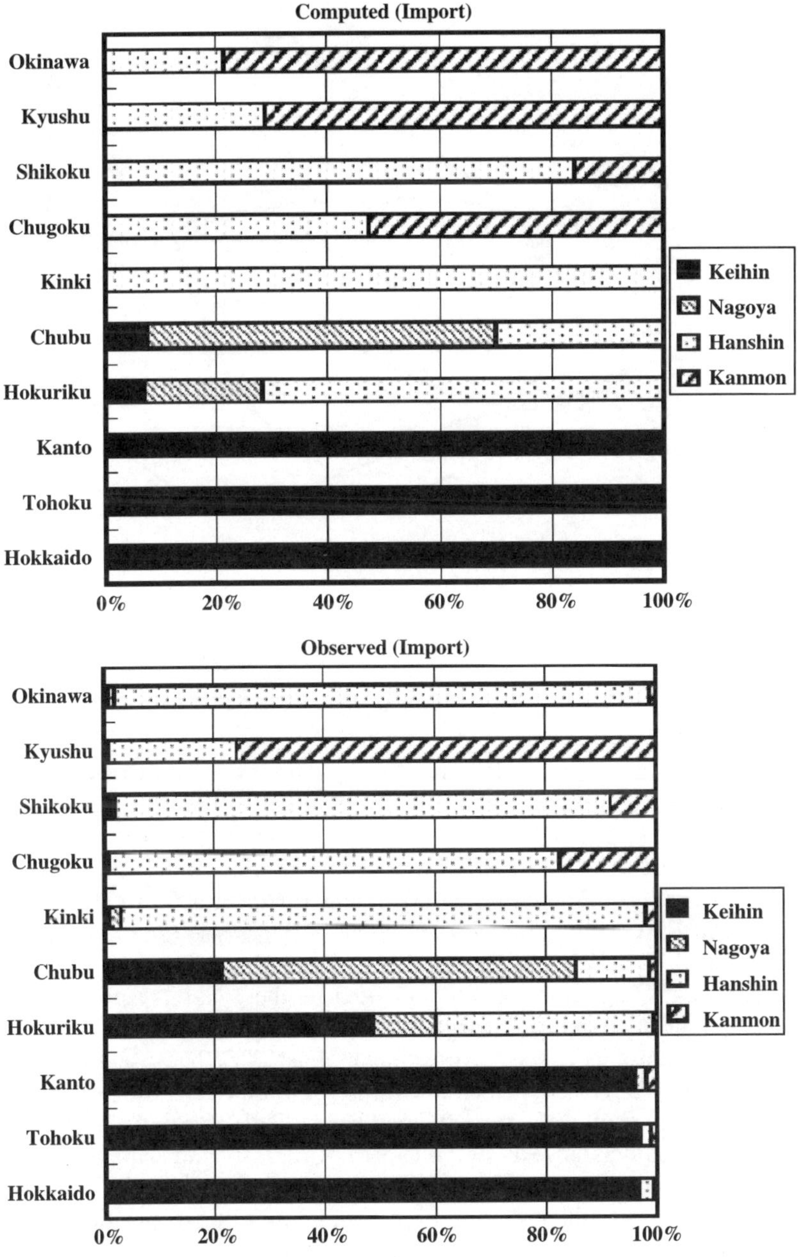

Fig. 6. Hinterland of Japanese Ports (Imports).

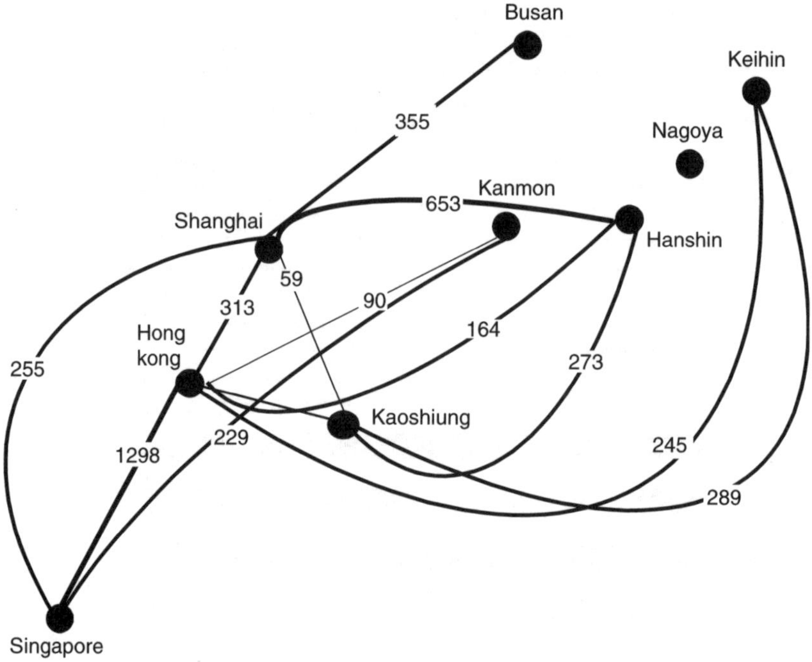

Fig. 7. Network of 1,000-TEU Vessel (Case 1).

for the North American trunk route and Singapore for the European trunk route. This is much more understandable if Figs. 9 and 10 are observed. Fig. 9 shows that Hong Kong invites more 1,000 TEU vessels than other ports, where as Fig. 10 shows that more than 97% of containers are Hong Kong's own cargo. Therefore, Hong Kong invites 1,000 TEU vessels to transport in East Asian countries and 3,000 TEU vessels for the trunk line such as North America and Europe.

Since Hong Kong is very much congested by these vessels from China, it cannot handle transshipped cargo of other countries. On the contrary, Kaoshiung maintains a special position in East Asia because Kaoshiung invites 3,000 TEU vessels, which accounts for about 50% of its shipping trade, and more than 20% transshipped cargo. This comes from being located almost equidistance between North American and Japanese ports. Kaoshiung's port and cargo handling charges are much cheaper than Japanese ports. Thus, as illustrated in Figs. 6 and 7, it keeps its position as gate port to North America in the East Asian region.

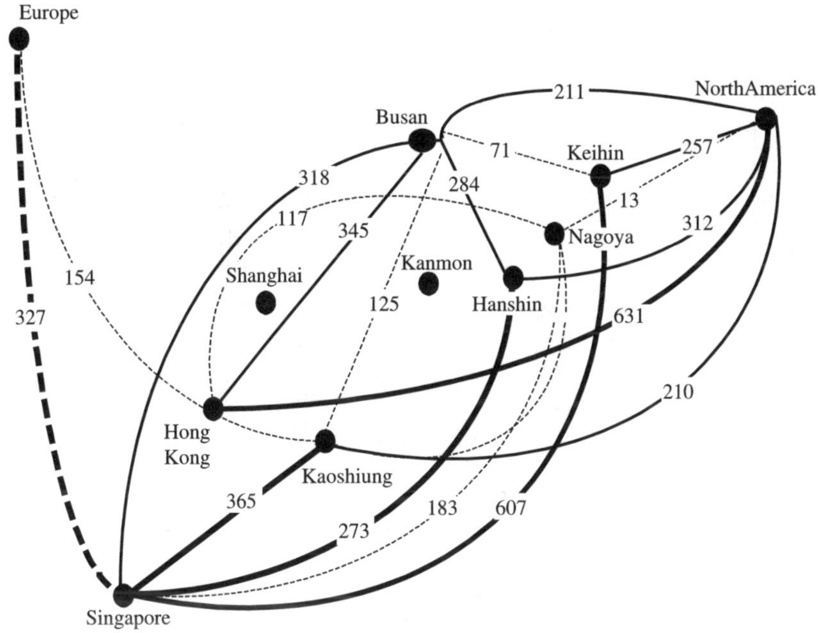

Fig. 8. Network of 3,0000-TEU Vessel (Case 1).

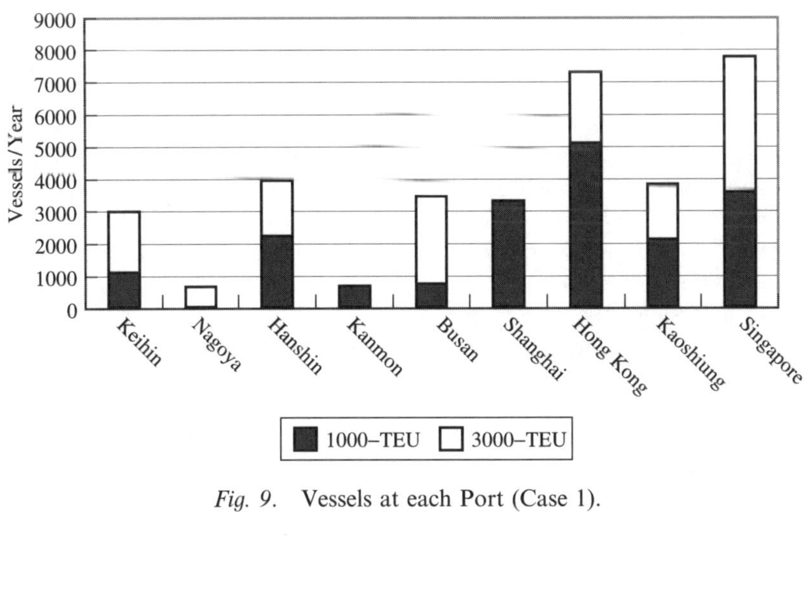

Fig. 9. Vessels at each Port (Case 1).

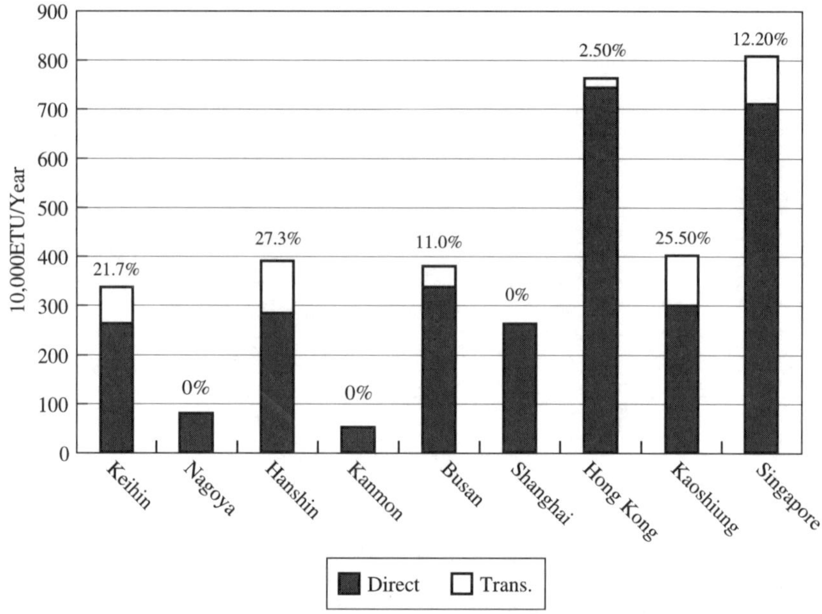

Fig. 10. Containers at Port (Case 1).

3.1.3. On Future Market in 2010 Including 6,000 TEU

It is anticipated that carriers will use huge container vessels such as 6,000 TEU in 2010 market to pursue economies of scale. Under such conditions, in a competitive market carriers may change routing strategies and marine tariffs. This is a big concern for shippers as well as for port administrators. In this section, numerical examples related to the future market including 6,000 TEU vessels are discussed and is represented as Case 2.

Figs. 11–13 show the routing network for each vessel size. This was compared with Figs. 6 and 7 and observed that 6,000 TEU vessels will replace 3,000 TEU vessels that are operating in between Kaoshiug to Los Angeles and Singapore to Rotterdam. This leads to a change in the routing networks of 1,000 and 3,000 TEU vessels. In particular, Hanshin to Los Angeles route will disappear and Shanghai to Busan route will be reopened for 1,000 TEU vessels.

A remarkable change is seen at port of Nagoya where 3,000 TEU vessel routes connecting with Busan, Hong Kong and Singapore will disappear. Such conclusions are understandable by Fig. 14, which shows vessels calling

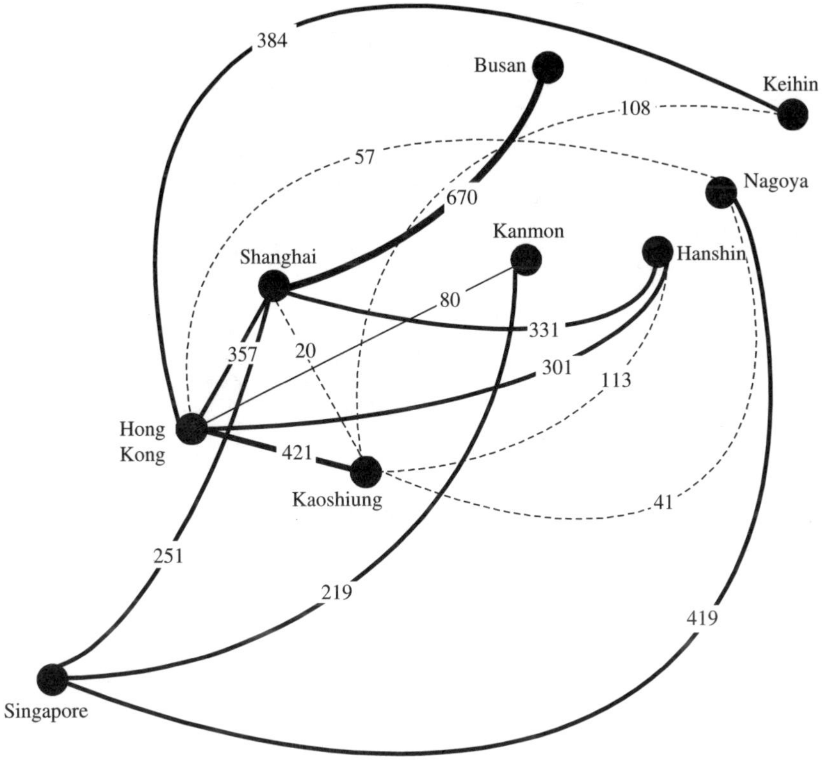

Fig. 11. Network of 1,000-TEU Vessels (Case 2).

at each port. Change in routing network causes a transformation of container's flow and is presented in Fig. 15. This change in routing network leads to the disappearance of transshipped cargo at Japanese ports, Kaoshiung port and big ports like Hong Kong and Busan. Drastic changes that concentrate vessels and cargo at Busan, Hong Kong and Singapore can be explained. Busan's gate port function for North America reflects Busan's port and cargo handling charges that are cheaper about 36% and 30% respectively than Japanese ports, even though port capacity is almost same. This situation invites an "economy of scale" for 6,000 TEU vessels. In fact, tariffs are drastically decreased on the Busan to North America route but not on the Hanshin to North America, which is explained later. In addition, at the same time Busan enjoys an "economy of density" that

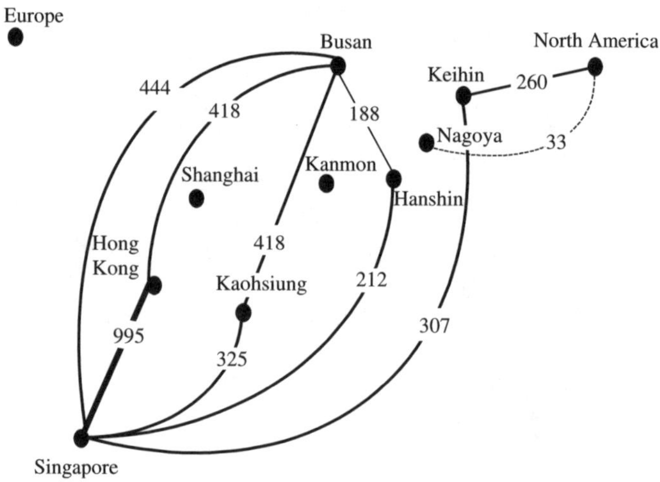

Fig. 12. Network of 3,000-TEU Vessel (Case 2).

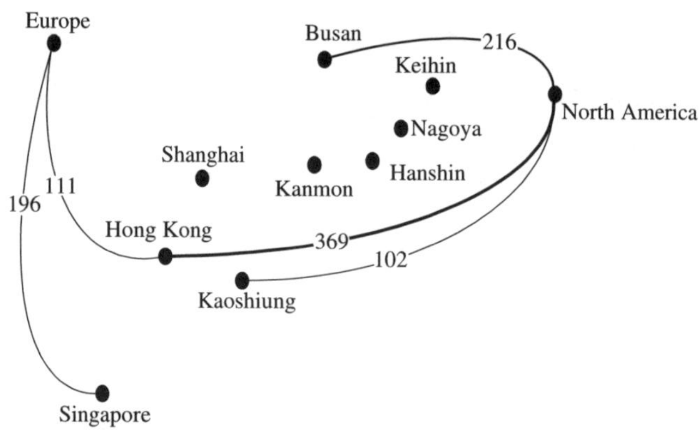

Fig. 13. Network of 6,000-TEU Vessels (Case 2).

results from the concentration of 1,000 and 3,000 TEU vessels for feeder transports.

On the other hand, for Europe's trunk route Hong Kong and Singapore will be the gate ports connected to the Busan, Shanghai, and Japanese ports. They enjoy economies of scale and density as explained above.

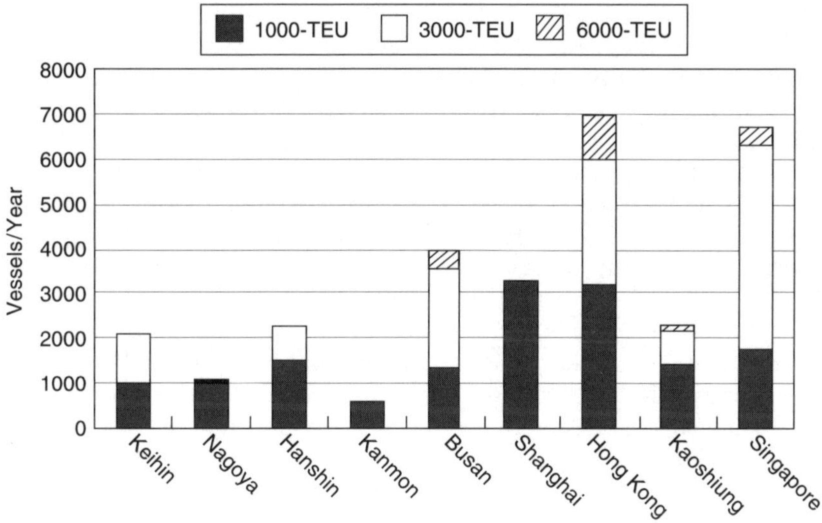

Fig. 14. Vessels at each Port (Case 2).

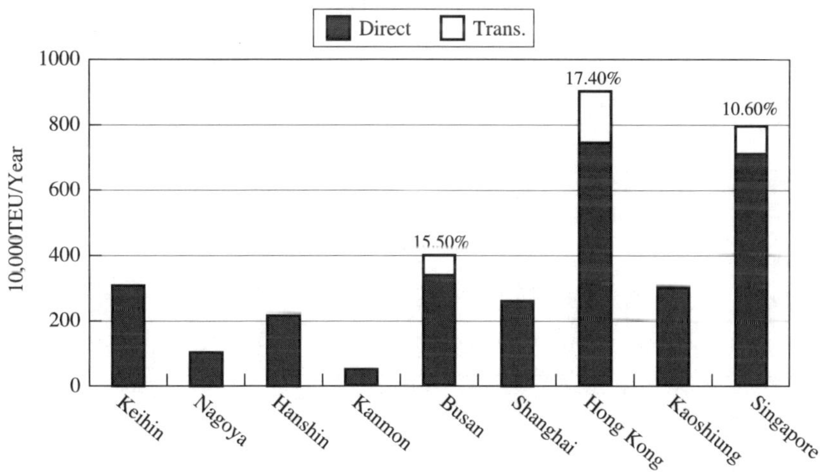

Fig. 15. Containers at each Port (Case 2).

The process that produces an "economy of scale" and "an economy of density" works in the market, i.e., at first employment of 6,000 TEU vessels invites "an economy of scale" that decreases transport cost per TEU. This results in decreases of 1,000 and 3,000 TEU vessels provided in the market, which dissolves congestion at ports. A decrease of TEU transport and congestion costs at a port may enable carriers to provide more vessels, but this will again invite congestion. However, the concentration of containers by feeder service at a gate port invites "an economy of density" that again decreases transport costs. Consequently, tariff and transportation times can be decreased. The process explained is considered as a "positive feedback" in economics. In fact, this process invites a big decrease in tariffs.

Tariffs of each route in Cases 1 and 2 are compared and presented in Figs. 16 and 17. Fig. 16 shows tariffs on North American and European routes are decreased except on Nagoya to North America route. Particularly, tariffs on routes from Busan and Hong Kong, and Nagoya and Hong Kong bound for North America and Europe are decreased remarkably. On the other hand, in intra-Asian routes there is a remarkable increase in tariff from/to to Shanghai.

Such tariff changes can be explained as cargo from Hanshin for North America is transported via Busan. Then, although the shipping distance is extended, the tariff itself is decreased because of the "positive feedback effect" at Busan, which was discussed previously. The tariff from Nagoya to North America rises due to increase of congestion cost for all vessels that are

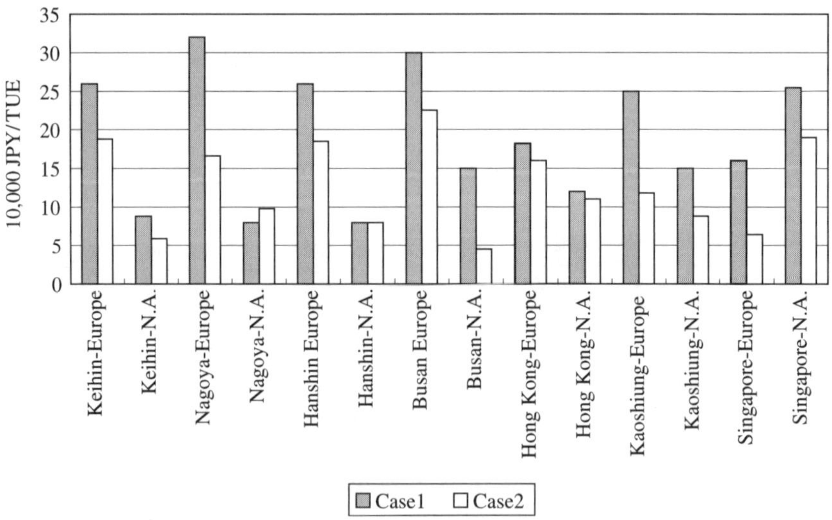

Fig. 16. Tariff charges at Routes for North America and Europe.

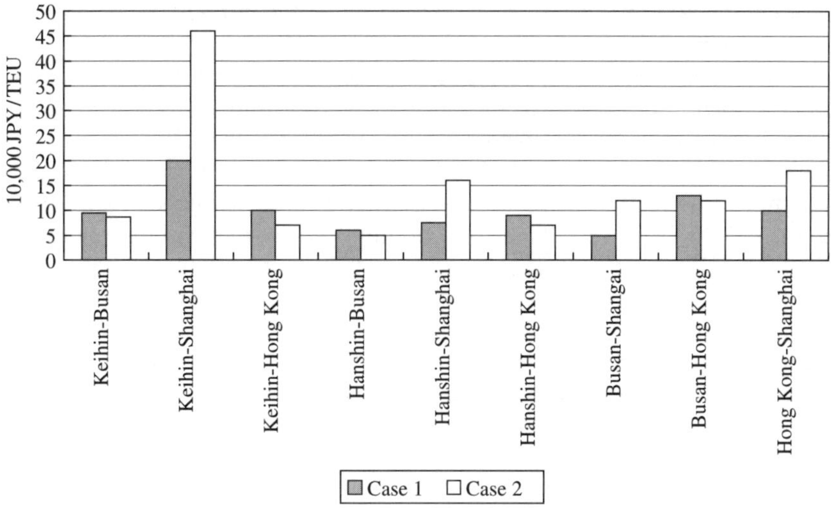

Fig. 17. Tariff Change at Intra-Asian Routes.

calling at Nagoya are changed from 3,000 to 1,000 TEU. But tariffs for European bound ships are decreased because Singapore enjoys an economy of scale and the resolution of congestion. Tariffs from Keihin to North America are reduced because of reduction in congestion, which occurred due to decrease in 3,000 TEU vessels and tariffs for European bound ships are remarkably reduced due to feeder service by 1,000 and 3,000 TEU vessels and transshipments to 6,000 TEU vessels at Singapore. On the contrary, Shanghai suffers due to increase in both tariffs on North America and Europe routes. Before the introduction of 6,000 TEU vessels, Shanghai was intensively provided feeder service to Hanshin. But after their employment, frequent feeder service that once concentrated on Hanshin was replaced by spread feeder service to Busan, Hong Kong, and Hanshin. This increases the costs using a redundant route instead of enjoying an economy of scale and density as Busan and Hong Kong.

4. CONCLUSIONS

The present paper proposes a tool to analyze the international marine transport market with few numerical examples to predict the influence of deploying 6,000 TEU class container vessels in the market, particularly focusing on the East Asian region.

Study results based on numerical examples suggests that the deployment of 6,000 TEU vessels in the marine transport market will drastically change carriers routing networks and tariffs as well as the volume of transshipped cargo handled at main ports in the region. The model assumes "perfectly competitive market", and concludes that: (i) before 6,000 TEU vessels are fully employed in the future market of 2010, Hanshin, Busan, Hong Kong, and Singapore will have equal power in marine transport market so the present routing network will not change so much; (ii) however, when 6,000 TEU vessels are fully employed in the market, the routing network will drastically change to a hub-and-spoke type and Busan, Hong Kong and Singapore will act as Hub ports; (iii) with 6,000 TEU vessels, Hub ports can have an advantage of "economy of scale" and "economy of density"; (iv) according to the changes in the network's shape, tariffs on almost all routes will be reduced, except on the Shanghai route due to positive feedback effect in economics; and (v) Japanese ports will lose their traditional role as gate ports to the East Asian region.

The above estimation assumes that the extension of some ports being planned will be completed and there won't be any change in port and cargo handling charges. However, the ports of Busan, Hong Kong, and Singapore may increase port tariffs according to the economic development and then the costs will approach those of Japan. In such a scenario, the above conclusions will change. In fact, another numerical scenario assumes the reduction of such charges and/or further capacity extension of Japanese ports, which paints a different future for the marine transportation market. However, discussion on those studies is omitted due to space limitations.

REFERENCES

Ieda, H., Shibasaki, R., & Naitoh, T. (1998a). Container flow model and application in Asian region. *Journal of Infrastructure Planning, JSCE, 16*, 469–480 (in Japanese).

Ieda, H., Naitoh, T., & Shibasaki, R. (1998b). A simulation model of international container shipping including domestic land and feeder transportation. *Proceedings of infrastructure planning, JSCE, 21*(2), 717–728 (in Japanese).

Imai, A., & Papadimitrious, S. (1996). Port choice for hubs in the container liner networks of Eastern Asia. *Proceedings of the 1st JSPS-NUS seminar on intergrated engineering, Kyoto University*, 115–124.

Kuroda, K., Yang, Z., & Takebayashi, M. (1997a). International container flow analysis considering feeder service network. *Proceedings of Infrastructure Planning, JSCE, 14*, 551–558 (in Japanese).

Kuroda, K., & Yang, Z. (1997b). Port management policy and its influence on behavior of container liner shipping company and shippers. *Journal of EASTS, 2*(2), 75–86.

Kuroda, K., Takebayashi, M., Tsuji, T., & Muto, M. (1999). Influence of port management policy on behavior of container liner ship company and shippers. *Journal of EASTS, 3*(2), 127–138.

Kuroda, K., Takebayashi, M., Muto, M., Okubo, T., & Tsuji, T. (2000). Modeling of international container flow and its application to East Asia. *Journal of JSCE, 653*(IV-48), 117–131.

Kuroda, K., Takebayashi, M., Muto, M., & Okubo, T. (2001). Impact study on introduction of post-panamax container cargo vessels to the Asian-pacific-European maritime transport market. *Journal of JSCE, 674*(IV-51), 35–48.

Osman, M. A., & Inamura, H. (1997). Port choice selection based on cargo physical distribution (Containerization) for export promotion. *Journal of EASTS, 2*(2), 129–139.